LABORATORY MANUAL

TEACHER EDITION

AUTHORS

Chris Kapicka
Biology Department
University of Nevada-Reno
Reno, NV

Alton Biggs
Allen High School
Allen, TX

Linda Lundgren
Bear Creek High School
Denver, CO

Albert Kaskel
Evanston Township High School
Evanston, IL

Contributing Authors

Diane Bynum
Belaire High School
Baton Rouge, LA

Rachel Hays
Heath Junior High School
Greeley, CO

Juliana Texley
Richmond Schools
Richmond, MI

Donald Emmeluth
Fulton-Montgomery
Community College
Johnstown, NY

Priscilla Lee
Venice High School
Los Angeles, CA

Maureen Wahl
Notre Dame-Cathedral
Latin School
Chardon, OH

Tom Russo
Chemistry Education
Consultant
Kemtec Educational Corp.
West Chester, OH

Consultants

Lucy Daniel
Rutherfordton County Schools
Spindale, NC

Lorena Farrar
Westwood Junior High School
Richardson, TX

Albert Kaskel
Evanston Township High School
Evanston, IL

Ouida Thomas
Rosenberg High School
Rosenberg, TX

GLENCOE

Macmillan/McGraw-Hill

New York, New York   Columbus, Ohio   Mission Hills, California   Peoria, Illinios

A GLENCOE PROGRAM

Biology
The Dynamics of Life

Student Edition
Teacher Wraparound Edition
Study Guide, SE and TE
Laboratory Manual, SE and TE
Biolab and Minilab Worksheets
Chapter Assessment
Videodisc Correlations
Science and Technology Videodisc Series Teacher Guide
Transparency Package
Concept Mapping
Biology Projects
Exploring Applications of Biology
Critical Thinking/Problem Solving
Great Developments in Biology
Spanish Resources
Lesson Plans
Computer Test Bank IBM/APPLE/MACINTOSH
English/Spanish Audiocassettes

Send all inquiries to:

GLENCOE DIVISION
Macmillan/McGraw-Hill
936 Eastwind Drive
Westerville, Ohio 43081

ISBN 0-02-826667-6

Printed in the United States of America.

4 5 6 7 8 9 10   POH   00 99 98 97 96

TABLE of CONTENTS

TO THE TEACHER

Biology: The Dynamics of Life, *Laboratory Manual* follows the chapter sequence and reinforces concepts presented in that text. However, the 72 activities in the manual are designed to be used with any high school biology text. The contents provide you with two to four experiments pertinent to each major topic covered in a one year comprehensive biology course. In this Teacher Edition, the chapters of **Biology: The Dynamics of Life** that correspond to each laboratory activity are listed in the table of contents.

Scientific literacy, scientific principles, and scientific inquiry are developed. Students increase their science vocabulary, learn how to handle laboratory equipment, use modern laboratory techniques, and acquire skill in working with tables and graphs. Scientific methods become most important as students perform each activity, collect and record data, and form conclusions based on analysis and interpretation of experimental results. Skills requiring proper and careful reading of experimental procedures, accurate data collection, data interpretation, and graphing are utilized throughout the manual.

The Teacher Edition for **Biology: The Dynamics of Life,** *Laboratory Manual* provides useful general information designed to aid you in the laboratory. Helpful teaching strategies and safety and disposal guidelines are outlined. Instructions for working with animals, cooperative learning, alternatives to dissection, a lab-by-lab materials list, and instructions for the preparation of solutions and reagents have been included to aid you in preparing for laboratory activities. Also found here are models to be used with three activities.

The Teacher Edition provides a variety of helpful information about each activity in the form of an Answers and Teacher Guide section. Teaching tips, helpful comments and suggestions, objectives, process skills, and time allotments, as well as answers to all questions are to be found in this section. Sample data, labeled diagrams, sources of materials, and alternative materials have been included.

USING THE LABORATORY MANUAL

Student Edition

The activities in this laboratory manual are of two types: Investigation or Exploration. In an Investigation, students are presented with a problem. Then, through use of scientific methods, they seek answers. Their conclusions are based on their observations, experimental data, and interpretation of these data and observations. In Explorations, students make observations but do not use other scientific methods to reach conclusions. The format each laboratory activity follows is outlined below.

Introduction: A brief introduction provides background information for each activity. Students may need to refer to the introduction for information that is important for completing an activity.

Objectives: Each statement listed in this section is a performance objective. These objectives should prove helpful to both you and the student. You may want to use them as a basis for evaluating student progress. Students can use them as a means for quickly determining what they will be doing and what is important in each activity.

Materials: Reagents, equipment, and supplies needed for each activity are listed here. Specific quantities of materials indicate minimum needs for each student or group. If you have very limited quantities of supplies and equipment, you may need to adjust the quantity of materials as listed or have students work in larger teams. Directions for preparing all reagents are found within this Teacher Edition. It is not necessary to prepare a fresh reagent each time it appears in a different activity unless directions state otherwise.

Procedure: Instructions are brief, yet complete, and often accompanied by diagrams for further student clarification. Emphasis is placed on developing student skill in carefully following directions and in observing, measuring, and recording data in an organized manner. Students often are asked to represent data by graphing.

Hypothesis: In Investigations, students will write a hypothesis statement to express their expectations of the results and as an answer to the problem statement. Explorations do not require students to make hypotheses.

Data and Observations: This section includes tables and space for students to record their data and observations.

Analysis: Major experimental concepts and objectives are drawn together in this section. Students are asked to answer questions that require analysis of experimental data.

Checking your Hypothesis: Students determine whether their hypotheses are valid, based on their data.

Further Investigations/Further Explorations: Further activities that students can do on their own are given here. The first activity is an enrichment; the second activity is a more difficult challenge.

In addition to the activities, the Student Edition contains several other features. A glossary provides definitions of all the important terms used in the activities. A description of how to write a laboratory report, a section on the care of living things, diagrams of laboratory equipment, and information on safety in the laboratory are also included. Three appendices give instructions on the use of laboratory equipment and SI units.

Teacher Edition

The Teacher Guide consists of a separate page for each activity. Each page provides answers to questions, sample data and labeled diagrams, helpful suggestions and teaching tips for preparing and teaching the activity, as well as a list of materials needed for the class with suggestions for size of student groups appropriate for the activity and recommended time allotments. The Teacher Guide also includes those critical thinking process skills students will use to develop the ability of thinking logically and abstractly.

The Teacher Edition also includes additional information designed to give you, the teacher, as much help as possible in equipping your classroom or laboratory and in preparing for each activity.

Teaching Strategy: This section gives ideas for modifying the activities to fit different time restrictions you may have. Methods for evaluating student performance are explained.

Cooperative Learning in the Laboratory: This section provides suggestions for teaching students how to work together cooperatively for the mutual benefit of all. Cooperative learning techniques, such as assigning roles to members of the group, are discussed.

Alternatives to Dissection: This section describes alternatives if you do not wish to teach dissection labs.

Working with Animals: The use of live animals in the classroom necessitates certain responsibilities on the parts of both teacher and student. Animals must be cared for properly. This section gives instructions on the care of animals in the classroom and how to set up a marine aquarium for class use.

Safety and Disposal Guidelines: This brief listing of safety and disposal procedures will help you make the laboratory a safe place to work and learn. Included here are basic procedures that should be followed by students and teacher while working in the biology laboratory. The material in this section should be stressed to students before any activities are undertaken.

Preparation of Solutions: Solutions used in this laboratory manual are listed in order by the number of the activity in which they are used. Preparation procedures, cautions, and amounts to make are indicated to aid you in the safe and economical use of reagents.

Course Materials List: All equipment, expendables, chemical supplies, and biological supplies needed to complete the 72 activities in this manual are listed for a class of 30 students. Using this list, all materials needed for the entire year may be assembled or ordered, thus eliminating the problem of improper or insufficient supplies.

Materials List Per Lab: This section lists the materials one student or group needs to complete each activity. Using this list, you can easily tell how much equipment you will need for each of your classes. If equipment is in short supply, you can use this list to increase the number of students you will need to assign to each team. This listing of materials should afford you the maximum amount of flexibility as well as quick access to all that is needed for each activity to aid in preplanning. Again, materials are listed as equipment, expendables, chemical supplies, and biological supplies to aid you in planning for the use and securing of materials.

Suppliers: A list of suppliers has been included for your use in ordering materials.

Student Models: For your convenience, all models and handouts needed to complete several activities throughout the Student Edition have been gathered together on pages T27 through T30 of the Teacher Edition.

DEVELOPING PROCESS SKILLS

Basic Process Skills

Observing: Students use one or more senses to increase their perceptions in order to learn more about objects and events.

Classifying: Students group objects or events based on common properties and/or categorize based on existing relationships among objects or events.

Inferring: Students propose interpretations, explanations, and causes from observed events and collected data.

Communicating: Students convey information verbally, both in oral and written forms, and visually through graphs, charts, pictures, and diagrams.

Recognizing and Using Spatial Relationships: Students estimate the relative positions of moving and non-moving objects to one another.

Measuring: Students identify and order length, area, volume, mass, and temperature to describe and quantify objects or events.

Predicting: Students propose possible results or outcomes of future events based on observations and inferences drawn from previous events.

Using Numbers: Students transfer or apply ordering, counting, adding, subtracting, multiplying, and dividing to quantify data where appropriate in Investigations or Explorations.

Complex Process Skills

Interpreting Data: Students explain the meaning of information gathered in scientific situations.

Forming Hypotheses: Students make an informed assumption in order to draw out and test its logical consequences.

Separating and Controlling Variables: Students recognize the many factors that affect the outcome of events and understand the relationship of the factors to one another so that one factor (variable) can be manipulated while the others are controlled.

Experimenting: Students test hypotheses or predictions under conditions where variables are both controlled and manipulated.

Formulating Models: Students construct mental, verbal, or physical representations of ideas, objects, or events. The models are then used to clarify explanations or to demonstrate relationships.

Defining Operationally: Students form a working definition that is based upon actual experience in which the student has participated.

TEACHING STRATEGY

The 72 activities in **Biology: The Dynamics of Life,** *Laboratory Manual* may exceed the number of activities needed for a one-year course. You have the option to choose those best suited for your students and time schedule. The activities are designed to be as flexible as possible. Most are completed in one laboratory period, although some may require starting one day and completing on another. If time is short, you may wish to omit parts of an activity. Or, you may want to reduce the number of trials or the time spent for each trial. Those activities or parts of activities requiring model building may be conducted by students on their own time outside of class. Using student teams may also allow for completion of specific activities that appear too lengthy for your school schedule. Each team may complete a different part of the three- or four-part activity and then share observed data to save time.

Student progress may be evaluated through written answers in the *Analysis* section or through the performance objectives in the *Objectives* section. For example, can a student properly record data from an activity so that it can be contributed towards class totals? Can a student follow dissection directions carefully so that organs can be observed and identified? You may wish to evaluate students directly using the performance objectives as guides.

COOPERATIVE LEARNING IN THE LABORATORY

The same principles apply to using cooperative learning techniques during laboratory activities as they do for classroom teaching. Some guidelines are laid out for you in **Biology: The Dynamics of Life,** *Teacher Wraparound Edition,* pages 32T–33T. Additional suggestions for cooperative learning strategies during lab activities are presented here.

At the beginning of the year, it is useful to design a lab called Biological Mysteries. Biological items such as skulls, preserved invertebrates, live insects, plants, fossils, microscopic specimens, photomicrographs, and pictures of unusual biological phenomena are set up with questions in the form of a lab demonstration around the room. Have the students move from station to station in their groups. Have one half of the class discuss their answers in their groups, arriving at a group consensus on a single paper. The other half of the class should work individually on their own papers. After the activity is completed, ask questions of each half of the class such as: What are the advantages of working in a group? What are the advantages of working alone?

When an Exploration or Investigation activity is assigned and you have explained the day's lesson, you may want to describe how to divide up the work. How you do this will depend on the available lab stations and equipment, as well as on students' individual needs. Each group should assign its members specific roles to carry out during the course of the lab activity. You will assign appropriate roles depending on whether the group is just beginning group work or has experience, and depending on the nature of the activity. One student can be assigned more than one role. The following are possible roles: The **reader** reads the group's material aloud. The **recorder** writes the group's consensus of the answers if you are requiring only one paper per group. The **materials handler** gets and puts away materials and equipment needed by the group. The **checker** checks to be sure that everyone agrees with the answer and from time to time may summarize the material learned. The **prober** tries to get students to elaborate and asks for other possible answers, rather than allowing the group to agree immediately. The **gatekeeper** may be an important role to assign in a group where one or two people dominate. The gatekeeper makes sure that everyone has a turn to speak or participate. During particular activities or in certain groups, you may need to assign a **task master** who makes sure the group stays on task.

The combination of roles needed will vary for each activity.

Before a lab activity begins, you might want students to get into the spirit of cooperation by having them work in their group on one sheet of paper. Ask them first to make a list of everything they already know about the topic. Then, ask them to read the introduction and write a list of things they think they might learn. They should take just a few minutes to complete this task. Emphasize that their evaluation on these questions will not be graded as right or wrong, but that their degree of concerted participation will be recognized. Request students to divide up the work and share results as the lab progresses, and remind them to come together again to do the analysis questions.

If background reading, library research, or textbook study is required to complete the lab activity, two team members might be assigned this responsibility while the other two team members carry out other steps of the activity. Some activities might be set up so each team member becomes an "expert" on a particular section of the activity by getting together with a group consisting of one member from each team in the class. When they have developed their needed expertise, they report back to their own group and share what they have learned.

The final stage of cooperative learning in a lab activity is processing, a group evaluation of cooperative learning skills used during the activity. The evaluation can be carried out for the team as a whole or for the specific roles students were asked to carry out. Three or four questions that students rate on a scale of one to ten may be enough for one activity. Examples are: How well did you perform your role? How well did your group work together? What things can your group do to improve their cooperative skills?

There are a variety of ways to grade cooperative group work. One paper or product can be selected at random. Students signatures indicate that each member of that team accepts the grade on that paper for their own grade. You might average individual member's scores, give bonus points based on the group average, or give individual scores plus bonus points if everyone in the group meets a preset criterion. The important point to remember in giving grades in a cooperative learning setting is that the students must perceive that they "sink or swim" together.

ALTERNATIVES TO DISSECTION

As a biology teacher, you probably wish to foster the same thrill and excitement with nature in your students as you continually experience yourself. You want your students to appreciate and respect life and the diversity of organisms. You would like them to develop an understanding of life processes and the interdependence of all living things. Above all, you would like to see your students take responsibility for the quality of their own lives as well as quality of life in our biosphere.

In order to achieve the above goals, **Biology: The Dynamics of Life,** *Laboratory Manual,* presents a large number of labs that provide students with the opportunity to observe life in its natural state. Although the authors of this lab manual have provided detailed procedures for dissections of an earthworm, a squid, a starfish, and a frog, we are aware that more and more biology teachers prefer to have the choice of an alternative to these dissections. When students have the opportunity to work with live animals by examining their behavior, growth, development, social interactions, or ecology, they learn the complexity of biological systems. They learn how an earthworm, a snail, a starfish, or a fish goes about its daily business. They learn important principles of behavior, physiology, and ecology. Students are encouraged to view living things, not as mere scientific subjects, expendable and unimportant, but as integral dynamic parts of the complex web of life.

Alternatives to dissection are important because students are taught that millions of species of life forms are being destroyed by habitat destruction and other human activities. Use of alternatives to dissection also satisfies concerns for animal welfare. Most important, current ethical issues including AIDS, the greenhouse effect, *in vitro* fertilization, euthanasia, acid rain, abortion, and drugs have a biological basis. Because biological problems are problems of living systems, it is important that students have an opportunity to study live specimens.

Other alternatives to dissection, in addition to the activities provided in this lab manual, include videos, films, transparencies, models, charts, posters, and computer simulations of dissections. Students can be taken on field trips to observe animals at zoos, aquaria, and wildlife parks or refuges.

When you decide to teach labs that are alternatives to dissection, you are taking on the responsibility of having animals in your classroom. You will need to provide their housing and feeding. You should make sure the animals are free of transmittable diseases and are obtained from a source where they have been maintained in good health. You should review the custodial care of your animals and supervise their care and handling by students. Plan who will be responsible for animal care on weekends, holidays and vacations, and who will be economically responsible for the animals' needs, such as food and veterinary care. Students could be assigned library research projects concerning the care of any of the species you plan to introduce to the classroom. The authors of **Biology: The Dynamics of Life** wish you success in achieving your goals for your students this year.

4. Never allow students to use a scalpel or other cutting device with more than one sharp edge. When dissecting specimens, be sure students use dissecting pans to support their specimens. Hand-holding a specimen is "asking for" cuts.
5. If your microscopes require a separate light source, be sure students use proper lamps. Using reflected sunlight can damage the eye.
6. Use extreme caution if you use a pressure cooker for sterilization purposes. Turn off the heat source, remove the cooker, and allow the pressure to return to normal before opening the cover.
7. Students should never point the open end of a heated test tube toward anyone.
8. Remove broken or chipped glassware from use immediately. Use a whisk broom and dust pan to pick up broken glass. Large wads of wet cotton should be used to pick up small pieces of glass. Also clean up any spills immediately that may occur. Dilute concentrated solutions with water before removing.
9. Be sure all glassware that is to be heated is a heat-treated type that will not shatter. If a gas flame is to be used as a heat source, the glassware should be protected from direct contact with the flame through the use of a wire gauze or asbestos-centered wire gauze.
10. Remind students that heated glassware looks cool several seconds after heating, but can still cause burns for several minutes.
11. Prohibit eating and drinking in the laboratory.

After the Activity
1. Be sure that the laboratory is clean. All work surfaces and equipment should be cleaned thoroughly after use.
2. Be sure students dispose of chemicals and broken glassware properly. Provide a container marked "Broken Glass."

3. Be sure all hot plates and burners are off before leaving the laboratory.

Disposal Guidelines
1. Bacterial and fungal cultures, used plastic petri dishes, cotton plugs, and contaminated growth medium should be autoclaved before disposal. Contaminated or used glassware should be autoclaved before being washed.
2. Be aware of local, state, and federal regulations for disposing of chemicals in municipal sewage systems or sanitary landfills.
3. Before disposing of acids and bases, neutralize them by adding dilute sodium hydroxide to acids and dilute hydrochloric acid to bases until pH paper indicates that they are no longer strongly acidic or basic. They may then be flushed down the drain with large amounts of water. Be aware that neutralization of strong acids and bases generates heat, so use caution by neutralizing slowly.
4. The organic chemicals used in this laboratory manual can be flushed down the drain with large quantities of water.
5. Solid wastes can be disposed of by placing them in a container suitable for disposal in a sanitary landfill. Be sure to follow applicable regulations.
6. Broken glass should be placed in a separate, well-marked container.

Remember, a positive attitude toward safety on the part of students is imperative in operating a safe laboratory. Student attitude often reflects the teacher's attitude. Therefore, it is most important that you, as the teacher, always have a positive attitude toward safety and set good safety examples when conducting demonstrations and experiments.

PREPARATION OF SOLUTIONS

Solutions used in the laboratory manual are listed in order by the number of the activity in which they are used. Preparation procedures, cautions, and amounts to make are also included. You may want to plan several weeks ahead so you will have all the solutions prepared.

Add solvents to the solutes. If a specific order of preparation is needed, it will be noted. Dissolve and mix thoroughly. Never add water directly to concentrated acid. Always add the acid to some of the water to be used and then continue diluting. Because the diluting produces heat, it is advised that you add the acid slowly down a stirring rod as you gently stir.

Unless directed otherwise, use distilled water in preparation of solutions requiring water. Using tap water may give erroneous results in the tests made by chemical solutions.

Mix solutions in a beaker or flask of greater capacity than the amount you are making. Usually a container 100 to 300 mL larger works well. It is better to make a little more than the exact amount needed. Students may spill or waste some solution.

Sometimes it is more economical to buy chemicals in large quantities. Many chemicals can be stored for several years. However, certain chemicals become extremely hazardous, even explosive, with age. Know the age limitations of stored chemicals and safe means of disposal of *all* chemicals in the laboratory. Flammable, volatile, and explosive chemicals should be stored in special secure areas and cabinets.

Solutions, once prepared, can be stored in large screw-cap or stoppered bottles. Glass is better than plastic as it reacts with fewer chemicals. These containers should be cleaned with a low sudsing detergent, and rinsed well in distilled water before use.

If possible, request your principal or department head to schedule some student laboratory assistants for your use. These should be qualified to assist in lab preparations and instruction. They should be A/B students, mature, responsible, and they should have taken a course in biology. These students should work directly under your supervision and they can be of tremendous value.

Lab	Solution	Preparation	Cautions
2-1	Solutions A and B	For 1 L: 1000 mL water, 10 g sodium hydroxide, 10 g glucose, 1 mL 1% methylene blue. (Dissolve 0.1 g methylene blue powder in 10 mL ethyl alcohol.)	Sodium hydroxide is very caustic. Avoid contact with skin and eyes. Flush with water.
4-1	Crystal violet stain	Add 1.5 g crystal violet to 10 mL denatured ethyl alcohol. Stir, allowing dye to dissolve. Add 90 mL distilled water.	Flammable. Eye irritant. Do not ingest. If body contact occurs, flush with water.
6-2	Liquid detergent solutions	1%: Add 1 part dishwashing liquid to 99 parts water. 10%: Add 10 parts dishwashing liquid to 90 parts water.	
7-1	Soluble starch solution	Place 4 g corn starch in 10 mL water. Stir to form a paste. Add to 1 L boiling water. Stir for 2 minutes and allow to cool.	
	Iodine solution	Dissolve 10 g potassium iodide and 3 g iodine in 1 L water. Store in brown bottle. Dispense in dropper bottles.	Iodine vapor is toxic or extremely irritating. Dust is hazardous when inhaled or touched. Causes burns, flush with water.
	2% gelatin solution	Add 2 g gelatin to 98 mL of boiling water. Stir for about 2 minutes and allow to cool.	
	Benedict's solution	Dissolve 200 g Na_2CO_3 (or 100 g anhydrous) and 173 g sodium or potassium citrate in 800 mL of water. Warm slightly to speed solution and filter. Dissolve 17.3 g copper sulfate ($CuSO_4$) in 100 mL of water and slowly pour into the first solution. Stir constantly, let cool, and add distilled water to make 1 L.	Avoid contact with skin and eyes. Flush with water. Caustic solution.
	Glucose solution	Place 2 to 3 spoonfuls of glucose in 500 mL of water. Exact amount is not important.	
	Biuret reagent	Add 160 g sodium hydroxide to 400 mL water. Dissolve slowly. Add 80 mL 1% copper sulfate to solution.	Sodium hydroxide is caustic. Avoid contact with eyes or skin. Flush with water if contact occurs.

Lab	Solution	Preparation	Caution
7-2	Benedict's solution	See Lab 7-1.	
	0.4% starch solution	Add 4 g cornstarch to 10 mL cold water to form a paste. Add to 1 L boiling water. Stir for 2 minutes; cool.	
	Iodine solution	See Lab 7-1.	See Lab 7-1.
	0.1% - 0.2% diastase solution	For 0.2% diastase solution, add 0.2 g to 99.8 mL water.	
8-2	Iodine stain	Dissolve 3 g potassium iodide and 0.6 g iodine in 2 L water. Store in a brown bottle. Fill dropping bottles.	Iodine is poisonous and can burn the skin.
9-1	6% salt solution	Dissolve 6 g sodium chloride (table salt) in 94 mL water.	
10-1	Chromatography solvent	Mix 135 mL petroleum ether, 15 mL ethyl alcohol, and 15 mL acetone. Store in a tightly closed bottle.	Use goggles and wear an apron. Solvents are highly flammable. Work under a hood or in a well ventilated room and away from flames.
10-2	Bromothymol blue solution	*Stock solution:* Mix 0.5 g bromothymol blue powder with 500 mL distilled water. Dilute 10 mL stock solution with 500 mL distilled water.	
	Molasses solutions	10%: Mix 100 mL molasses with 900 mL water. 20%: Mix 200 mL molasses with 800 mL water. 50%: Mix 500 mL molasses with 500 mL water.	
11-2	Feulgen stain	Purchase from a biological supply house.	Do not inhale. Flush with water.
	Methanol-acetic acid fixative	Add 1 part glacial acetic acid to 3 parts 100% methanol.	Prepare under hood, wearing goggles and rubber gloves. Poisonous, corrosive. Flush with water.
	45 % acetic acid	Add 45 mL acetic acid to 55 mL water.	Poisonous, corrosive irritant. Vapor may be harmful. Avoid spilling. Flush with water if contact occurs.
	3% hydrochloric acid	Carefully pour 3 mL concentrated hydrochloric acid into 97 mL distilled water.	Corrosive irritant. Flush with water if contact occurs. Work under a hood wearing gloves, goggles, and apron.
13-1	0.7% saline solution	Dissolve 0.7 g of sodium chloride in 93 mL of distilled water.	
	18% hydrochloric acid solution	Carefully pour 18 mL concentrated hydrochloric acid into 82 mL of distilled water.	See Lab 11-2.
	Aceto-orcein stain	The stain can be purchased from a biological supply house or prepared as follows: Heat 45 mL of acetic acid. (The acid acts as a fixative.) When the acid is hot, add 2 grams of orcein. Allow the solution to cool and dilute with 55 mL of distilled water.	Poisonous, corrosive irritant. Vapor may be harmful. Flush with water if contact occurs.
22-1	Iodine stain	See Lab 8-2.	Iodine is poisonous and can burn the skin.
22-2	Congo red	Add 3 g Congo red dye to 100 mL distilled water.	May cause severe allergic reaction.
24-1	Iodine stain	See Lab 8-2.	Iodine is poisonous and can burn the skin.

Lab	Solution	Preparation	Cautions
28-1	Congo red	See Lab 22-2.	May cause severe allergic reaction.
29-1	Acetic acid solution	For 2% solution, add 2 mL acetic acid to 98 mL water.	Poisonous, corrosive irritant. Vapor may be harmful. Avoid spilling or body contact. Flush with water if contact occurs.
38-3	Iodine solution	See Lab 7-1.	See Lab 7-1.
40-1	1% Acetic acid solution	Add 10 mL acetic acid to 990 mL water.	
	0.1% Quinine sulfate	Add 1 g quinine sulfate to 1000 mL water.	
	Sodium chloride	Add 100 g NaCl to 900 mL water.	
	Sucrose solution	Add 100 g sucrose (table sugar) to 2 L water.	
42-1	Disinfectant solution	Add 250 mL bleach to 4 L water.	Avoid contact with skin and eyes. Flush with water.
	Lactose agar	Add 2 g non-nutrient agar and 5 g skim milk powder to 100 mL water. Dispense into tubes and autoclave. Refrigerate.	
	Sterile nutrient agar	For 10 plates: Add 5 g peptone, 3 g beef extract, and 15 g agar to 1 L warm water in a flask. Stir constantly and bring to a boil. After liquid appears clear, pour agar into individual tubes. Plug with cotton or autoclavable caps. Autoclave for 15 minutes. Refrigerate tubes until ready to use. To use, heat tubes in a hot water bath until agar is melted. Pour into sterile petri dishes while it is still liquid. Invert when hard and refrigerate. Agar may also be stored in bottles. When needed, heat bottles in a pan of water until agar has melted. Agar may be kept on shelf for several years without contamination.	
	Tincture of iodine	Purchase from pharmacy.	
43-1	Phenolphthalein solution	Add 1 g phenolphthalein to 50 mL denatured ethyl alcohol. Add 50 mL distilled water. Put in dropping bottles.	Flammable. Eye irritant. Do not ingest. If body contact occurs, flush with water.
	Sodium hydroxide (0.4%)	Add 0.4 g to 99.6 mL water.	Sodium hydroxide is caustic. Avoid contact with eyes or skin. Flush with water if contact occurs.
	Solution A	Add 240 g $MnSO_4 \cdot H_2O$ to 500 mL distilled water. Keeps indefinitely.	
	Solution B	Add 350 g KOH to 500 mL distilled water, then add 75 g KI.	Extremely caustic and corrosive. Do not ingest. Flush spillage or body contact with water immediately.
	Solution D	Mix 2 g starch with 10 mL water to form a paste. Add to 100 mL boiling water. Boil for 2 minutes while stirring. Keeps 2–3 days.	
	Solution E	Add 3.1 g sodium thiosulfate to 1 L water. Keeps 1–2 weeks.	

MICROCHEMISTRY

MICROCHEMISTRY TECHNIQUES

Using chemicals for teaching chemistry or biochemistry at the high school level is presently facing several problems at once. Concern for student safety, environmental questions, cost of materials, and the necessity of adhering to a prescribed curriculum have all worked to make the laboratory program difficult to carry out. Still, the "lab" is the most tangible, best remembered, and most visible aspect of science instruction.

When you have limited time and a limited budget, microchemistry provides a safe, inexpensive, and time-efficient means to conduct a biochemical laboratory activity.

Laboratory activities and experiments that use micro-amounts of chemicals provide a way to involve students in observation and manipulation of some substances that might otherwise be regarded as hazardous. Because drastically reduced amounts of chemicals are used, safety is greatly increased. Likewise, the expense of running a laboratory program is cut. As with traditional laboratories using chemicals, all processes should be conducted with safety goggles and protective clothing.

WASTE DISPOSAL IN MICROCHEMISTRY

Collect all chemical solutions, precipitates, and rinse solutions in a polyethylene dishpan or similar container devoted to that purpose. Retain the solutions until the end of the period or day. In the fume hood, set up a hot plate with a 1-L or 2-L beaker. Pour the collected solutions into the beaker. Turn the hot plate on low and allow the beaker to heat with the hood running and the hood door closed. The liquids and volatiles in the mixture will evaporate, leaving dried chemicals. Allow the beaker to cool. Continue to add solutions and waste until the beaker is 2/3 full. Treat the waste as heavy metal waste. *Dispose of the beaker and its dry contents in an approved manner. Review the Disposal Guidelines on page T11.*

MICROCHEMISTRY

Microchemistry uses smaller amounts of chemicals than do other chemistry methods. The hazards of glass have been minimized by the use of plastic labware. If a chemical reaction must be heated, hot water will provide the needed heat. Open flames or burners are seldom used in microchemistry techniques. By using microchemistry, you will be able to do more experiments and have a safer environment in which to work.

Microchemistry uses two basic tools, the microplate and the plastic pipette.

The Microplate

The first is a sturdy plastic tray called a microplate. The tray has shallow wells arranged in rows (running across) and columns (running up and down). These wells are used instead of test tubes, flasks, and beakers. Some microplates have 96 wells, and other microplates have 24 larger wells.

The Plastic Pipette

Microchemistry uses a pipette made of a form of plastic that is soft and very flexible. The most useful property of the pipette is the fact that the stem can be stretched without heating into a thin tube. If the stem is stretched and then cut with scissors (Figure 1.), the small tip will deliver a tiny drop of reagent. You may also use a pipette called a microtip pipette which has been pre-stretched at the factory. It is not necessary to stretch a microtip pipette.

The pipette can be used over and over again simply by rinsing the stem and bulb between reagents. The plastic inside the pipette is non-wetting and does not hold water or solutions the way glass does.

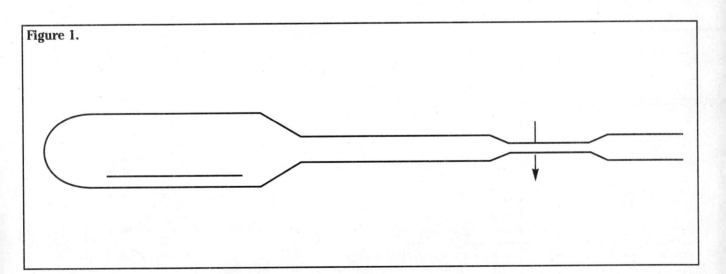

Figure 1.

COURSE MATERIALS LIST

Equipment

anesthetic wands: 8
antibiotic disks: 100
aquarium: 2
aquarium fish nets: 8
autoclave or pressure cooker: 1
baby food jars: 30
balances: 15
beakers—50 mL: 15, 100 mL: 15, 150 mL: 12, 250 mL: 12, 600 mL: 12, 1 liter: 2
brachiopod molds, plastic: 10 sets
Bunsen burners: 15
cages: 5
camel's hair brushes: 8
capillary tubes: 1 box
celluloid, colored: 8 pieces
chart of *Tenebrio* life cycle: 1
coverslips: 10 boxes
culture dishes
    20-cm diameter: 8, medium: 50, small: 50
culture vials: 30
    50 mL with stoppers: 30
dissecting pans: 30
dissecting pins: 250
dissecting probes: 30
droppers: 120
feeding containers with lids, plastic: 15
fermentation tubes, graduated: 32
field guide to birds: 10-15
flashlights: 10
flasks, Erlenmeyer, 250 mL: 30 with stoppers
flowerpots, 10 cm: 30
foam plugs: 32
forceps: 30
funnels: 30
glass jars with lids: 60
glass plates: 30
glass stirring rods: 30
graduated cylinders, 10 mL: 18, 25 mL: 15, 50 mL: 15, 100 mL: 32, 500 mL: 8
hand hole punches: 8
hand lenses: 30
hot plates: 15
incandescent lamps: 8, 25-watt bulbs: 8, 100-watt bulbs: 8
incubator: 1
laboratory aprons: 30
microplates: 30
microscopes: 30
microscope slides: 4 boxes
mirrors: 30
microscope slides, depression: 30
mortar and pestle: 15
nets, insect: 3
pails, 5 L: 3
pans, 30 cm × 50 cm × 20 cm: 8
petri dishes: 100
pins, #2 insect: 100
pins, straight: 1 box
pipettes, Pasteur 22-cm: 8
pipettes, plastic: 45
planting flats: 6
pruning shears: 8
razor blades: single-edged: 30
rubber surgical tubing, 5-mm bore: 80 cm
rulers, metric: 30
safety goggles: 30
scalpels: 30
scissors: 30
scoops, small: 8
specimen jars with lids: 30
spoons, measuring plastic, 1/2 teaspoon: 10
spot plates, plastic: 10
stereomicroscopes: 30
strikers: 15
stopwatches: 15
terrarium: 2
test-tube brushes: 30
test-tube holders: 30
test-tube racks: 30
test-tube stoppers: 1 box
test tubes, 100 × 12 cm: 200, 150 × 19 cm: 200
thermal mitts: 8
thermometer tubes, plastic, with caps: 30
thermometers, Celsius: 30
vials, brown: 30
vials, plastic: 30
vials, 50 mL with stoppers: 24
watchglasses: 30
watches with second hand: 30
water baths: 8

Expendables

absorbent cotton: 1 roll; 60 5-cm squares
cardboard boxes, small: 48
cardboard display boxes: 8
cardboard from a box: 15 pieces
cardboard, corrugated: 30 sheets
cards, white index: 1 package
celluloid, colored: 1 piece
cheese cloth: 2 m
cotton balls, sterile: 1 box
duct tape: 1 roll
fabric: 8 pieces, 80 cm × 80 cm, brightly colored floral pattern
facial tissue: 2 boxes
film canisters, plastic: 30
filter paper: 3 packages
glue, white: 30 bottles
graph paper: 1 ream
labels, white adhesive: 100
lens paper: 4 packages
magazine pages: 30
marking pens, permanent: 30
masking tape: 10 rolls
newspaper: 200 sheets
paper bags: 10
paper cups: 200
paper disks, sterile: 200
paper towels: 15 rolls
paper, construction: 1 package each of 10 colors and black, brown wrapping: 8 pieces, tissue: 2 packages, tracing: 1 package, white: 3 reams
pencils, colored: 30 sets, lead with erasers: 30
petroleum jelly: 8 jars
polystyrene foam pieces: 16 (25 cm × 3 cm)
poster board, large: 8
receptacles, garbage: 4
rubber cement: 30 bottles
rubber bands: 1 box
sand, clean: 1 small bag, coarse: 1 bag
sandpaper, medium: 8 sheets

sterile cotton swabs: 3 boxes
straws, plastic: 30
tape, cellophane: 30 rolls
toothpicks: 3 boxes
thread, black: 1 spool, white: 1 spool
vermiculite: 1 bag
wax marking pencils: 30
wood chips, bedding material: 3 bags

Chemical Supplies for Solutions

acetic acid: 400 mL
acetone: 1 bottle
aceto-orcein stain: 400 mL
agar: 56 g
alcohol, ethyl: 1 bottle
alcohol, isopropyl: 1 bottle
anesthetic: 1 bottle
beef extract: 1 tube
bleach: 1 bottle
bromothymol blue: 1 bottle
calcium chloride: 0.5 kg
Congo red dye: 30 mL
copper sulfate: 25 g
corn starch: 1 box
crystal violet: 10 g
diastase: 1 bottle
ethanol, 95%: 1 bottle
fertilizer: 1 bottle or box
fingernail polish remover (ethyl acetate): 1 bottle
ethyl alcohol, denatured: 1 bottle
food coloring: 1 box
gelatin: 1 box
glucose: 100 g
glycerin: 200 mL
hydrogen peroxide: 1 bottle
hydrochloric acid: 400 mL
liquid detergent: 1 bottle
manganous sulfate: 240 g
methanol-acetic acid fixative: 160 mL
methyl cellulose: 10 g
molasses: 1 bottle
moth balls: 1 box
peptone: 5 g
petroleum ether: 1 bottle
phenolphthalein: 5 g
potassium hydroxide: 100 g
potassium iodide: 5 g
quinine sulfate: 100 g
salt, coarse: 1 box
soda lime: 100 g
sodium bicarbonate: 60 g
sodium chloride: 700 g
sodium carbonate: 160 g
sodium citrate: 200 g
sodium hydroxide: 450 g
sodium thiosulfate: 100 g
sucrose: 200 g
sulfuric acid: 1 bottle
vegetable oil: 1 bottle
vinegar: 1 bottle
water, aquarium: 20 L
water distilled: 5 bottles
water samples, 2 different sources: 5 L each

Biological Supplies

beans, red: 1 lb, black: 1 lb.
bones:

leg bones, duck: 8; leg bones, turkey: 8; rib bones, beef: 16; rib bones, pork: 16; wing bones, duck: 8; wing bones, turkey: 8
bread: 15 slices
chick eggs, fertilized
    at 0 hours development: 15
    at 24 hours development: 15
    at 48 hours development: 15
    at 72 hours development: 15
Cladonia (reindeer moss): 30
coffee, instant: 2 g
Cultures: bread mold: 30; *Daphnia*: 15; *Escherichia coli*: 8; *Escherichia coli* nutrient broth: 15; fruit flies, ebony-bodied, long-winged: 15; vestigial winged: 15; fungi: 10; *Hydra* culture: 15; *Paramecium*: 15, *Planaria*: 15; rotifers: 10; vinegar eels: 10
dicot branches with leaves: 8
Drosophila medium, instant: 1 bag
earthworms, live: 30
Elodea: several sprigs
fern plants, mature: 8
fish food: 1 box
grass
guinea pigs, live: 5
guinea pig pellets: 1 case
human skeleton or pictures of the skeleton: 15
insect collections from Exploration 1-1
juice, citrus samples
lettuce, large-leaf: 1 head
lichen samples, three types: 30 each
living specimens of: crayfish, goldfish, lizards, grasshoppers, flies, *Closterium*, *Oedogonium*, *Scenedesmus*, *Synedra*, *Ulothrix*, *Volvox*: 30 each
mealworms: (eggs, larva, pupa, adult): 8 samples of each
mealworm pupae (of the same age): 30
milk, whole: 2 gallons
onions: 40
parsnip root: 30 cross sections
plant extracts
potatoes: 8
potato flacks, dehydrated: 1 box
prepared slides, adrenal gland: 30; compact bones: 30; gills: 30; starfish: 8; lily anthers: 30; pedicellaria, starfish: 8; thyroid gland: 30
preserved specimens-crayfish: 30; earthworms: 8; frogs: 8; grasshoppers: 30; insect legs: 30; spiders: 30; starfish: 8; sand dollar, sea urchin, brittle star, jellyfish, butterfly or moth, sea anemone, squid: 10
prothalli, mature: 30
raisins: 1 box
seeds: barley: 300, each of 5 different kinds; bean: 300; corn: 60; radish: 900
snails, live: 60
spinach leaves, dried: 1 package
squid, fresh or thawed: 8
starfish, live: 8
sterile nutrient agar: 15 tubes
yeast: 1 package

MATERIALS LIST PER LAB

Lab	Equipment	Expendables	Chemical Supplies	Biological Supplies
1-1	#2 insect pins (160-240) insect nets (8) (optional) 10X hand lenses (8) small vials (50-mL) with stoppers (16-24) glass jars (1-L) with lids (16) scissors (8)	white glue (8 bottles) white paper (8 pieces) white index cards (unlined) (16) facial tissues (32) masking tape (8 rolls) cardboard display boxes (8) plastic film canisters (8) polystyrene foam pieces (25 cm × 25 cm × 3 cm) (16) wax marking pencils (8) small cardboard boxes (48)	ethyl acetate (1 bottle) moth balls, crystals (1 box) 95% ethanol (1 bottle) fingernail polish remover (ethyl acetate) (1 bottle)	
2-1	Erlenmeyer flasks (30) clocks or watches with second hand (15) stoppers, to fit flasks (30) beakers (15) laboratory aprons (30) goggles (30)			
2-2	balances (several) 10-cm flowerpots (60) beakers (60) metric rulers (15) 50-mL graduated cylinders (15)	vermiculite (1 bag) masking tape (15 rolls) colored pencils (15 each of 4 colors) newspapers wax marking pens (15)	water fertilizer solutions: full-strength, double strength, half-strength (15L each)	bean seeds (180)
3-1	balances (several) specimen jars with lids (30) beakers (30) scoops (10) metric rulers (10) pins (30)	20-cm cloth squares (30) masking tape (10 rolls)	water soil samples (30)	
3-2	green and red colored pencils (30 of each)			
4-1	droppers (30) microscope slides (90) coverslips (60) Bunsen burners (several) strikers (several) laboratory aprons (30) 100-mL graduated cylinders (10) microscopes (30) small beakers or paper cups (30) funnels (10) 250-mL beakers (10)	spring-type clothes pins (30) filter paper (10 pieces)	water (1 L) pond water (1 L) bean water (tap water in which beans have been soaked for several days) (10 mL) crystal violet stain (10 mL)	

Lab	Equipment	Expendables	Chemical Supplies	Biological Supplies
5–1	metric rulers (15) scissors (15)	index cards (15) 20 cm × 20 cm pieces of paper (30) uncooked white rice grains (1 bag) uncooked rice grains dyed red with food coloring (1125) graph paper (30 sheets) colored pencils (15 each of 6 colors)		
6-1	metric rulers (30)	graph paper (30 sheets) colored pencils (30 each red, green, blue) garbage receptacles (2) bags of selected garbage (8) newspaper (30-40 large sheets) electric bills (30)		
6-2	scissors (10) petri dishes (30) 50-mL graduated cylinders (10) metric rulers (10)	masking tape (10 rolls) wax marking pencils (10) toothpicks (10) paper towels (10) colored pencils (30) graph paper (30 sheets)	distilled water (200 mL) 1% liquid dishwashing solution (200 mL) 10% liquid dishwashing solution (200 mL)	radish seeds (900)
6-3	planting flats (6) metric rulers (6) laboratory aprons (30)	masking tape (6 rolls) plastic ties (6) potting soil (1 bag) colored pencils (30) graph paper (60 pieces) plastic bags (transparent) (6)		five kinds of irradiated barley seeds (6 packets of each kind of seed)
7-1	droppers (36) test tubes (88) test-tube racks (8) test-tube holders (8) test-tube stoppers (16) test-tube brushes (8) hot plates (8) water baths (8) laboratory aprons (30) safety goggles (30)	brown paper (8 pieces) wax marking pencils (8)	vegetable oil (1 bottle) biuret reagent glucose solution Benedict's solution soluble starch solution iodine solution 2% gelatin solution 95% ethanol (50 mL)	
7-2	droppers (24) test tubes (42) test-tube racks (6) test-tube holders (6) hot plates (6) water baths (6) mortars and pestles (6) 100-mL beakers (6) 50-mL beakers (6) stirring rods (12) scoops (6) plastic spot plates (6) 10-mL graduated cylinders (18) laboratory aprons (30) safety goggles (30) clock	clean sand (1 small bag) white paper (6 pieces)	Benedict's solution (100 mL) 0.4% starch solution (100 mL) iodine solution (25 mL) distilled water (300 mL) 0.1-0.2% diastase solution (20 mL)	germinating barley seeds, 5 days old (150)

Lab	Equipment	Expendables	Chemical Supplies	Biological Supplies
8–1	microscopes (30) lamps (if needed) (30) microscope slides (30) coverslips (30) forceps (30) droppers (30)	lens paper (1 book)	water	preserved insect legs (30)
8-2	microscopes (30) microscope slides (30) coverslips (30) scissors (30) laboratory aprons (30) single-edged razor blades (30) droppers (30) forceps (30)	absorbent cotton (1 ball) black thread (60 cm) white thread (60 cm) lens paper (1 package) magazine pages (30) tissue paper (30 small pieces)	iodine solution (10 mL) pond water (10 mL)	potato, peeled (2-3)
9-1	microscopes (30) microscope slides (30) coverslips (30) droppers (30) forceps (15)	paper towels (1 roll)	water 6% salt solution (100 mL)	*Elodea* (several sprigs)
10-1	200-mL brown bottle mortars and pestles (30) thin layer chromatography slides (30) funnels (30) small dark-colored bottles or vials (30) capillary tubes (30) metric rulers (30) laboratory aprons (30) safety goggles (30)	baby food jars with lids (30) cheesecloth (2 yards) pencils (30)	petroleum ether (135 mL) acetone (15 mL) ethyl alcohol (75 mL)	spinach leaves, dried (1 package)
10-2	graduated fermentation tubes (32) laboratory aprons (30) stopwatches (8)	colored pencils (32) graph paper (30 sheets) wax marking pencils (8)	molasses (1 bottle) bromothymol blue solution (40 mL) distilled water	dry yeast (1 package)
11-1	scissors (30) balances (several) small scoops (several)	photocopies of 3 cell models (15) white glue (15 bottles) coarse sand (1 bag)		
11-2	150-mL glass jars (32) metric rulers (8) scalpels (8) slides (32) coverslips (32) forceps (8) microscopes (8) 25-mL graduated cylinders (16) test tubes (64) test-tube holders (8) test-tube racks (8) thermometers (8) hot plates (8) water baths (8) laboratory aprons (30) goggles (30)	toothpicks (2 boxes) wax marking pencils (8) paper towels (1 roll)	Feulgen stain methanol-acetic acid fixative (160 mL) 3% hydrochloric acid (160 mL) 45% acetic acid in dropper bottles (8) coffee solutions (See page T51.) distilled water	onion bulbs (32)

Lab	Equipment	Expendables	Chemical Supplies	Biological Supplies
12–1	microscopes (30)	drawing paper (30 pieces) pencils, colored (30)		prepared slides of lily anthers (30)
13-1	microscope slides (30) coverslips (30) dissecting probes (30) forceps, fine-nosed (30) dissecting microscope or hand lenses (30) compound microscopes (30) metric rulers (30) laboratory aprons (30) goggles (30)	paper towels (1 roll)	0.7% saline solution 18% hydrochloric acid solution 45% acetic acid solution aceto-orcein stain	cultures of fruit fly larvae
14-1	culture vials with medium and foam plugs (16) anesthetic wands (8) camel-hair brushes (8) hand lenses or stereo microscopes (8)	white index cards (8) wax marking pencils (8)	alcohol (8 vials) anesthetic (8 vials)	cultures of vestigial-winged, normal-bodied fruit flies (8) cultures of long-winged, ebony-bodied fruit flies (8) instant *Drosophila* medium (1 bag)
15-1	mirrors (30)	PTC test paper (30 strips) untreated taste paper (30 strips)		
15-2	scissors (30)	rubber cement (30 bottles) photocopies of metaphase chromosomes from six insects (30 sets of 2 pages)		
16-1	scissors (8) meter sticks (8) plastic trays or rectangular lids, small (32)	colored pencils (32) large poster boards (8) tape (1 roll) wax marking pencils (8) markers (8) photocopies of 6 DNA strands (32 sets of 1 page)		
17-1	plastic sheets of brachiopod molds (10 sets) metric rulers (30)	colored pencils (60) graph paper (30 sheets)		
18-1	hand hole punches (8)	colored paper (purple, brown, blue, green, tan, black, orange, red, yellow, white) (8 sheets of each color) plastic film canisters or petri dishes (80) pieces of brightly colored, floral fabric (80 cm × 80 cm) (8) graph paper (60 sheets)		

Lab	Equipment	Expendables	Chemical Supplies	Biological Supplies
18-2	plastic feeding containers (20 cm in width) with perforated lids (15) scissors (15) stereomicroscopes or hand lenses (15)	graph paper (15 sheets) wax marking pencils (15)		land snails (15) plant extracts large leaf lettuce or Chinese cabbage (1 head)
19-1	plastic food containers, tall (40) blindfolds (8)	wax marking pencils (8)		beans: red (600), black (200)
21-1	scissors (15)	glue (15 bottles) photocopies of viruses and virus parts (15)		
21-2	hot plates (8) water baths (8) thermometers (8) test tubes (40) test-tube holders (8) test-tube rack (8) sterile pipettes (8) incubator (at 37°C) 250-mL beakers (8) 25-mL graduated cylinders (8) laboratory aprons (30) goggles (30)	sterile cotton swabs (10) wax marking pencils (8) paper towels (1 roll) tape, masking (1 roll)	alcohol (800 mL)	broth cultures of *Escherichia coli* (8) sterile petri dishes containing hardened bacto-methylene blue agar (10) milk, whole refrigerated (3.8 L)
22-1	microscopes (30) microscope slides (180) coverslips (180) droppers (6) laboratory aprons (30)	paper towels (1 roll)	iodine solution	living specimens of: *Closterium* *Oedogonium* *Scenedesmus* *Synedra* *Ulothrix* *Volvox*
22-2	small beakers (5) microscope slides (30) coverslips (30) droppers (45) clock or stopwatches (15) microscopes (30)	wood splints (5) toothpicks (30)	cream or whole milk (300 mL) Congo red indicator petroleum jelly (1 jar) methyl cellulose	culture of *Paramecium*
23-1	microscopes (30) microscope slides (120)	cellophane tape (1 roll)		cultures of fungi (4)
23-2	hand lenses or stereomicroscopes (30) petri dishes (15) small jars with covers (90)	labels or wax marking pencils (15) cardboard from a box (15 pieces) cotton swabs (90) dehydrated potato flakes (1 box) raisins (1 package)	water	living bread mold (enough for 15 petri dishes)
23-3	microscopes (30) microscope slides (30) coverslips (30) droppers (30)		water	*Cladonia* (reindeer moss) lichen samples, three types (30 each)

Lab	Equipment	Expendables	Chemical Supplies	Biological Supplies
24–1	single-edged razor blades (30) microscope slides (30) droppers (30) coverslips (30) microscopes (30) laboratory aprons (30)		iodine stain (10 mL)	cross-sections of parsnip root (30)
24-2	prepared single-leaf slides of various species of gymnosperm, such as: hemlock leaf, Japanese cedar leaf, redwood leaf, yew leaf, juniper leaf, spruce leaf (10 sets) compound microscopes (10) calculators (10) metric rulers, transparent (10) field guide to conifers			
25-1	microscopes (30) microscope slides (120) coverslips (120) droppers (2) scalpels (15)	wax marking pencils (15) toothpicks (30) pencils with erasers (15)	water petroleum jelly (1 jar) glycerin (10 mL)	mature fern plants (1 or 2) fern-spore culture mature prothalli (30)
26-1	stop watches (8) Pasteur pipettes, 22-cm (8) pans, 30 cm × 50 cm × 20 cm (8) pruning shears (8) metric rulers (8) rubber surgical tubing 5 mm bore 10 cm pieces (8) incandescent lamps (8) 25-watt bulbs (8) 100-watt bulbs (8) thermal mitts (8) laboratory aprons	permanent markers (8)	petroleum jelly (8 jars)	dicot branches with leaves (8)
27-1	test tubes (30) metric rulers (10) beakers, small (10) measuring half-teaspoons (10) laboratory aprons (30)	absorbent cotton (1 roll) masking tape (10 rolls) rubber bands (10) marking pens (10)	soda lime (100 grams) water, colored (350 mL)	bean seeds, soaked (50) bean seeds, dry (50)
27-2	bowls or shallow dishes (15) staplers (8) metric rulers (15) straight pins (60)	corrugated cardboard (15 pieces) paper towels (1 roll) plastic bags (15) masking tape (8 rolls)	water	soaked corn seeds (60)

Lab	Equipment	Expendables	Chemical Supplies	Biological Supplies
28–1	aquarium or fish bowl terrarium watchglasses (10) compound microscopes (10) pipettes (10) hand lenses or dissecting microscopes (10) depression slides (10) coverslips (10) glass jars, large (20)		spring, pond, or aquarium water (2–4 L)	museum mount of a bird preserved or dried specimens of a sand dollar, sea urchin, brittle star, jellyfish, butterfly or moth, sea anemone, and squid, live crayfish, goldfish, lizards, nd grasshoppers cultures of *Daphnia*, *Planaria*, *Hydra*, and rotifers grass beef liver living flies lettuce guinea pig pellets fish food yeast boiled in Congo red
29-1	stereomicroscopes (15) depression slides (30) small beakers (15) droppers (45) dissecting probes (15)		dilute acetic acid (25 mL) aquarium water (1 L)	*Hydra* culture *Daphnia* culture
30-1	dissecting pins (80-120) dissecting pans (8) hand lenses or stereo-microscopes (8) single-edged razor blades or scalpels (8) laboratory aprons (30)	colored pencils (32)		earthworms, preserved or freshly anesthetized (8)
30-2	hand lenses or stereo-microscopes (6) glass jars (1 pint) or 250 mL beakers (6) plastic trays with lids, 15 cm × 20 cm × 5 cm (6) metric rulers (30) stopwatches (6) or clock camel-hair brushes (6)	white paper (30 pieces) pencils (6) paper towels (1 roll) cardboard, approx. 15 cm × 10 cm (6 pieces)	moist soil (1 small bag) vinegar or lemon juice (1 bottle)	lettuce (6 leaves) live snails (30)
30-3	scissors (8) dissecting pins (40-80) dissecting pans (8) hand lenses or stereo-microscopes (8) laboratory aprons (30)			squid (fresh or thawed) (8)
31-1	stereomicroscopes (8) chart of *Tenebrio* life cycle plastic vials (32) foam plugs (32) incubator (at 30°C) thermometer	wax marking pencils (8)		samples of mealworms (eggs, larva, pupa, adult) (8 of each) mealworm pupae (of the same age) (32)
31-2	hand lenses or stereo-microscopes (several)	white adhesive labels		insect collections from Exploration 1-1
31-3	dissecting probes (16) dissecting pans (8) hand lenses or stereo-microscopes (8)			preserved spiders (8) preserved crayfish (8) preserved grasshoppers (8)

Lab	Equipment	Expendables	Chemical Supplies	Biological Supplies
32-1	glass containers for starfish (8) glass plates (8) colored celluloid (8 pieces)	drawing paper (8 pieces)	sea water	living starfish (8)
32-2	stereomicroscope scissors (8) dissecting trays (8) dissecting probes (8) laboratory aprons (30)			preserved starfish (8) prepared slide-gills and pedicellarias
32-3				specimens, ,models, or photographs of starfish, sea cucumber, sea urchin or sand dollar, brittle star, sea lily (8 of each)
33-1	scissors (8) dissecting pins (80) dissecting pans (8) stereomicroscopes (8) dissecting probes (8) forceps (8) microscope slides (16) coverslips (8) microscopes (8) laboratory aprons (30)			frogs (preserved) (8)
33-2	petri-dish halves (30) aquarium net glass slides (60) microscopes (30) droppers (30) transparent plastic metric rulers (30)	absorbent cotton (60 5-cm squares)	aquarium water	goldfish in aquarium (30)
34-1	field guides to birds (15)			
34-2	graduated cylinders, 500-mL (8) graduated cylinders, 100-mL (8) dissecting probes (8) balances (several)	pencils (8)		rib bones, beef (16) rib bones, pork (16) leg bones, turkey (8) leg bones, duck (8) wing bones, turkey (8) wing bones, duck (8)
35-2	5-L pails (3) 100-mL graduated cylinders (32) thermometers, 30 cm long (8) stirring rods (8) clock or stopwatches (8)	rubber bands (32) food storage size plastic bags (32) newspapers (160 pages) colored pencils (32) graph paper (30 sheets) masking tape (8 rolls)	hot tap water (45°-55°C) ice cubes (72 cubes)	
36-1	droppers (20) flashlights (10) plastic thermometer tubes with caps (30) test-tube racks (10) rulers (10) stopwatches (10) hand lenses (10) laboratory aprons (30)	black construction paper (10 pieces) duct tape (1 roll) wax marking pencils (10) filter paper (30 pieces)	coarse salt (1 box)	living cultures of *Planaria* vinegar eels (*Turbatrix aceti*) *Daphnia*

Lab	Equipment	Expendables	Chemical Supplies	Biological Supplies
36-2	cages (5) bells, whistles or other sources of sound	wood chips or other bedding material	guinea pig pellets or fresh food	guinea pigs (5)
37-1	microscopes (30)			prepared slides of compact bone (30) human skeleton or pictures of the skeleton
38-1	microscopes (30)			prepared slides of thyroid gland (30) prepared slides of adrenal gland (30)
38-2	food tables (with Calories and nutrients listed) (15)			
38-3	plastic or microtip pipettes (45) microplate, 96-well (15) microplate, 24-well (15) scissors or scalpels (15) laboratory aprons (30) safety goggles (30)	paper towels (1 roll) toothpicks (30) tape (1 roll) plastic straws (30) wax marking pencil (15)	vitamin C solution 1% starch solution distilled water iodine solution citrus juices (including fresh orange juice)	
39-1	stopwatches(15) or clock with second hand visible to all students			
40-1	25-mL graduated cylinders (15) stirring rods (15) permanent marking pens (15)	paper cups (210) plastic wrap (1 roll) cotton swabs (630) paper towels (15) paper bags (15) toothpicks (270) potato, apple, onion (90 pieces 0.5 cm thick)	sweet solution (5.0% sucrose solution) sour solution (1.0% acetic acid solution) bitter solution (0.1% quinine sulfate solution) salty solution (10% sodium chloride solution)	
41-1	incubator (38° C) dissecting probes (15) small dishes (15) scissors (15) forceps (15) stereomicroscopes (15) metric rulers (15)	graph paper (60 sheets) absorbent cotton (1 roll)		fertilized chick eggs at 0 hrs development (15), at 24 hrs development (15), at 48 hrs development (15), and at 72 hrs development (15)
41-2	metric rulers (15)			
42-1	laboratory aprons (30) forceps (8) Bunsen burners (8) test-tubes (48) test-tube racks (8) strikers (8) metric rulers (8)	paper towels (1 roll) sterile cotton swabs (32) wax marking pencils (8) sterile filter paper disks (112) masking tape (8 rolls) antibiotic disks (16)	disinfectant solution isopropyl alcohol (100 mL) household bleach (100 mL) tincture of iodine (100 mL) 3% hydrogen peroxide (100 mL) mouthwash (100 mL) disinfectant (100 mL)	sterile petri dishes containing lactose agar (16) sterile petri dishes containing nutrient agar (16) sour milk (200 mL) *Escherichia coli* cultures (8)

Lab	Equipment	Expendables	Chemical Supplies	Biological Supplies
43-1	small flasks or beakers (30) droppers (105) safety goggles (30) laboratory aprons (30)	masking tape (15 rolls)	water samples from 2 sources (3 L of each) solutions of: phenolphthalein, sodium hydroxide, manganous sulfate, potassium hydroxide— potassium iodine solution, concentrated sulfuric acid, starch, sodium thiosulfate	

SUPPLIERS

Cambosco Scientific Company Inc.
342 Western Avenue
Brighton Station, Boston, MA 02135

Carolina Biological Supply Company
2700 York Road
Burlington, North Carolina 27215

Central Scientific Company
2600 South Kostner Avenue
Chicago, IL 60623

Connecticut Valley Biological Supply Company Inc.
82 Valley Road
Southhampton, MA 01073

Edmund Scientific Company
101 E. Glouchester Pike
Barrington, NJ 08007

Frey Scientific Company
905 Hickory Lane
Mansfield, Ohio 44905

Nasco
901 Janesville Avenue
Fort Atkinson, Wisconsin 53538

Nebraska Scientific
A Division of Cygus Company Inc.
3823 Leavenworth Street
Omaha, Nebraska 68105

Sargent Welch Scientific Company
35 Stern Avenue
Springfield, NJ 07081

Ward's Natural Science Establishment Inc.
5100 West Henrietta Road
Rochester, NY 14692

per models for activities 15–2, 16–1, and 21–1 can be found in this teacher edition. In order for students mplete these activities, you must photocopy or mimeograph copies of the models for your students. that each team will need four copies of page T29 for Exploration 16–1, "DNA Sequencing."

**METAPHASE CHROMOSOMES FOR INVESTIGATION 15–2,
"HOW CAN KARYOTYPE ANALYSIS EXPLAIN GENETIC DISORDERS?"**

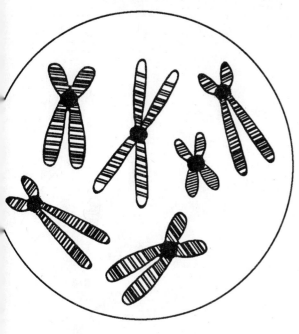

Metaphase chromosomes Insect 1

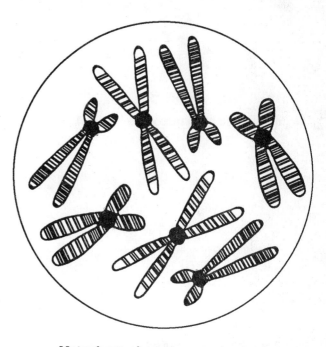

Metaphase chromosomes Insect 2

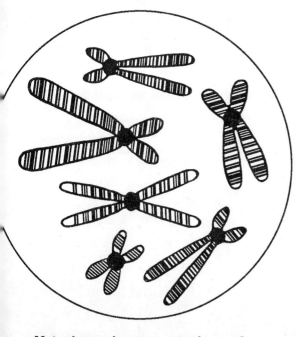

Metaphase chromosomes Insect 3

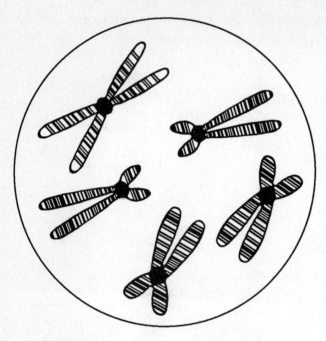

Metaphase chromosomes Insect 4

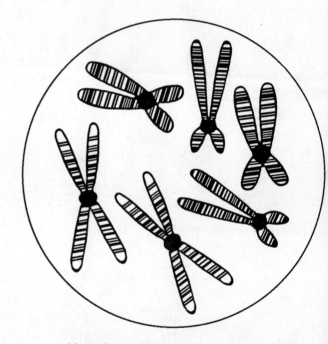

Metaphase chromosomes Insect 5

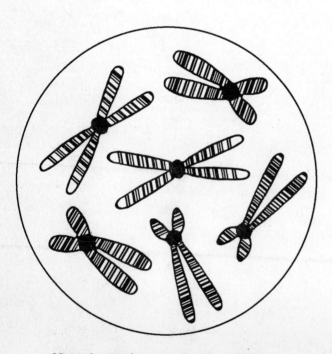

Metaphase chromosomes Insect 6

ACGGACTACCATGGGCCTTA

ACGGACTACCATGGGCCTTA

ACGGACTACCATGGGCCTTA

ACGGACTACCATGGGCCTTA

ACGGACTACCATGGGCCTTA

ACGGACTACCATGGGCCTTA

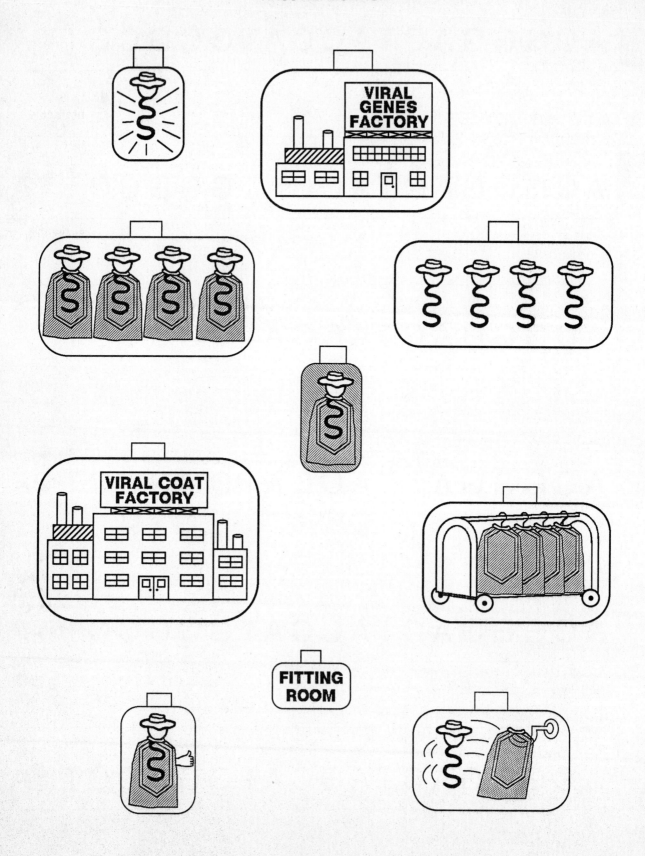

The purpose of the experiment was to test the effect of overcrowding on the growth of plants in a limited area.

materials needed for the experiment include 16 bean seeds, three containers, potting soil, water, metric ruler, and graph paper

1. Fill three containers with equal amounts of potting soil.

2. Plant one bean seed in Container 1, five bean seeds in Container 2, and 10 bean seeds in Container 3. Plant all seeds at the same depth.

3. Water all three containers with equal amounts of water. Place all three containers in a well-lit room.

4. For the next two weeks, water each container once a day. Use an equal amount of water in each container.

5. Measure the average height of the plants in each container every day for the next two weeks. Record these measurements in a table.

6. After two weeks, plot the data in your table on a graph.

The least amount of growth was observed in Container 3. Living space in Container 3 had to be divided among 10 growing plants. As a result, no one plant had enough space to grow well.

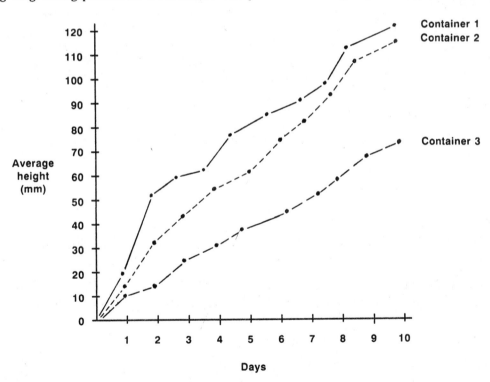

1-1 EXPLORATION

LAB

Preparing an Insect Collection

OBJECTIVES
- Collect and prepare an insect collection.
- Recognize and become familiar with the features used to identify insects.

PROCESS SKILLS: observing, classifying, communicating, recognizing and using spatial relationships

TIME ALLOTMENT: 1 class period

MATERIALS

scissors (8)
white glue (8 bottles)
white paper (8 pieces)
#2 insect pins (160–240)
fingernail polish remover (ethyl acetate) (1 bottle)
white index cards (un-lined) (16)
facial tissue (32)
insect nets (optional) (8)
masking tape (8 rolls)
cardboard display box (8)
plastic film cannisters (8)

moth balls (crystals) (1 box)
95% ethanol (1 bottle)
10X hand lenses (8)
small vials (50 mL) with stoppers (16-24)
clean glass jars (approx. 1-L size) with lids (16)
polystyrene foam pieces (25 cm × 25 cm × 3 cm) (16)
wax marking pencils (8)
small cardboard boxes (approx. 48)

ALTERNATE MATERIALS
- Any sturdy cardboard box can be used for storing insects. School boxes store well. They can be reused year after year.

PREPARATION
- Assemble all materials prior to beginning this Exploration. Students can bring in some of the materials, such as jars and boxes, from home.

TEACHING THE LAB
- Have students work in groups of four.
- Inform students that the jar is reinforced with tape in case of breakage. The tape will bind most pieces of glass, making it easy to clean up.
- Explain to the students that moth crystals are placed in the collection to prevent dermestid beetles from destroying the collection. Without these crystals, dermestid beetles can turn a collection into dust.
- Students can preserve an additional insect in alcohol for use in Exploration 8–1.
- **Troubleshooting:** Beetles are pinned through the thorax like other insects, but because the outer wing covers are on top of the thorax, students may have difficulty knowing where to place the pin.

- **Troubleshooting:** Students who are allergic to insect stings should not attempt to make an insect collection without adequate protection.

DATA AND OBSERVATIONS
1. Answers may include three body regions, segmented bodies, 6 legs, or trachea for breathing.
2. Answers may include chewing (grasshopper), piercing-sucking (true bugs), siphoning (moths), lapping (bee), or sponging (house fly).
3. Answers may include pincers, claspers, egg depositors, springs, or stingers.

ANALYSIS
1. Some wings are thin and transparent, others are not. Some have visible veins, scales, or hairs. Some insects have four wings while others have two wings. Wings of an insect may be of different sizes or shapes.
2. Yes; answers will vary, but may include the observations that some are club-shaped, cigar-shaped, short, long, smooth or hairy.
3. Type of mouthparts, type of appendages or type of antennae could be used to separate insects into different groups.
4. No; the same insect can be found in various habitats and different insects can be found in the same habitat.

Can Scientific Methods Be Used to Solve a Problem?

OBJECTIVES
- Use scientific methods to decide whether two liquids are similar or different.
- Make careful observations.
- Hypothesize whether the two liquids are the same or different.
- Record accurate experimental results.

PROCESS SKILLS: observing, forming hypotheses, experimenting, inferring, communicating, predicting, interpreting data

TIME ALLOTMENT: 1 class period

MATERIALS
Erlenmeyer flasks containing liquids, labeled A and B (15 of each)
clock or watches with second hand (15)
stoppers to fit flasks (30)
beakers (15)
laboratory aprons (30)
goggles (30)

PREPARATION
- Prepare and label flasks for students before class. Flask A is half-filled and stoppered. Flask B is filled to the top with no air space below the stopper. See page T12 for the preparation of the solution. Prepare a fresh solution each day. Add more 1% methylene blue (several drops) to the flasks during the day if color formation is weak. Solution will be blue when first prepared; allow 10 minutes for it to clear.

TEACHING THE LAB
- Have students work in teams of two. One student can shake the flasks, and the other can keep time and record results.
- The questions posed in the Procedure are designed to promote critical thinking in the scientific methods. Answers will vary.
- **Troubleshooting:** Cork stoppers are preferable to rubber; rubber is slippery when wet and the stoppers may pop out.

HYPOTHESIS
The liquids in the flasks are the same.

DATA AND OBSERVATIONS
Table 1.

First Observations	
Similarities	**Differences**
1. Liquid in both	1. More liquid in B
2. Colorless	2. Liquid in B touches stopper.
3. Clear	3. Flask A has space at top.
4. White film on bottom	

Table 2.

Results of Experiment 1	
Similarities	**Differences**
1. Liquid in both	1. Half-filled flask A turns blue when shaken, then becomes colorless again.
	2. Flask B does not change color.

Table 3.

Results of Experiment 2	
Similarities	**Differences**
1. Liquid in both	Intensity of blue color may vary.
2. Volume of liquid the same	
3. Blue color in both after shaking, then blue disappears.	

Table 4.

Three Trials of Experiment 3									
	Time in seconds to return to original condition								
	1 shake			2 shakes			3 shakes		
Trial	1	2	3	1	2	3	1	2	3
Flask A	40	44	40	62	65	59	73	88	82
Flask B	43	50	39	57	64	58	80	91	84

ANALYSIS
1. no
2. no
3. Experiments 2 and 3; they showed that the liquids behaved in the same way.
4. air (some students may say oxygen)
5. No air was in the flask.
6. Air must mix with the liquid in order for the color change to occur.
7. Each added shake mixes more air into the liquid and the blue color stays longer.
8. a. yes
   b. Experimentation provides evidence and data with which a problem can be solved.
   c. using observation, experimentation, interpretation, and hypothesis formation to solve a problem

CHECKING YOUR HYPOTHESIS
Answers may vary. Some students may say that their hypotheses were supported; others may say that more tests would be needed. Students who predicted that the two liquids were the same should say that their hypotheses were supported.

ing SI Units

BJECTIVES
Measure the mass of bean seeds.
Measure, record, and graph the growth of bean seeds.

ROCESS SKILLS: observing, measuring, using numbers, interpreting data, communicating

ME ALLOTMENT: 1 class period, then 5 minutes daily; 30 minutes on days 6 and 9 after germination

ATERIALS

an seeds (180)	beakers (60)
asking tape (15 rolls)	colored pencils (15 each of 4 colors)
lances (several)	metric rulers (15)
-cm flower pots (60)	newspapers
rmiculite (1 bag)	wax marking pencils (15)
ter	50-mL graduated cylinders (15)
rtilizer solutions: full-strength, double-strength, half-strength (15 L each)	

REPARATION
Either clay or plastic pots can be used. Foam drinking cups, with a drainage hole punched in the bottom, can also be used.
Prepare the three commercial plant fertilizer solutions. Refer to the instructions on the product. The double-strength solution should be prepared with half the water specified. The half-strength solution is prepared with twice the water specified.
Practice use of the balance with the students before beginning the lab.
Remind students that to find an average, they add a group of numbers and then divide the sum by the number of numbers added.

EACHING THE LAB
The students may work in pairs.
This lab will be a review of the SI system for most of your students. If some of your students are not familiar with SI or metric, they may need extra help.
Have 1L containers of plain water available for the students to rinse the roots.
Review with students how to multiply and divide by 10's by simply moving decimal points.

- A wide variety of results will probably occur, caused by students not following directions, by differences in light and temperature, and by variations in amounts of water. The different results could lead to a discussion of scientific methods, the need to be consistent in procedures, and the need to make sure there is only one variable in an experiment.
- Review the unit factor method of problem solving, showing students that units will cancel out as they do the calculations. For example,

$$1.5 \, \cancel{g} \times \frac{1 \, kg}{1000 \, \cancel{g}} = 0.0015 \, kg$$

DATA AND OBSERVATIONS
Student results will vary greatly. In general, plants with the recommended fertilizer amounts will yield the greatest growth. It is possible that half the recommended water (double-strength fertilizer) will produce the greatest growth. However, this much fertilizer could also harm the young plants.

ANALYSIS
1. milligram, gram, kilogram; millimeter, centimeter, meter, kilometer
2. **a.** Divide by 1000.
   b. Multiply by 1000.
   c. Multiply by 10.
   d. Divide by 100.
3. 1000
4. Multiply by 1000.
5. to avoid introducing another variable—the amount of liquid
6. Answers will vary. Ideally, Pot A will have least growth because no fertilizer was used, and Pot B will have greatest growth because it had the optimum amount of fertilizer. Pot D should have less growth than Pot B because it used less fertilizer than Pot B. The position of Pot C depends on whether or not the excess fertilizer helped or harmed the plants.
7. Answers will vary. These results may parallel the growth in height since the longer plants would have more mass. However, some shorter plants may have more lateral growth, which could result in greater mass.
8. Answers will vary. See answer to question 6.
9. Answers will vary. See answer to question 6.

3–1 EXPLORATION

Physical Factors of Soil

OBJECTIVES
- Determine the amounts of various particle types in three soil samples.
- Calculate the water contents and water-holding capacities of three soil samples.

PROCESS SKILLS: observing, inferring, communicating, measuring, using numbers, interpreting data, experimenting

TIME ALLOTMENT: 2 class periods

MATERIALS
soil samples (30)	scoops (10)
20-cm cloth squares (30)	water
balances (several)	metric rulers (10)
specimen jars with lids (30)	pins (30)
beakers (30)	masking tape (10 rolls)

PREPARATION
- Have each student bring a soil sample in a plastic bag. Explain to the students that soil water will not be lost if samples are sealed in plastic bags.

TEACHING THE LAB
- Have students work in groups of three, with each student collecting data on one sample.

- If 100-mL graduated cylinders are available, add 50 mL of loose soil to the cylinder. Add 50 mL of water and shake as directed. The amount of various mineral particles can be determined by direct reading.
- Soil samples can be dried in a warm oven for several hours or in an incubator overnight.
- In Parts B and C, students actually measure the water-holding capacity and water content of the soil and the cloth, but the water content and water-holding capacity of the cloth is small and so it can be ignored.
- To keep the pan of the balance dry when massing a wet soil sample, have students place the sample in a cup made from aluminum foil. The mass of the aluminum foil will be negligible compared with the soil and can be ignored.
- **Troubleshooting:** Students may have trouble with the calculations. Go over them in class.

DATA AND OBSERVATIONS
See the tables below.

ANALYSIS
1. a., b. Answers will vary with soil sample used.
2. a. clay and silt
   b. sand and gravel
3. Loosely packed soil allows water to drain through it. Closely packed soil does not drain as well.

Table 1.

Soil Particle Size Data					
	Amount of each particle type (in mm)				
Soil location	**Gravel**	**Coarse sand**	**Fine sand**	**Silt**	**Clay**
1. Oak forest	20	10	8	40	12
2. Garden soil	46	11	21	16	12
3. Cow pasture	20	12	9	35	10

Table 2.

Water Content and Water-holding Capacity					
Soil location	**Mass of soil and cloth**	**Mass of dried soil and cloth**	**Mass of saturated soil and cloth**	**Percentage water content**	**Percentage water-holding capacity**
1. Oak forest	125 g	95 g	250 g	31.6%	163%
2. Garden soil	150 g	105 g	175 g	42.9%	66.7%
3. Cow pasture	135 g	120 g	145 g	12.5%	20.8%

The Lesson of the Kaibab

OBJECTIVES
- Graph data on the Kaibab deer population of Arizona from 1905 to 1939.
- Analyze the methods responsible for the changes in the deer population.
- Propose a management plan for the Kaibab deer population.

PROCESS SKILLS: observing, communicating, interpreting data, formulating models

TIME ALLOTMENT: 1 class period

MATERIALS
green and red colored pencils (30 of each)

TEACHING THE LAB
- This Exploration offers a good opportunity for open-ended discussions since it touches on emotional issues. Have students discuss their answers to the Analysis. Students may raise pros and cons to the various management possibilities.

DATA AND OBSERVATIONS

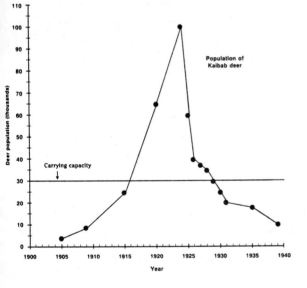

ANALYSIS
1. They stopped hunting and eliminated predators.
2. More than 8724 animals were removed.
3. In 1915, the population was near the carrying capacity.
   In 1920, the population was twice the carrying capacity.
   In 1924, the population was more than three times the carrying capacity.
4. Yes, the deer population was increasing.
5. The deer population was over the carrying capacity of the range. As a result, the deer were starving to death because there was not enough food for all of them on the range.
6. Yes, the carrying capacity decreased. The large number of deer had severely deteriorated the condition of the range.
7. The population may have been at a low point of its natural population cycle. Perhaps the predator population was unusually high at the time.
8. The deer population probably would have increased to a point nearer the natural carrying capacity of the range.
9. Several lessons were learned from the Kaibab study, including the following: (1) predators provide a natural balance in a healthy population, and (2) eliminating hunting and predators from a populations allow the population to grow unchecked, leading to starvation of the population.
10. Answers will vary. Possible answers include the following:
    In 1915, begin allowing hunting, allow predators back into the area.
    In 1923, begin systematic removal of deer to reduce the population, take steps to improve the state of the vegetation, stop all grazing by cattle, sheep, and horses.
    In 1939, with population at a reasonable number, monitor the predator populations, maintain a reasonable hunting program.
11. Answers will vary. Students answers should be similar to answers in question 10.

INVESTIGATION

What Organisms Make Up a Microcommunity?

OBJECTIVES
- Observe and identify the organisms found in the microcommunities in bean water and pond water.
- Determine if each organism is motile or sessile.
- Identify each organism as a producer or a consumer.
- Make a hypothesis about the relationship of an organism as a consumer or producer to its motility.

PROCESS SKILLS: observing, classifying, inferring, communicating, predicting, forming hypotheses

TIME ALLOTMENT: 2 class periods

MATERIALS
water
pond water (1L)
droppers (30)
spring-type clothespins (30)
bean water (tap water in which beans have soaked for several days) (10 mL)
microscope slides (90)
coverslips (60)
Bunsen burners (several)

strikers (several)
laboratory aprons (30)
crystal violet stain (10 mL)
100-mL graduated cylinders (10)
filter paper (10 pieces)
microscopes (30)
small beakers or paper cups (30)
funnels (10)
250-mL beakers

PREPARATION
- Obtain water from a variety of pond environments and if possible from a variety of depths, particularly bottom sludge or sediment. Label all containers as to the source or depth of water.
- If pond water is not available, scrapings from the side or sediment from the bottom of a fish tank may be used.
- Prepare bean water 3 to 4 days in advance of class. Add 10 to 15 bean seeds (any species) to 100 mL of water. Do not cover. Add water as needed.
- See page T12 for preparation of crystal violet stain.

TEACHING THE LAB
- Students should prepare their own slides and make their own observations, but several students can share a Bunsen burner. At least three students can make slides from one filter containing pond water organisms.
- Using the clothespin will prevent students from burning their fingers or staining their hands.

HYPOTHESIS
Most consumers must be motile in order to find food. Producers make their own food and do not need to be motile.

DATA AND OBSERVATIONS
See the table below.

ANALYSIS
1. air, bean seeds, water
2. a. consumers
   b. The organisms were not green; they lived on nutrients from the beans.
3. Producers are shades of green; they contain chlorophyll or other pigments.
4. Yes; motility is an adaptation that aids in food-getting.

CHECKING YOUR HYPOTHESIS
Answers will vary. If students said that most consumers are motile, their data will support their hypotheses.

Table 1.

Organisms in Microcommunities			
Name of Organism	Community (bean water or pond water)	Motile or sessile	Producer or consumer
bacteria	bean water	motile	consumer
Tubifex	pond water	motile	consumer
Fragilaria	pond water	sessile	producer
	Answers will vary, depending on what organisms were present in the microcommunities.		

How Does the Environment Affect an Eagle Population?

OBJECTIVES
- Make a model to show how abiotic and biotic factors affect a bald eagle population.
- Hypothesize how biotic factors affect a bald eagle population.

PROCESS SKILLS: observing, communicating, predicting, interpreting data, forming hypotheses, formulating models

TIME ALLOTMENT: 1 class period

MATERIALS
index cards (15)
uncooked rice grains dyed red with food coloring (1125)
20 cm × 20 cm pieces of paper (30)
uncooked white rice grains (1 bag)

metric rulers (15)
scissors (15)
graph paper (30 sheets)
colored pencils (15 each of 6 colors)

PREPARATION
- The white rice can be dyed red by putting it in a plastic bag and shaking it with several drops of food coloring until the desired shade is obtained. Allow time for the rice to dry, spread out on paper towels. It might be easier for you to have the sheets of paper and squares of index card ready for the students.

TEACHING THE LAB
- Have the students work in pairs.
- You may want to have the students use forceps to pick up the rice.
- The students should drop the "eagle" carefully and count the "fish" accurately if the Investigation is to work. If the "eagle" does not land on the "lake," simply drop it over again.
- Since this is a model and involves random sampling, different students may get different results and may not even get the "desired" results. The data students collect may show the fish population to be increasing when it "should" decrease.
- If you wish to have students do all six factors in Part B, have them make additional tables on a separate piece of paper.

HYPOTHESIS
Students will probably hypothesize that there will be fewer fish to go around and one or more eagles might die.

DATA AND OBSERVATIONS
The students will fill in the tables with the number of rice grains that appear on the grid square. Chances are, due to the size of the rice grains and the size of the grid, there will be no more than 5 or 6 grains per square.

ANALYSIS
1. Answers may vary, but students should note that the fish population decreases over time due to eagle predation.
2. A small-scale decrease in the fish population should not affect the eagle population.
3. Answers may vary, but students should note that competition makes the fish population decrease more quickly. It limits food availability for the eagles.
4. A climate change does not necessarily affect a population. The degree to which a climate change affects a population is often related to the severity of the change or how the change affects needed resources.
5. An increase in the fish population increases food availability but does not necessarily cause an increase in the eagle population.
6. A seasonal change can cause the eagles to leave the territory in search of food.
7. The insecticide has no effect on the fish population. The insecticide prevents the eagle population from increasing by preventing the eggs from hatching.
8. Yes.
9. Pollution can cause algal blooms, which cause a decrease in the fish population and thus limit food availability.
10. An increase in the eagle population makes the fish population decrease more quickly. This is the same effect as competition.

CHECKING YOUR HYPOTHESIS
Answers will vary. Students who hypothesized that there will be fewer fish and that one or more eagles might die, will say their hypotheses were supported.

Recycling Garbage

OBJECTIVES
- Identify the major components of a bag of selected garbage.
- Determine which types of garbage can be recycled.
- Construct a graph that compares a bag of selected garbage with the garbage generated by the average United States household.
- Calculate how many aluminum cans would have to be recycled to pay a typical electric bill.

PROCESS SKILLS: observing, classifying, communicating, using numbers, interpreting data

TIME ALLOTMENT: 1 class period

MATERIALS
individual bags of selected garbage (8)	household electric bill (30 copies)
metric rulers (30)	receptacles for recyclable and non-recyclable materials (2)
graph paper (30 sheets)	
colored pencils (30 each red, blue, green)	
newspapers (30–40 large sheets)	

PREPARATION
- It is best to bring in "selected" garbage from your own home or from the school so that no dangerous items, such as broken glass, are brought to class. Carefully cleaning and preparing the garbage will provide an accident, injury-free activity. Make each bag different so that the pooled class percentages will be different from the individual bag percentages.
- Have two receptacles labeled and ready to receive the garbage.

TEACHING THE LAB
- Have students work in groups of four. If enough bags of garbage are available, they can work in smaller groups. Each student should make a graph and do the electric-bill calculations.
- Instruct the students to dispose of the garbage in the proper containers. Try to recycle as much garbage as possible.
- This Exploration would make a good take-home activity. Have students classify and count their own garbage items, using one bag of garbage the night before it is collected. Data can be pooled and graphs made the following day in class.
- Calculators can be brought to class to speed up the calculations for Table 1.

- **Troubleshooting:** Students may have trouble with the electric bill calculations. You may want to do them in class. See the answer to question 6 in the Analysis.

DATA AND OBSERVATIONS
Table 1 will vary, depending on the components of the bags. The percentages of the class totals will usually be close to the national percentages, however.

ANALYSIS
1. Answers will vary, depending on the components of the individual bag.
2. Answers will vary, but class data should be close to the national percentages.
3. Answers will vary. It should be the total of the percentages of paper, glass, metal and plastic.
4. Answers will vary. It should be equal to the total of the percentages of food, rubber, wood and other.
5. The advantages should include the availability of the plastic for use as a new product rather than being thrown away. There would also be the reduction in landfill space needs and the environmental damage due to landfills.
6. Answers will vary depending on the number of kilowatt-hours on the bill. First determine the number of kilowatt-hours saved by recycling each can. Multiply 3.5 hours by 100 watts (the wattage of the light bulb) and divide by 1000 (watts per kilowatt). Thus, 0.35 kilowatt-hours of electricity is saved by recycling each can. By dividing the number of kilowatt-hours in the electric bill by the number of kilowatt-hours saved per can (0.35), the number of recycled cans required to pay the bill can be calculated.

How Does Detergent Affect Seed Germination?

OBJECTIVES
- Hypothesize what effect different concentrations of detergents labeled *phosphate-free* and *biodegradable* will have on seed germination.
- Determine the effect of different concentrations of liquid detergent on seed germination.
- Graph the data collected in the experiment.

PROCESS SKILLS: observing, inferring, communicating, measuring, predicting, using numbers, interpreting data, forming hypotheses, separating and controlling variables, experimenting

TIME ALLOTMENT: 25 minutes on the first day, then 10 minutes a day for five days

MATERIALS
masking tape (10 rolls)
scissors (10)
petri dishes (30)
wax marking pencils (10)
toothpicks (10)
paper towels (10)
colored pencils (30)
radish seeds (900)
distilled water (200 mL)
50-mL graduated cylinders (10)
graph paper (30 sheets)
1% liquid dishwashing detergent solution (200 mL)
10% liquid dishwashing detergent solution (200 mL)
metric rulers (10)

ALTERNATE MATERIALS
- Liquid laundry detergent can be used instead of dishwashing liquid.

PREPARATION
- Prepare a 1% detergent solution by adding 1 part detergent to 99 parts distilled water. Prepare a 10% detergent solution by adding 10 parts detergent to 90 parts distilled water.
- Purchase radish seeds. There are about 500 radish seed in a non-name brand seed packet for the home gardener.
- There is little difference in results among different brands of detergent. Select a liquid dishwashing detergent that is labeled phosphate-free and biodegradable.

TEACHING THE LAB
- Have students work in groups of three. Each student can be responsible for one petri dish.
- Brown school paper towels work well because the germinating seeds show up well against the contrasting background.
- Keep petri dishes in a warm place. Cooler temperatures will cause slower growth.
- Emphasize to students that they should make sure their fingers and graduated cylinders are detergent free when handling seeds, water, and paper toweling for the control dish.
- Begin this Investigation on a Monday so daily observations can be made before the weekend.

HYPOTHESIS
Detergent affects the number of seeds that germinate and causes abnormal growth of seedlings. The higher the detergent concentration, the greater the effect on germination and growth.

DATA AND OBSERVATIONS
Table 1.

Number of Seeds Germinated			
Day	Control	1% detergent solution	10% detergent solution
1	0	0	0
2	8	1	0
3	15	9	0
4	22	11	0
5	24	14	0

Table 2.

Growth of Germinating Seedlings			
Day	Control	1%	10%
	Roots have many root hairs and are 6 to 12 mm long after 2 days.	Roots have fewer root hairs and are shorter than the control seedlings. Length of roots after 2 days will be about 3–6 mm.	Few seeds will germinate. They may crack open and the point of the root may show, but they will probably not germinate normally.

ANALYSIS
1. Answers will vary, but approximately 24, 14 and 0 will germinate in water, 1%, and 10% detergent.
2. Detergent inhibited seed germination.
3. The control provided a basis for comparison for seeds soaked in detergent.
4. Seedlings growing in detergent have fewer root hairs and shorter roots.
5. Crop yield would probably be lower because fewer seeds would germinate to produce plants. Once the seeds germinated, their growth would be inhibited because they would have fewer root hairs and shorter roots.
6. No, because biodegradable simply means that a substance can be broken down by bacteria.

CHECKING YOUR HYPOTHESIS
Answers will vary. Students who predicted that seeds would not germinate in 10% detergent and that germination would be inhibited and growth abnormal in 1% detergent will say that their hypotheses were supported.

How Does Ionizing Radiation Affect Plant Growth?

OBJECTIVES
- Hypothesize how different amounts of ionizing radiation affect germination and growth of barley seeds.
- Observe the effects of different amounts of ionizing radiation on the germination of barley seeds.
- Observe the effects of different amounts of ionizing radiation on the size and structure of barley plants.
- Compare percent germination, size, and structural changes among control and experimental groups.

PROCESS SKILLS: observing, inferring, measuring, predicting, using numbers, interpreting data

TIME ALLOTMENT: 1 class period, then 20 minutes each week for 3 weeks

MATERIALS
planting flats (6)
masking tape (6 rolls)
plastic ties (6)
potting soil,
   moistened (1 bag)
metric rulers (6)
colored pencils (30)
graph paper (60 pieces)
laboratory aprons (30)
plastic bags (transparent,
   to enclose planting flat) (6)
five kinds of barley seeds
   (6 packets of each kind
   of seed)

PREPARATION
- Purchase irradiated seeds from a supply house.
- Divide each type of seed into 6 groups so that each team of students receives five groups of seeds.

TEACHING THE LAB
- Have students work in groups of five.
- Plant seeds on a Monday so that soil can be checked daily for proper moisture during germination.
- To pool the class data write Table 1 on the chalkboard or on a piece of posterboard. Have each group record its data. Calculate averages as a class activity.
- **Troubleshooting:** If mold begins to develop on the soil, have students remove the plastic bag.

HYPOTHESIS
Germination and growth of barley seeds are affected by ionizing radiation. The higher the dose of radiation, the more damage there will be to the seeds and resulting seedlings.

DATA AND OBSERVATIONS
See the table below.

ANALYSIS
1. Nonirradiated seeds serve as a control to which the radiation treatments can be compared.
2. The number and percentages of germinating seeds decreases.
3. As the amount of radiation increases, the average height of the seedlings decreases. At the higher doses of radiation the seedlings develop curled and scorched leaves and sometimes a reduction of chlorophyll.
4. Ionizing radiation can cause the destruction of base pairs and improper pairing. This can result in damaged or non-functional proteins and improper growth and development.
5. Seeds have fewer meristems so radiation damage to a small number of them will have a large impact on the plant.
6. At higher altitudes and near the equator and poles, there is not as much protective atmosphere. Organisms there are exposed to higher levels of ionizing radiation. As the ozone layer degrades, more radiation will reach Earth's surface.
7. The bacteria and fungi that cause decomposition of food can be killed by ionizing radiation. Insects and harmful parasites also can be killed and viruses can be deactivated.

CHECKING YOUR HYPOTHESIS
Students who predicted that the amount of damage increases with increasing radiation dosage will report that their hypotheses were supported.

Table 1.

Effects of Ionizing Radiation											
Radiation dosage (RADs)	Number of seeds at start	Germ. date	Number of seeds germinated			Percent germination			Average height (cm)		
			Week			Week			Week		
			1	2	3	1	2	3	1	2	3
No radiation (control)	99				87			87.9			19.5
20 000	120				86			71.7			17.8
30 000	132				76			57.6			14.2
40 000	110				48			43.6			10.5
50 000	90				13			14.4			3.3

Tests for Organic Compounds

OBJECTIVES
- Determine the presence of starch by a chemical test.
- Analyze a glucose solution for the presence of simple reducing sugars.
- Analyze a sample of vegetable oil for the presence of lipids.
- Analyze a sample of gelatin for the presence of protein.

PROCESS SKILLS: observing, classifying, inferring, communicating, interpreting data, experimenting

TIME ALLOTMENT: 1 class period

MATERIALS
droppers (36)
test tubes (88)
test-tube racks (8)
test-tube holders (8)
test-tube stoppers (16)
test-tube brushes (8)
brown paper (8 pieces)
wax marking pencils (8)
vegetable oil (1 bottle)
biuret reagent

glucose solution
Benedict's solution
hot plates (8)
water baths (8)
soluble starch solution
95% ethanol (50 mL)
2% gelatin solution
iodine solution
laboratory aprons (30)
safety goggles (30)

PREPARATION
- Make up all solutions prior to beginning this Exploration. See page T12 for directions.

- Prepare a tray of solutions, reagents, and equipment for each group of students to avoid congestion and confusion.

TEACHING THE LAB
- Have students work in groups of four, with one student doing the starch test, one doing the sugar test, a third student doing both lipid tests, and the fourth student doing the protein test. All students should be responsible for observing all results and filling in all five tables.
- Contents of all test tubes can be flushed down the drain with large amounts of water.
- If hot plates are in short supply, more than one group can share a water bath. Be sure tubes are labeled with group names.

DATA AND OBSERVATIONS
See the tables below.

ANALYSIS
1. iodine solution
2. The substance turns blue-black in the presence of iodine.
3. Benedict's solution
4. The substance changes color from blue to green, yellow, orange, or red.
5. Water was used as a control for comparison with the other substances.
6. biuret reagent
7. The substance will turn lavender to violet.
8. Skin contains protein.
9. a. Lipids are not soluble in water.
   b. Alcohol or other non-polar solvents will remove a greasy food stain because the lipid will dissolve in these solvents.

Table 1.

Test for Starch				
Test tube	Substance	Color at start	Color after adding iodine	Starch present (+/–)
1	Starch	Cloudy white	Blue-black	+
2	Glucose	Clear	Light brown	–
3	Water	Clear	Light brown	–

Table 2.

Test for Simple Reducing Sugars				
Test tube	Substance	Color at start	Color after adding Benedict's solution	Reducing sugar present (+/–)
1	Starch	Cloudy white	Blue	–
2	Glucose	Clear	Yellow, orange, or red	+
3	Water	Clear	Blue	–

Table 3.

Brown Paper Test for Lipids		
Substance	Translucent on brown paper?	Lipids present (+/–)
Water	No	–
Oil	Yes	+

Table 4.

Solubility Test for Lipids		
Substance	Dissolves?	Lipids present (+/–)
Oil in ethanol	Yes	+
Oil in water	No	+

Table 5.

Test for Proteins				
Test tube	Substance	Color at start	Color after biuret reagent	Protein present (+/–)
1	Gelatin	Clear	Violet	+
2	Glucose	Clear	Blue	–
3	Water	Clear	Blue	–

What is the Action of Diastase?

OBJECTIVES
- Prepare an extract of germinating barley seeds.
- Analyze a known solution of diastase for the presence of starch and sugars.
- Hypothesize what changes will occur when barley extract is mixed with starch solution.
- Determine the action of barley extract on starch solution over a period of time.

PROCESS SKILLS: observing, inferring, communicating, measuring, predicting, interpreting data, forming hypotheses, separating and controlling variables, experimenting

TIME ALLOTMENT: 1 class period

MATERIALS
droppers (24)
test tubes (42)
test-tube racks (6)
germinating barley
  seeds, 5 days old (150)
hot plates (6)
water baths (6)
Benedict's solution
  (100 mL)
0.4% starch solution
  (100 mL)
iodine solution (25 mL)
distilled water (300 mL)
0.1–0.2% diastase
  solution (20 mL)

mortars and pestles (6)
100-mL beakers (6)
50-mL beakers (6)
clean sand (1 small bag)
stirring rods (12)
scoops (6)
plastic spot plates (6)
test-tube holders (6)
10-mL graduated
  cylinders (18)
white paper (6 pieces)
goggles (30)
laboratory aprons (30)
clock

ALTERNATE MATERIALS
- Watchglasses can be used instead of spot plates.

PREPARATION
- Prepare starch solution, Benedict's solution, iodine solution, and diastase solution according to directions on page T12. Care must be taken to avoid vigorous boiling when making the starch solution. Excess heat will break down the starch, and it will then yield a positive Benedict's test.
- Start germinating barley seeds five days prior to beginning this Investigation.

TEACHING THE LAB
- Have students work in groups of five. Each student can run the tests for one time period.
- **Troubleshooting:** Starch may give a positive Benedict's test if some of the starch has broken down to sugar.

HYPOTHESIS
Diastase in barley extract will break down starch to form maltose. This reaction will take place within 12 minutes of mixing barley extract with starch.

DATA AND OBSERVATIONS
Table 1.

Testing for Substances		
Substance	Iodine test Starch (+/–)	Benedict's test Sugars (+/–)
Starch solution	+	–
Diastase solution	–	–

Table 2.

Testing for Enzyme Action		
Time (Minutes)	Iodine test Starch (+/–)	Benedict's test Sugars (+/–)
0	+	+
3	+	+
6	–	+
9	–	+
12	–	+

Answers may vary for Benedict's test at 0, 3, and 6 minutes depending on concentration of the enzyme. The Benedict's test is usually positive at time 0 because the enzyme breaks down starch faster than students can run the test. By 12 minutes, all the starch should be broken down.

ANALYSIS
1. the presence of starch
2. Starch was present.
3. No starch was present.
4. sugars
5. Benedict's test on starch solution should be negative, indicating that sugars are absent. Answers can vary because of some breakdown of starch to sugar during preparation of the starch solution.
6. the absence of sugar
7. They are controls.
8. Both starch and sugar were present.
9. Starch was broken down to maltose. Eventually all starch was converted to maltose. The Benedict's test indicated that sugar was being produced.
10. It was all converted to maltose.

CHECKING YOUR HYPOTHESIS
Answers will vary. Students who hypothesized that the starch will be converted to maltose by barley extract will say that their hypotheses were supported.

Use of the Light Microscope

OBJECTIVES
• Practice proper handling and use of the light microscope.
• Identify the parts of a light microscope.
• Locate objectives under low- and high-power magnification.
• Prepare a wet mount of an insect leg.

PROCESS SKILLS: observing, recognizing and using spatial relationships

TIME ALLOTMENT: 1 class period

MATERIALS
light microscopes (30)	droppers (30)
lamps (if needed) (30)	preserved insect legs
microscope slides (30)	(30)
coverslips (30)	lens paper (1 book)
forceps (30)	water

PREPARATION
• Preserved insects are available from biological supply houses. Students can also prepare their own insects by dropping an insect into a jar of ethanol. You may wish to have students collect an additional insect in Exploration 1–1 and keep it in ethanol until needed for this Exploration.
• Check microscopes to determine whether the stage moves instead of the objective when the coarse wheel adjustment is turned. Modify Part B, questions 5a and b accordingly.

TEACHING THE LAB
• Students should work individually in this Exploration.
• If permanent slides are used for Part C, students should still become familiar with steps needed to prepare wet mounts.
• Directions for cleaning slides and coverslips may be given when Part C has been completed.
• In Part D, failure to observe the insect leg or to see it clearly may be due to improper centering of slide over stage opening, objective raised too far, objective not raised high enough, improper lighting, low-power objective not clicked into position, dirty eyepiece or objective lens, dirt or water on coverslip, or water on lens of objective.
• If permanent mounts are available, students may want to extend their experience using the microscope with other samples.
• Instruct students to use only a small drop of water when making a wet mount. Water must not flow onto the top surface of the coverslip.

• Answers: **Part A. 3a., b.** Answers will vary depending on the type of microscope used.
Part B. 5. a. up
  b. down

DATA AND OBSERVATIONS
1. Answers will vary. Most microscopes will have a total magnification under low power of 100X.
2. Answers will vary. Most microscopes will have a total magnification under high power of 400X to 450X.

ANALYSIS
1. **c** diaphragm
   a stage opening
   g mirror or lamp
   i eyepiece
   d low-power objective
   e high-power objective
   h revolving nosepiece
   b coarse wheel adjustment
   j fine wheel adjustment
   f stage
2. **a.** False
   b. True
   c. True
   d. True

INVESTIGATION

How Can a Microscope Be Used in the Laboratory?

OBJECTIVES
- Compare the position of an object when viewed through a light microscope with its position on the microscope stage.
- Use stains to aid in viewing objects.
- Hypothesize how the field of view using low power compares with that under high power.
- Compare the depth of field under low and high powers.

PROCESS SKILLS: observing, recognizing and using spatial relationships, predicting, forming hypotheses

TIME ALLOTMENT: 2 class periods

MATERIALS

light microscopes (30)	white thread (60 cm)
microscope slides (30)	droppers (30)
coverslips (30)	lens paper (1 package)
scissors (30)	magazine pages (30)
laboratory aprons (30)	forceps (30)
absorbent cotton (1 ball)	iodine solution (10 mL)
single-edged razor blades (30)	tissue paper (30 small pieces)
black thread (60 cm)	peeled potato (2–3)
	pond water (10 mL)

ALTERNATE MATERIALS
- If pond water is not available, mix yeast with 100 mL of tap water. No movement will be detected.

PREPARATION
- Scrapings from the sides or debris from the bottom of a fish tank will supply a variety of organisms for viewing.
- See page T13 for preparation of an iodine solution.
- Peel potatoes and obtain pond water.

TEACHING THE LAB
- Students should work individually.
- All objects viewed under the microscope will require adjustment of light. Explain to the students that many problems associated with microscopic observations can be overcome by adjusting the diaphragm for proper lighting.
- Have students remove all excess paper surrounding the "e."
- Use lens or tissue paper to remove excess iodine solution from the top of the coverslip.
- Iodine solution may be added directly by lifting the coverslip and dropping iodine onto the starch grains.

- The questions posed in the Procedure are designed to promote careful observation. Answers follow.
  - **Part A.** 3. backwards and upside down
    - 4. right to left
    - 5. away from you
  - **Part B.** 2. little light
    - 3. maximum light
  - **Part C.** 3. yes
    - 4. no
  - **Part F.** Answers will vary, but students should observe about 16 times more grains under low power than under high power.

HYPOTHESIS
When using low power on a microscope, the total area of the field of view is greater than when using high power. The difference is determined by the difference in the magnifications of the two objectives.

DATA AND OBSERVATIONS

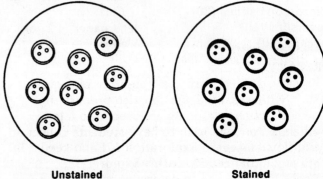

Unstained   Stained

ANALYSIS
1. True
2. False
3. True
4. False
5. False
6. True
7. True
8. False
9. True

CHECKING YOUR HYPOTHESIS
Answers will vary. Students who hypothesized that when using low power, the total area of the field of view is greater than when using high power, will say that their hypothesis was supported.

EXPLORATION

Normal and Plasmolyzed Cells

OBJECTIVES
- Prepare a wet mount of an *Elodea* leaf.
- Observe plasmolysis in the cell as salt solution is added to the wet mount.
- Observe the reversal of plasmolysis as the salt solution is diluted.
- Compare and diagram normal cells in tap water with plasmolyzed cells in salt solution.

PROCESS SKILLS: observing, experimenting, defining operationally

TIME ALLOTMENT: 1 class period

MATERIALS
microscopes (30)
microscope slides (30)
paper towels (1 roll)
coverslips (30)
forceps (15)

Elodea (several sprigs)
droppers (30)
tap water
6% salt solution (100 mL)

ALTERNATE MATERIALS
- A filamentous alga may be used in place of *Elodea*.

PREPARATION
- Prepare the 6% salt solution before class begins. Put a few milliliters into each of 15 small beakers, test tubes, or dropper bottles.

TEACHING THE LAB
- Have students work individually, but they can share supplies, such as forceps and droppers.
- Caution students to remove the paper towel from the edge of the coverslip before all water is removed.
- It will take a few minutes for plasmolysis to occur after salt solution is added.

DATA AND OBSERVATIONS
See the diagrams below.

ANALYSIS
1. The chloroplasts are evenly distributed throughout the cell.
2. The chloroplasts are bunched together in the center of the cell.
3. The cell membrane pulls away from the cell wall; the cell contents bunch together in the center of the cell.
4. out of the cell into the surrounding salt solution
5. The cell contents returned to their normal positions.
6. from surrounding solution into the cell
7. Plasmolysis is the "shrinking" of cell contents due to the loss of water from a cell. Water moves out of the cell when the concentration of water outside the cell is less than that inside the cell.

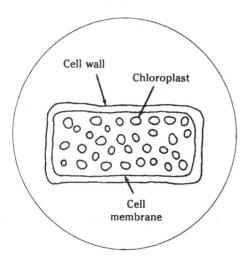

Normal cell

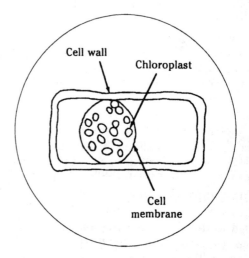

Plasmolyzed cell

Chloroplast Pigment Analysis

OBJECTIVES
- Extract a mixture of plant photosynthetic pigments.
- Separate pigments of spinach leaves by thin layer chromatography.
- Prepare and analyze a silica gel chromatogram.
- Calculate the R_f values for various photosynthetic pigments.

PROCESS SKILLS: observing, classifying, inferring, recognizing and using spatial relationships, measuring, using numbers, interpreting data

TIME ALLOTMENT: 1 class period

MATERIALS
baby food jars with lids (30)
spinach leaves, dried (1 package)
petroleum ether (135 mL)
acetone (15 mL)
200-mL brown bottle
thin layer chromatography slides (30)
funnels (30)
cheesecloth (2 yards)
small dark-colored bottles or vials (30)
goggles (30)
capillary tubes (30)
metric rulers (30)
pencils (30)
ethyl alcohol (75 mL)
mortars and pestles (30)
laboratory aprons (30)

ALTERNATE MATERIALS
- Any dried, dark green leaves can be used.
- You can make your own capillary tubes by drawing out glass tubing.
- To make your own chromatography slides, mix 20 grams of silica gel H with 70 mL of acetone in a 150-mL beaker or jar to form a silica gel slurry. A baby food jar works well. This amount of slurry will coat at least 20 slides. If the slurry becomes thick due to evaporation of the acetone, add a few more mL and stir. Stir the slurry well with a glass stirring rod. Place two glass microscope slides back to back with one slightly higher than the other. Dip the slides into the slurry and remove them slowly, touching the bottom of the slides to the edge of the container to remove excess slurry. Separate the slides and set them aside to dry. Drying will take only a few seconds.

PREPARATION
- All materials should be on hand when class begins. Make up the solvent shortly beforehand.
- Fresh or frozen leaves must be dried for 48 hours in a warm oven or 37°C incubator.

TEACHING THE LAB
- Have each student make a chromatogram.

- You may wish to compare the students' R_f values for specific pigments. They should get similar values for specific pigments. Students should realize that the R_f value for each pigment is a constant, regardless of how far the solvent front moved.
- **Troubleshooting:** It will take only 3 or 4 minutes for the solvent front to near the top of the slide. Caution students to remove the slide from the solvent before the front reaches the top.

DATA AND OBSERVATIONS
Table 1.

Chromatography Data		
Substance	**Distance from original spot (mm)**	**R_f value**
Solvent front	50 mm	
Carotenes	45	0.90
Lutein	43	0.86
Chlorophyll *a*	25	0.50
Xanthophylls	12	0.24
Chlorophyll *b*	12	0.24

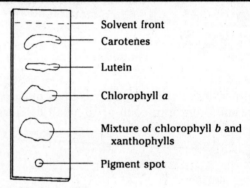

- Solvent front
- Carotenes
- Lutein
- Chlorophyll *a*
- Mixture of chlorophyll *b* and xanthophylls
- Pigment spot

Your thin layer slide

ANALYSIS
1. Students should be able to identify chlorophylls *a* and *b*, as well as the xanthophylls, carotenes, and lutein.
2. Chlorophyll *a* appears to be most abundant.
3. Carotenes usually travel the fastest. Chlorophyll *b* usually travels most slowly.
4. Carotenes have the highest R_f value.
5. The higher the R_f of the pigment, the faster the rate of travel.
6. The pigments have different weights and electrical charges that result in different attractions to the solvents carrying them. Heavier molecules travel more slowly than lighter molecules.
7. Students should expect to get similar results because all green leaves have similar pigments.
8. There is usually much more chlorophyll present in leaves and it masks the other pigments.
9. Once chlorophylls *a* and *b* are broken down, other pigments, present in smaller quantities, can be seen.

How Does Concentration of Sugar Affect Fermentation?

OBJECTIVES
• Prepare four different growth environments for yeast and measure the rate of respiration in each.
• Make a hypothesis that describes what effect an increase in carbohydrates will have on yeast respiration.
• Draw a graph that compares the rates of respiration for yeast in the four environments.

PROCESS SKILLS: observing, inferring, communicating, measuring, predicting, forming hypotheses, experimenting

TIME ALLOTMENT: 1 class period, then 20 minutes the next day.

MATERIALS
colored pencils (32)
molasses (1 bottle)
graduated fermentation tubes (32)
bromothymol blue solution (40 mL)
laboratory aprons (30)
graph paper (30 sheets)
wax marking pencils (8)
distilled water
dry yeast (1 package)
stopwatches (8)

ALTERNATE MATERIALS
• If fermentation tubes are unavailable, substitute by filling a test tube with the liquid, as in the Procedure. Invert the test tube into a beaker containing some of the liquid.

PREPARATION
• Purchase bromothymol blue solution from a supply house.
• Make up the different molasses concentrations before class. See page T13.
• Make the yeast suspension shortly before the beginning of class. Mix 1 packet of dry yeast with 50 mL of warm, not hot, water.

TEACHING THE LAB
• Have students work in groups of four, with each student working with one fermentation tube.
• Caution students not to get bubbles trapped in their fermentation tubes when they set up the tubes.
• If fermentation tubes are not graduated, students can measure with a metric ruler the linear amount of gas produced. By measuring the diameter of the tube and calculating the radius, the volume of gas can be calculated using the formula $V = \pi r^2 h$.

HYPOTHESIS
Increasing the concentration of carbohydrates increases the rate of respiration in yeast.

DATA AND OBSERVATIONS
Table 1.

	Time elapsed	Color of solution	Odor of ethyl alcohol (+/–, weak, strong)	Gas production (Volume in mL)
	0 min.	blue	none	none
	10 min.	blue	none	none
Tube 1	20 min.	blue	none	none
	30 min.	blue	none	none
	40 min.	blue	none	none
	24 hrs.	blue	none	none
	0 min.	blue	none	none
	10 min.	blue	none	2 bubbles
Tube 2	20 min.	yellow	none	0.5 mL
	30 min.	yellow	slight	1 mL
	40 min.	yellow	slight	1.5 mL
	24 hrs.	orange	slight	10 mL
	0 min.	blue	none	none
	10 min.	blues	none	4 bubble
Tube 3	20 min.	yellow	none	0.6 mL
	30 min.	yellow	slight	1.1 mL
	40 min.	yellow	slight	1.7 mL
	24 hrs.	orange	strong	12 mL
	0 min.	blue	none	none
	10 min.	yellow	slight	0.6 mL
Tube 4	20 min.	yellow	slight	1.2 mL
	30 min.	yellow	slight	1.7 mL
	40 min.	orange	strong	2.5 mL
	24 hrs.	orange	strong	25 mL

Graphs will vary depending upon the data obtained.

ANALYSIS
1. The fluid in tubes 2, 3, and 4 changed from blue to yellow. These changes took place at different times according to concentration of molasses—the higher the concentration, the sooner the color changed.
2. The odor of ethyl alcohol could be detected in all of the tubes except tube 1. The odor became noticeable first in the tube with the highest molasses concentration.
3. Carbon dioxide was released and could be demonstrated by placing a glowing splint into the gas that collects in the tubes.
4. Little or no change was detected in tube 1 because there was no source of carbohydrate. Tube 1 served as a control.
5. The most gas was produced in tube 4 because it had the greatest concentration of molasses. With more carbohydrate available, more fermentation could occur and more gas could be produced.

CHECKING YOUR HYPOTHESIS
Answers will vary. Students who hypothesized that the rate of respiration increases with an increase in carbohydrate concentration will say that their hypothesis is supported.

Why Don't Cells Grow Indefinitely?

OBJECTIVES
- Make a hypothesis that describes the relationship among surface area, volume, and mass of a cell.
- Determine the relationship between surface area and volume of a model cell.
- Determine the relationship between surface area and mass of a model cell.
- Apply these mathematical relationships to living cells.

PROCESS SKILLS: observing, recognizing and using spatial relationships, measuring, predicting, using numbers, interpreting data, forming hypotheses, experimenting, formulating models

TIME ALLOTMENT: 1 class period

MATERIALS
photocopies of 3 cell models (15)
white glue (15 bottles)
scissors (30)
balances (several)
coarse sand (1 bag)
small scoops (several)

PREPARATION
- Make the photocopies of the cell models on heavy paper. The heavier the paper, the more sturdy the models will be and the less likely they will be to break when filled with sand.

TEACHING THE LAB
- Students may find it helpful to use calculators for some of the ratio calculations.
- **Troubleshooting:** Make sure students realize that they should add an imaginary sixth side to the cell models when they do their calculations.

HYPOTHESIS
The surface area-to-volume and surface area-to-mass ratios both decrease as a cell grows larger.

DATA AND OBSERVATIONS
See the tables below.

ANALYSIS
1. total surface area
2. volume and mass
3. It will need more cell membrane in order to get more materials into and out of the cell.
4. It gets smaller.
5. It gets smaller.
6. the cell with sides of one unit
7. a. They will have difficulty getting necessary materials such as oxygen and food into the cell and wastes out of the cell fast enough.
   b. the smallest cell (s=1)
8. 27
9. 27 s=1 cells
10. The large cell can divide into two smaller cells.
11. It also doubles.

CHECKING YOUR HYPOTHESIS
Answers will vary. Students who predicted that the ratios would decrease as the size of the cell increases will report that their hypotheses were supported.

Table 1.

Measurements of Cell Models					
Formulas	s^2	$6s^2$	s^3	$\frac{1}{2}s$	
Cell size (Length of one side)	Area of one face (Square units)	Total surface area of cell (Square units)	Volume of cell (Cubic units)	Distance from center to edge (Units)	Mass of cell (Grams)
1	1	6	1	1/2	Answers
2	4	24	8	1	will
4	16	96	64	2	vary.

Table 2.

Ratios of Cell-Model Measurements		
Cell size (Length of one side)	Total surface area to volume	Total surface area to mass
1	6:1	Answers will vary.
2	3:1	Surface area-to-mass ratio will
4	1.5:1	decrease as cell size increases.

INVESTIGATION

How Does the Environment Affect Mitosis?

OBJECTIVES
- Prepare squashes of onion root tips to observe mitosis.
- Make a hypothesis to describe the effect of caffeine on mitosis.
- Compare growth of onion roots in water and caffeine.

PROCESS SKILLS: observing, inferring, measuring, hypothesizing, experimenting

TIME ALLOTMENT: 2–3 class periods

MATERIALS
onion bulbs (32)	3% hydrochloric acid
toothpicks (2 boxes)	45% acetic acid in a
150-mL glass jars (32)	dropper bottles (8)
concentrations of	forceps (8)
caffeine (see below)	microscopes (8)
metric rulers (8)	25-mL graduated
wax marking pencils (8)	cylinders (16)
scalpels (8)	test-tubes (64)
paper towels (1 roll)	test-tube holders (8)
distilled water	test-tube racks (8)
slides (32)	thermometers (8)
coverslips (32)	hot plates (8)
Feulgen stain	water baths (8)
methanol-acetic acid	goggles (30)
fixative	laboratory aprons (30)

PREPARATION
- Buy onion bulbs from a garden center. To save lab time, do steps 2 and 3 of Part A for students. It should take about 2 days for the roots to reach 1 cm. To save more time, do steps 1 and 3 of Part B the day before the lab. Leaving the tips in the fixative for 24 hours eliminates the need for the water bath in step 4.
- For more complete results, provide each team with 11 sprouted bulbs: 1 bulb in 132 mL of distilled water and 10 bulbs in progressively stronger concentrations of caffeine. The stock caffeine solution is made by adding 238 mL of water to 2 grams of instant coffee to produce a 6.6% solution of caffeine. Dilute the stock solution as follows: 0.1% caffeine: 130 mL water + 2 mL stock solution; 0.2%: 128 mL water + 4 mL stock; 0.3%: 126 mL water + 6 mL; 0.4%: 124 mL water + 8 mL stock; 0.5%: 122 mL water + 10 mL stock; 0.6%: 120 mL water + 12 mL stock; 0.7%: 118 mL water + 14 mL stock; 0.8% 116 mL water + 16 mL stock; 0.9%: 114 mL water + 18 mL stock; 1.0%: 112 mL water + 20 mL stock.

TEACHING THE LAB
- Have students work in groups of 4.

HYPOTHESIS
Students who observed that root growth slowed down in the caffeine treatments probably said caffeine will delay or arrest mitotic division.

DATA AND OBSERVATIONS
Data for Table 1 will vary, but students will probably find that roots grown in pure water will reach 1 cm in 2–3 days. The table below is equivalent to Day 3 in Table 2 of the lab, using 10 different concentrations of caffeine.

% CAFFEINE	# ROOTS	AVG LENGTH (mm)
0.0 (pure water)	41	21
0.1	35	11
0.2	27	12
0.3	26	9
0.4	26	6
0.5	25	6
0.6	25	6
0.7	25	6
0.8	22	4
0.9	20	4
1.0	20	5

Table 3. There should be a graded decrease in the number of cells in metaphase, anaphase, and telophase from the lowest to the highest percent of caffeine. In water-treated roots, about 80% of cells are found in interphase.

ANALYSIS
1. The control is the bulb treated in plain water; the variable is the concentration of caffeine.
2. Answers will vary, partly depending on students' technique in preparing the squashes. In general, students will find that the rate of growth decreases with higher concentrations of caffeine.
3. In cells from the water treatment, students will probably observe all events in mitosis. The caffeine-treated roots will probably show the highest percentage of metaphase cells. With increased concentrations of caffeine, there will be fewer cells completing cell division and probably no cells in anaphase or telophase.
4. Answers will vary, but students will probably find a direct correlation. The higher the concentration of caffeine, the slower the rate of root growth and the fewer cells found in metaphase, anaphase, and telophase.
5. Answers may include ultraviolet light, pH, temperature, and various pollutants.

CHECKING YOUR HYPOTHESIS
Students who said that increasing the concentration of caffeine would alter the number of mitotic stages in the root tips will say their hypotheses were supported.

Observation of Meiosis

OBJECTIVES
- Observe the stages of meiosis in lily anthers.
- Draw and label the stages of meiosis from lily anthers.

PROCESS SKILLS: observing, classifying , defining operationally

TIME ALLOTMENT: 1 class period

MATERIALS:

microscopes (30)	drawing paper (optional)
prepared slides of	(30 pieces)
lily anthers (30)	pencils (colored pencils
	if desired) (30)

PREPARATION
- Check all microscopes to see that lamps are not burned out and lenses focus properly.

TEACHING THE LAB
- Students should work individually if there is a sufficient number of microscopes.
- Go over the phases of meiosis with the class using Figure 1 as a guide.
- Encourage students to move the slide around on the stage in order to observe additional stages of meiosis.

DATA AND OBSERVATIONS
Student drawings should resemble those in Figure 1. All stages of meiosis should be observed and drawn by students.

ANALYSIS
1. Answers will vary. Meiosis occurs in great frequency in anthers. Members of the lily family have few and very large chromosomes.
2. Answers will vary. Metaphase I, metaphase II, and anaphase II are usually abundant.
3. Chromosomes should appear as colored, threadlike objects within the cells.
4. Meiosis produces pollen grains containing half the number of chromosomes found in body cells. The pollen grains will eventually produce the male reproductive cells.

EXPLORATION

Chromosome Extraction and Analysis

OBJECTIVES
• Extract chromosomes from a fruit fly larva.
• Stain and examine the chromosomes under a microscope.

PROCESS SKILLS: observing, recognizing and using spatial relationships, communicating

TIME ALLOTMENT: 1 class period

MATERIALS

cultures of fruit fly larvae (15)
0.7% saline solution
18% hydrochloric acid solution
45% acetic acid solution
aceto-orcein stain
microscope slides and coverslips (30)

paper towels (1 roll)
dissecting probes (30)
forceps, fine-nosed (30)
dissecting microscopes or hand lenses (30)
compound microscopes (30)
laboratory aprons (30)
goggles (30)

PREPARATION
• Cultures of wild-type fruit flies, instant *Drosophila* media, culture tubes, and foam plugs should be ordered from a biological supply house.
• Anesthetize the original culture of flies and separate males from females. (See Lab 14–1 and p. T54.) Place two males and two females in the new culture, and cap the tubes. Keep the new cultures at 20–25 °C. Each female should lay about 500 eggs. The larvae should be ready to use within 4–5 days after mating the flies. Remove adult flies prior to giving students the culture tubes. A minimum of 5 larvae should be available to each student.
• See p. T13 for preparing solutions.

TEACHING THE LAB
• Students can share chemicals and microscopes, though each student should make his or her own slide preparations and sketches.
• Show the students how to hold the larva and pull on the head region. When the posterior end is held with the forceps, the anterior mouth area pops out. Students should place the dissecting probe firmly behind the protruding mouth area and pull gently but firmly across the slide so the head does not break off. Students may have to try over and over. If the dissecting probe does not work satisfactorily, have the students use a second forceps to grasp the larva behind the protruding head region.
• **Troubleshooting:** The slide has to be squashed for the cells to release the chromosomes. Gently tap the coverslip if the squash does not work the first time.

DATA AND OBSERVATIONS
• See the diagram of polytene chromosomes below. Expect student sketches to be much rougher and incomplete.

ANALYSIS
1. Answers will vary. Students may say that each pair appears as a single structure because each chromosome is tightly attached to the other member of the pair.
2. Increased gene activity would produce the large quantity of proteins needed during a period of growth of the larva and metamorphosis of the larva into the adult fruit fly.
3. No, because the banding is genetic and thus unaffected by nutrients.
4. The polytene chromosomes did not separate after they replicated, unlike chromosomes in typical mitotic division.

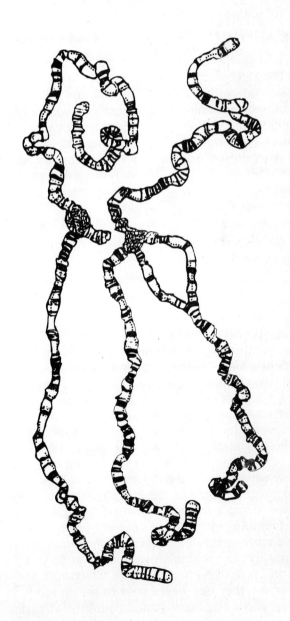

What Phenotypic Ratio Is Seen in a Dihybrid Cross?

OBJECTIVES
- Learn to care for and raise two generations of fruit flies.
- Develop hypotheses to describe the results of two dihybrid crosses.
- Determine the results of a cross between two flies showing recessive traits.
- Construct Punnett squares for two dihybrid crosses.

PROCESS SKILLS: observing, classifying, inferring, communicating, predicting, using numbers, interpreting data, forming hypotheses, experimenting

TIME ALLOTMENT: 3 class periods, each two weeks apart

MATERIALS
culture vials with medium and foam plugs (16)
cultures of vestigial-winged, normal-bodied fruit flies (8)
cultures of long-winged, ebony-bodied fruit flies (8)
vials of alcohol (8)

anesthetic (8 vials)
anesthetic wands (8)
white index cards (8)
camel-hair brushes (8)
wax marking pencils (8)
stereomicroscopes or hand lenses (8)
instant *Drosophila* medium (1 bag)

PREPARATION
- *Drosophila* cultures and medium should be ordered at least three weeks in advance of lab.

TEACHING THE LAB
- Have students work in groups of four.
- To save time, you could order custom F_1 cultures of *Drosophila*. There is little added expense in this and it saves one lab period.
- Virgin flies can be obtained for about six hours after new flies begin to emerge. All adults must be removed from the vials prior to emergence. Adult males will mate with newly-emerged females.
- Review the life cycle of *Drosophila* with students. Show what the pupal and larval stages look like.
- The traits involved in this Investigation are easy to see. Using a stereomicroscope or hand lens will aid identification. Pooling data will probably produce more accurate results than individual data.
- A fruit fly life cycle takes two weeks at 20°C. This lab will take three periods—one to set up the F_1 cross, another two weeks later to set up the F_2 cross, and a third to sort and count the offspring of the F_2 generation and complete the questions.

- Fruit flies, vials, foam plugs, and instant media can be obtained from most biological supply houses. FlyNap™, a safer and more effective anesthetic than ether, can be obtained, with instructions, from Carolina Biological Supply.

HYPOTHESES
All F_1 offspring will be long winged and have normal body color. The F_2 offspring will be produced in the ratio of 9 long-winged, normal-bodied: 3 long-winged, ebony-bodied: 3 vestigial-winged, normal-bodied: 1 vestigial-winged, ebony-bodied.

DATA AND OBSERVATIONS
Student data will vary. See sample data below.

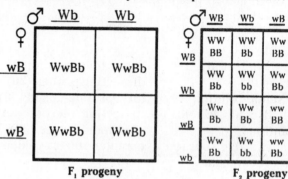

F₁ progeny   F₂ progeny

Table 1.

Results of Dihybrid Crosses				
Generation	Long-winged normal-bodied	Vestigial-winged normal-bodied	Long-winged ebony-bodied	Vestigial-winged ebony bodied
Parental males			5	
Parental females		5		
F_1 males	135			
F_1 females	120			
F_2 males	261	95	92	27
F_2 females	275	80	90	33

Total number of F_1 individuals __255__   Total number of F_2 individuals __536__

ANALYSIS
1. All individuals in the F_1 generation were long-winged and normal-bodied flies.
2. The F_2 generation had individuals that were long-winged and normal-bodied, long-winged and ebony-bodied, vestigial-winged and normal-bodied, and vestigial-winged and ebony-bodied.
3. 9:3:3:1; observed phenotypic ratios will vary but should be close to this.
4. Crosses must be perfectly performed, a large number of offspring must be produced, and virgin females must be used.
5. Mendel's Law of Segregation and Law of Independent Assortment.
6. It is in Metaphase I that the homologous chromosomes pair up and sort independently.
7. A female can store enough sperm from one mating to fertilize all of the eggs she produces in her lifetime. Only virgin females are guaranteed not to carry sperm from a previous mating. Males do not pick up and store gametes.

EXPLORATION

Determination of Genotypes from Phenotypes in Humans

OBJECTIVES
- Determine your phenotype for nine different characteristics.
- Determine your possible genotypes for the nine different characteristics.
- Compare your phenotypes and genotypes with those of other students in the class.
- Evaluate your uniqueness as an individual.

PROCESS SKILLS: observing, classifying, inferring, communicating

TIME ALLOTMENT: 1 class period

MATERIALS
PTC taste paper (30 strips)
untreated taste paper (30 strips)
mirrors (30)

PREPARATION
- It is easier to order PTC taste papers than to make them up yourself. The papers can be made by soaking strips of paper in a 1% solution of phenylthiocarbamide and allowing the papers to dry.

TEACHING THE LAB
- Review with students the differences between simple Mendelian inheritance and incomplete dominance.
- Go over each trait with the class before students begin this Exploration.
- The practice with incompletely dominant traits is to use all capital letters when illustrating genotypes, with different letters for each allele. The use of this technique reminds the student that this trait is incompletely dominant and that neither gene is dominant over the other.

DATA AND OBSERVATIONS
Table 1.
Answers will vary.

ANALYSIS
1. Students' answers will vary but should be the same as in his or her data table.
2. Students' answers will vary but should be the same as in his or her data table.
3. Students' answers will vary but should be the same as in his or her data table.
4. Answers will vary. Students should note that they share some, but not all, traits of many of their classmates.
5. Answers will vary. Students should note that they share some, but not all, traits of many of their classmates.
6. If two classmates shared all of the same traits in the list, they could prove uniqueness by listing other traits that are not shared. These will vary.
7. The genes inherited from parents determine the traits of the offspring.
8. Genotypes that are recessive can be determined because a person must have two alleles for the recessive trait. Genotypes can also be determined for incompletely dominant alleles because there is a different phenotype for each genotype. Genotypes cannot be determined for dominant traits because the homozygous dominant and heterozygous conditions appear as the same phenotypic trait.
9. The untreated paper was used as a control to prove to the taster that the paper itself did not taste bitter.

How Can Karyotype Analysis Explain Genetic Disorders?

OBJECTIVES
- Construct a karyotype from the metaphase chromosomes of a fictitious insect.
- Analyze prepared karyotypes for chromosome abnormalities.
- Identify the genetic disorders of six fictitious insects by using the insects' karyotypes.
- Hypothesize how karyotype analysis can be used to explain the presence of a genetic disorder.

PROCESS SKILLS: observing, classifying, inferring, communicating, recognizing and using spatial relationships, predicting, forming hypotheses, formulating models

TIME ALLOTMENT: 1 class period

MATERIALS
photocopies of metaphase chromosomes from six insects (30 sets of 2 pages)

scissors (30)
rubber cement (30 bottles)

PREPARATION
- Prepare all materials before class begins. Photocopy chromosomes from pages T27 and T28.

TEACHING THE LAB
- If you have enough materials, students should work alone.
- Demonstrate how to prepare a karyotype as an example.
- Set expectations for analysis by informing students to follow the directions closely.

HYPOTHESIS
Karyotype analysis can be used to explain abnormal phenotypes by identifying abnormal chromosomes and errors that have occurred during chromosome separation in cell division.

DATA AND OBSERVATIONS
See the diagrams below.

ANALYSIS
1.

	Sex	Genetic disorder	Chromosome error
Insect 1	Male	Normal	None
Insect 2	Female	Clear wing disorder	Trisomy 2
Insect 3	Male	Duplication disorder	Duplication 1
Insect 4	Female	Size reduction disorder	Monosomy 3
Insect 5	Female	Normal	None
Insect 6	Female	Unsegmented disorder	Deletion 3

2. addition or deletion of a short section of chromosome; addition or deletion of entire chromosomes; the larger changes are more obvious

3. The duplicated segment must have genes that control pigment production in the wings, and segmentation of the body and head.

4. Prior studies must have established that specific changes in the visible structures of chromosomes are associated with specific genetic disorders.

CHECKING YOUR HYPOTHESIS
Answers will vary. Students who said that karyotype analysis can detect the presence of genetic disorders by identifying abnormalities will say that their hypothesis was supported.

Insect 1

Insect 2

Insect 3

Insect 4

Insect 5

Insect 6

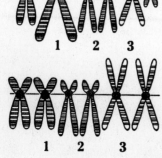

EXPLORATION

DNA Sequencing

OBJECTIVES
- Learn to read a gel electrophoresis.
- Model the process of DNA sequencing through gel electrophoresis.
- Convert the DNA sequence into a protein by using an mRNA codon chart.

PROCESS SKILLS: interpreting data, analyzing, communicating,

TIME ALLOTMENT: 1–2 class period

MATERIALS
photocopies of strands of DNA
colored pencils (32)
scissors (8)
meter sticks (8)
large poster boards (approx. 24" × 36") (8)

tape (1 roll)
plastic trays or rectangular lids, small (32)
wax marking pencils (8)
markers (8)

PREPARATION
- Photocopy the student model on p. T29 to make 4 copies, 24 strands, per team. You could use different colors of paper for each treatment.
- Plastic cafeteria trays, freezer containers, or shoe boxes could be used as the "test tubes."

TEACHING THE LAB
- Have students work individually on Part A and in groups of 4 on Part B.
- If the photocopies are on white paper, students can color the radioactive probe with a different color for each chemical treatment. This will help keep the segments separate.
- Stress that only the segments with the radioactive probe attached will show up on the X ray film of the gel, so students should not use the segments that do not have a probe. There will be 5 segments for nitrogen base A, 6 for C, 6 for G, and 4 for T. Segments will be of different lengths, each starting with the probe and ending with a "destroyed" base.
- To determine on which line to place a segment, it is helpful to count the number of bases on the segment, including the "destroyed" base, backwards from 21. Thus, a segment with 12 bases would be taped on line 10.

DATA AND OBSERVATIONS
Part A.
DNA sequence of Figure 1: TGT GTC CTA TAA AAG

Part B.
DNA sequence of your gel:
ACG GGA CTA CCA TGG GCC TTA

Figure 4.

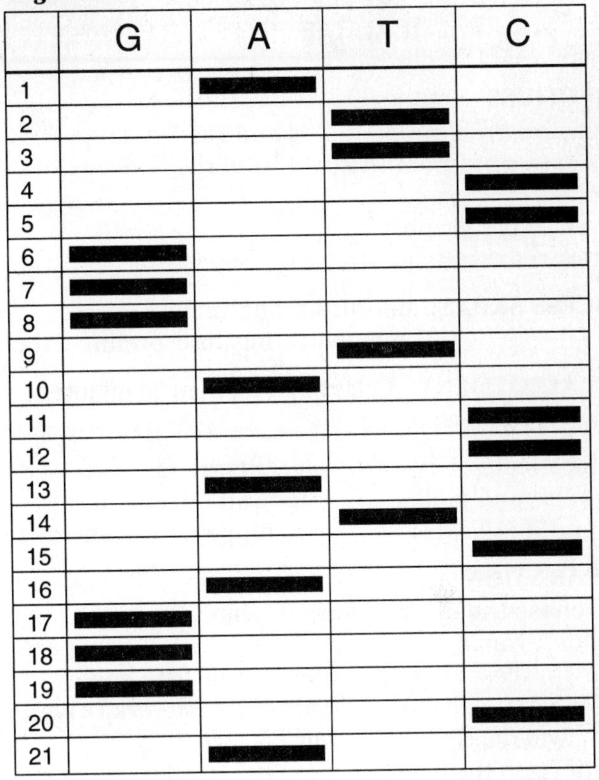

	G	A	T	C
1		▄		
2			▄	
3			▄	
4				▄
5				▄
6	▄			
7	▄			
8	▄			
9			▄	
10		▄		
11				▄
12				▄
13		▄		
14			▄	
15				▄
16		▄		
17	▄			
18	▄			
19	▄			
20				▄
21		▄		

ANALYSIS
1. ACA CAG GAU AUU UUC
2. threonine-glutamine-aspartic acid-isoleucine-phenylalanine
3. Most proteins contain several hundred amino acids. Other more complex proteins have several thousand amino acids in their structure.
4. UGC CCU GAU GGU ACC CGG AAU
5. cysteine-proline-aspartic acid-glycine-threonine-arginine-asparagine
6. Answers may vary. Students may suggest that an actual gene could consist of as many as 30 000 bases.

Analyzing Fossil Molds

OBJECTIVES
- Measure brachiopod molds and record data.
- Prepare a graph of the data from the two sets of molds.
- Analyze the graphs.
- Interpret results relative to geologic time.

PROCESS SKILLS: measuring, analyzing, communicating, interpreting data

TIME ALLOTMENT: 1 class period, then 30 minutes

MATERIALS
plastic sheets of brachiopod molds, A and B (10 sets)
metric rulers (30)
colored pencils (60)
graph paper (30 sheets)

PREPARATION
- Purchase the plastic "Variation and Evolution Molds" from a supply company. The following are two sources: Frey Scientific (see p. T26) and Science Kit and Boreal Labs, 777 East Park Drive, Tonawanda, NY 14150-6784.
- Distinguish the two sets of molds, A and B, by marking them with a permanent marker. The molds represent cast reproductions of two rocks, each containing approximately 50 fossil brachiopods.

TEACHING THE LAB
- Have students work in groups of 3.
- Make sure each team member is measuring the molds and recording data the same way. One person could measure sheet A while the other measures sheet B. The third person could record the data for the other two students.
- Make sure students graph the data correctly. The x axis is for the lengths of brachiopods in mm and the y axis for frequency. For each measurement on the x axis, there should be two sets of data, one for mold A and one for mold B.
- Devise a mechanism for keeping track of which molds have already been measured. Colored dots or other types of stickers would work.

VARIATIONS TO THE LAB
- Measure and graph the width of the fossil in addition to the length.
- Run a "t" statistical test between population A and population B. The null hypothesis would be that there is no statistical difference between the two populations of brachiopods, using length as an indicator. The results should reject that hypothesis, showing that the two populations are indeed statistically different and therefore are two separate populations of brachiopods living at two different time periods.
- Purchase some present day preserved brachiopods. A jar of 10 brachiopods from the Philippines can be purchased from Ward's (see p. T26).
- Have students measure each mold in mm and record the data. Then have students group the data in measurement sequences, such as 0-15 mm, and identify the number of molds for that sequence.

DATA AND OBSERVATIONS

SHEET "A" BRACHIOPODS		SHEET "B" BRACHIOPODS	
Length in mm	Number of molds	Length in mm	Number of molds
0–15	0	0–15	0
16–20	8	16–20	0
21–25	14	21–25	2
26–30	8	26–30	9
31–35	0	31–35	11
		36–40	8
		41–45	1

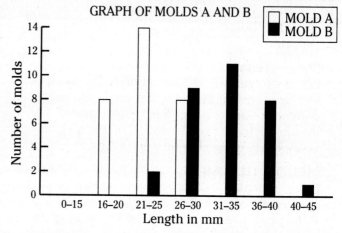

GRAPH OF MOLDS A AND B

ANALYSIS
1. Specific answers will vary but students should find that the means differ for the two populations.
2. Students may find that a line graph (or a line over the bar graph) will form a bell-shaped curve, which reflects a normal distribution.
3. Answers will vary. If the two sets of molds show a different mean and a different curve, which they do in the sample data, students may say that they appear to be two different populations.
4. Answers will vary. If the shape of the line graphs is different as shown in the sample, students may say that the molds came from different time periods. It would be difficult to know which organism evolved first, since evolution has not always favored an increase in size.
5. Students may suggest that the set of molds with the greater mean size would have been found in a more recent layer of rocks. In the sample data, that is sheet B.

How Is Camouflage an Adaptive Advantage?

OBJECTIVES
- Use an artificial environment to demonstrate the concept of natural selection.
- Construct bar graphs to show the results of the Investigation.
- Hypothesize what will happen if natural selection acts on organisms exhibiting camouflage over a period of four generations.
- Compare this example of artificial selection to natural selection.

PROCESS SKILLS: observing, inferring, communicating, predicting, interpreting data, forming hypotheses, experimenting, formulating models, defining operationally

TIME ALLOTMENT: 1 class period

MATERIALS
hand hole punches (8)
colored paper (purple, brown, blue, green, tan, black, orange, red, yellow, and white) (8 sheets of each color)
plastic film cannisters or petri dishes (80)

pieces of brightly colored, floral fabric (80 cm × 80 cm) (8)
graph paper (60 sheets)

PREPARATION
- Assemble all materials before beginning this Investigation.

TEACHING THE LAB
- Once the dots have been punched, they can be saved from year to year in the petri dishes or film cannisters.
- To save time, you can punch out all the dots prior to class. Have a hole punch and sheets of colored paper available, though, because students may need additional dots of some colors.
- Each team will have different results, depending on the colors present on their cloths. You can purchase the same color and pattern cloth for each group or have different colors and patterns. If all groups have the same kind of cloth, the groups can pool their results. Graphs can then be made of the class data.
- The various colors in the cloths could be used to demonstrate the fact that Earth has various biomes with different colored features.

HYPOTHESIS
Dots of the predominant colors in the cloth will become more numerous with each succeeding generation. Dots that contrast with the colors in the cloth will become less numerous with each succeeding generation.

DATA AND OBSERVATIONS
Data will vary depending on the colors and patterns of the cloths.

ANALYSIS
1. Answers will vary. The colors that stand out against the background of the cloth will be picked up.
2. Answers will vary. The colors that blend into the background of the cloth will not be picked up because they will not be seen as easily.
3. The predator will not see the food.
4. Answers will vary. Some examples are the peppered moth, chameleons, and walking stick insects.
5. Those colors that blend in with the background increase in number, while colors that stand out decrease in number.
6. The pattern of survival would change. Colors that would normally stand out against their background would blend in if the predator were color blind.
7. If those dots were bad-tasting, stung, or possessed other characteristics that the predator found undesirable, it would learn to associate the color with these characteristics and would avoid those dots.
8. Snakes have skin patterns that blend with their background; polar bears are white to blend with their snowy backgrounds. Accept all reasonable answers.

CHECKING YOUR HYPOTHESIS
Answers will vary. Students who predicted that dots of predominant fabric colors would become more numerous and that dots of colors that contrast with the fabric would become less numerous will say that their hypotheses were supported.

Plant Survival Exploration

OBJECTIVES
- Prepare a feeding tray containing plant tissues soaked in various chemicals.
- Observe the feeding behavior of a herbivore.
- Analyze the effect of different chemical defenses on a land snail.

PROCESS SKILLS: observing, inferring, measuring, analyzing, interpreting data

TIME ALLOTMENT: 2 class periods

MATERIALS

land snails (15)	graph paper
plant extracts prepared by your teacher	scissors (15)
large leaf lettuce (firm) or Chinese cabbage	stereomicroscopes or hand lenses (15)
plastic feeding containers (approx. 20 cm in width) with perforated lids (15)	wax marking pencil (15)

PREPARATION
- Prepare plastic containers for students by perforating the lids.
- Prepare a variety of plant extracts, both noxious and edible, such as nicotine from cigarettes, tannic acid from tea, alkaloids from milkweed, caffeine from coffee, oxalic acid from onions, and juice from beet leaves. Use plant sources from the local region as much as possible. Blend each food source in distilled water, using the same concentration of plant material per mL of water. Weigh the plant material in grams. Determine the percent concentration of plant material. (1 g of water = 1 mL water). Label each container.
- Land snails could be collected by students or can be purchased from biological supply houses. They should not be fed for two days prior to the lab. You may wish to keep the snails in a terrarium for use in Lab 30–2.

- You may wish to make a plastic transparency of graph paper for students to use as a template for cutting out the lettuce or Chinese cabbage.

TEACHING THE LAB
- Have students work in pairs.
- Show students how to pick up the snails by the shell and gently place them into a new environment. Stress humane treatment of the snails.
- Keep the snail containers in a shaded environment that does not exceed 21°C. Snails live in cool, shaded areas of the forest, so you would want to simulate the condition.
- Have students place the snail equidistant from each food source to avoid preference by proximity. If food is placed in a circle around the snail, there is a problem with the snail headed initially in whatever direction it is placed. Leaving the snail with the food for 24 hours avoids the preference by direction alone.

DATA AND OBSERVATIONS
See the table below. If 3.5 cm = 10 squares, the total number of squares would be 100 in the leaf soaked in extract.

ANALYSIS
1. Answers will vary based on the extracts chosen. Snails should be repelled by strong extracts such as tannins, caffeine, toxins, and alkaloids. They should eat the lettuce entirely, as well as typical vegetables, such as beet greens, which are edible by humans.
2. Those plants that, through mutation, developed chemical defenses are the plants that survived the onslaught of herbivores and were able to reproduce. Thus, chemical defenses, like other types of plant defenses, were an adaptive advantage and were favored by natural selection.
3. Answers may include camouflage, thorns, spines, tough leaves, sticky hairs, and prickles.

Type of Extract	Number of squares consumed	Percent of squares consumed
1. water	100	100%
2. beet greens	90	90%
3. onions	65	65%
4. nicotine	10	10%

EXPLORATION

Gene Frequencies and Sickle Cell Anemia

OBJECTIVES
- Determine the change in allelic frequencies of the Hemoglobin A and S genes.
- Explain the process of natural selection as a force affecting allelic frequencies.

PROCESS SKILLS: observing, communicating, predicting, using numbers, interpreting data

TIME ALLOTMENT: 1 class period

MATERIALS
75 red and 25 black beans per team
plastic food containers, tall (40)
wax marking pencils (8)
blindfolds (8)

PREPARATION
- Purchase enough red and black beans for a class set.
- Set out the 5 containers for each team.

TEACHING THE LAB
- Have students work in teams of four.
- Be sure students understand the procedure before starting the lab.
- Review allelic frequencies and shift of allelic frequencies over time as well as mechanisms of natural selection.
- Point out that the student calling out "malaria" should not be looking at the beans that are being selected in order to ensure that the calls are random.
- Remind students that AS individuals with malaria will live, even though they are circled (contract malaria). These alleles should be counted when calculating allelic frequencies.

DATA AND OBSERVATIONS
Sample data shown below.

Table 1.

Second Generation		
AA genotypes	AS genotypes	SS genotypes
/ / / / /①/ / / / ⓪⓪⓪⓪⓪/ /⓪⓪⓪ ⓪⓪⓪⓪⓪/ / / /	/ / / / /⓪⓪⓪⓪⓪ ⓪/ /⓪/⓪⓪	⓪/ /⓪
29	17	4

a. 47
b. 17
c. 64
d. 73%
e. 27%

Table 2.

Third Generation		
AA genotypes	AS genotypes	SS genotypes
/ /⓪⓪⓪/⓪⓪⓪/ ⓪⓪/ / /⓪	⓪/⓪⓪/⓪/ /⓪/ ⓪/⓪/ /	/
16	15	1

a. 29
b. 15
c. 44
d. 66%
e. 34%

ANALYSIS
1. a. 75%
   b. 25%
   c. 73%
   d. 27%
   e. 66%
   f. 34%
   g. The frequency of the A allele decreased from 75% to 66%, while the frequency of the S allele increased from 25% to 34% in two generations. The presence of malaria was the factor that selected for people who were heterozygous for the sickle-cell trait.

2. Selection favors the sickle-cell allele because it protects heterozygous individuals from malaria and thus balances the tendency of selection to eliminate the allele.

3. The sickle cell allele confers no special advantage on any population of individuals. Therefore selection moves the frequency of the allele in the direction of its elimination.

4. The malaria vaccine would give protection to all genotypes equally. Therefore, there would be no selective advantage for the AS genotype and the frequency of the sickle-cell allele would slowly decline.

5. No. The mutation occurs too infrequently to have a major effect.

INVESTIGATION

Can a Key Be Used to Identify Organisms?

OBJECTIVES
- Use a key to identify fourteen shark families.
- Examine the method used in making statements for a key.
- Construct your own key that will identify another group of organisms.
- Hypothesize how organisms can be identified with a key.

PROCESS SKILLS: observing, classifying, communicating, recognizing and using spatial relationships, predicting, forming hypotheses

TIME ALLOTMENT: 1 class period

MATERIALS
none

PREPARATION
none

TEACHING THE LAB
- Help the students with step 3 of the Procedure. They should understand the principle of making choices on which the key is based.
- Emphasize the importance of making the correct decision at each step, since one wrong choice will lead in the wrong direction.

HYPOTHESIS
Organisms can be identified when a key subdivides a group into smaller and smaller groups based on distinguishing features.

DATA AND OBSERVATIONS
1. Rajidae
2. Alopiidae
3. Pristophoridae
4. Carcharhinidae
5. Scyliorhinidae
6. Rhinocodontidae
7. Isuridae
8. Squalidae
9. Dasyatidae
10. Scapanorhynchidae
11. Pseudotriakidae
12. Hexanchidae
13. Sphyrnidae
14. Mobulidae

ANALYSIS
1. A key is a listing of specific characteristics of a group of organisms. The group is divided into smaller and smaller groups until each organism is identified.

2. body shape, nose shape, head shape, presence or absence of fins (Answers will vary.)

3. **a.** presence of anal fin
   b. shape of caudal fin

4. Answers will vary. There is no right or wrong grouping as long as those fish in a group fit the description of that group. A sample key is shown.
   1. A. Tubelike body I
      B. Body not tubelike . . . Go to statement 2
   2. A. One dorsal fin . . . . . Go to statement 3
      B. Two dorsal fins . . . . Go to statement 4
   3. A. Small mouth II
      B. Large mouth III
   4. A. Large mouth IV
      B. Small mouth V

CHECKING YOUR HYPOTHESIS
Answers will vary. Students who identified all the sharks correctly will say that the method was correct.

Virus Replication

OBJECTIVES
- Trace the steps of viral reproduction in cells.
- Describe each step of viral reproduction.
- Construct an analogy for viral replication in cells.

PROCESS SKILLS: observing, recognizing and using spatial relationships, formulating models.

TIME ALLOTMENT: 1 class period

MATERIALS
glue (15 bottles)
scissors (15)
photocopies of viruses and virus parts (15)

PREPARATION
- Photocopy the student model page from page T30.

TEACHING THE LAB
- Have students work in pairs.
- Emphasize the use of the key in Figure 3 before beginning. Encourage students to go on to the next step if they can't figure out which cutout matches a particular number.
- **Troubleshooting:** If students are having problems, reinforce the idea that each cutout matches a number in the sequence and that it also matches a label on the cell diagram. The sequence goes from top to bottom on the cell diagram.

DATA AND OBSERVATIONS
See the figure below.

ANALYSIS
1. Viral replication can take place only inside a living cell and uses the cell's machinery to carry this out. Cells reproduce by mitosis or binary fission.
2. In replication, viral genes are produced; in translation, viral coats are produced.
3. The new viral genes are inserted into the new viral coats.
4. The cell may be weakened and die, it may break open and die, or it may recover.
5. Answers will vary. One could block attachment, block coat removal so cell machinery is not taken over, block production of either coats or viral genes so new viruses could not be formed, or block escape so the new viruses could not invade other cells.

6. Answers will vary. Viral reproduction involves complex steps regulated by enzymes. Immunity of the body is at work blocking the process. Factories operate on supply and demand and need to make a profit. Factories usually don't become weakened or destroyed making products.

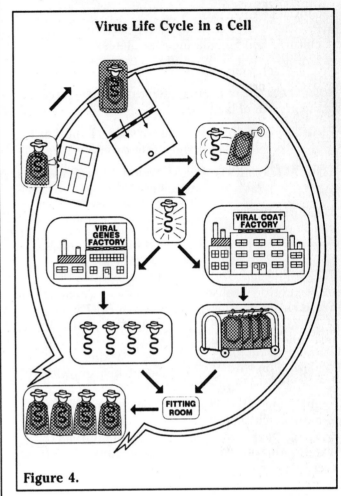

Virus Life Cycle in a Cell

VIRAL GENES FACTORY

VIRAL COAT FACTORY

FITTING ROOM

Figure 4.

How Are Bacteria Affected by Heat?

OBJECTIVES
- Practice sterile techniques for handling bacterial cultures and inoculating agar plates.
- Hypothesize how heat affects the growth of bacteria.
- Compare the effects of different temperatures on the growth of bacteria.

PROCESS SKILLS: observing, forming a hypothesis, experimenting, measuring

TIME ALLOTMENT: 1 class period, then 30 minutes after 24 hours

MATERIALS

sterile petri dishes containing hardened bacto-methylene blue agar (10)
sterile cotton swabs (10)
broth cultures of *Escherichia coli* (8)
test tubes containing 10 mL of *E. coli* culture (8)
test tubes with 10 mL of milk inoculated and autoclaved (8)
hot plates (8)
water baths (8)
thermometers (8)

test tubes (24)
test-tube holders (8)
test-tube racks (8)
sterile pipettes (8)
wax marking pencils (8)
incubator (37°C)
refrigerated milk (3.8 L)
paper towels (1 roll)
tape, masking (1 roll)
250-mL beakers with 100 mL of alcohol (8)
25-mL graduated cylinders (8)
laboratory aprons (30)
goggles (30)

PREPARATION
- Prepare agar plates (see p. T14 under Lab 42-1) or purchase sterile agar plates from a biological supply house.
- Obtain individually-wrapped sterile cotton swabs from a biological supply house or hospital supply store or sterilize them yourself by wrapping swabs in paper and autoclaving them for 15 minutes at 15 pounds pressure.
- Order a non-pathenogenic strain of *E. coli*.
- Have sufficient quantities of disinfectant solution to clean work surfaces.
- Purchase whole milk. Each container should have the same expiration date.

TEACHING THE LAB
- Have students work in groups of 4.
- You may wish to demonstrate the lab techniques used in this Investigation.

- Point out to students that meat products can also become contaminated by *E. coli* through unsanitary processing of animal products.
- Clarify that the entire section of agar must be streaked with bacteria for the proper results to occur. Given the manipulations involved in this investigation and the small size of each agar section, you may wish to use a separate agar plate for each of the treatments and assign each group of students to one of the experimental plates.
- Emphasize the importance of washing hands and disinfecting work surfaces even with non-pathogenic strains.

HYPOTHESIS
Hypotheses will vary. Many students will probably say that heating will inhibit bacterial growth.

DATA AND OBSERVATIONS
Students' results will vary depending on inoculation and sterilization techniques and whether or not the purchased milk already shows some contamination. However, both the control and the autoclaved milk should show no bacterial growth and the pure *E. coli* culture will probably result in a lawn of bacterial growth.

Treatment	Observations
#1	no growth
#2	no growth
#3	significant growth—colonies
#4	no bacterial growth
#5	lawn of bacteria
#6	no *E. coli* colonies

ANALYSIS
1. It is necessary to have a control area where no treatment is done with which to compare experimental variables.
2. Its function was to test whether the milk purchased in the store contained any *E. coli* bacteria prior to the experiment.
3. Its function was to show how a highly contaminated culture would appear.
4. The milk inoculated with *E. coli* should show some colonies on the agar but not a lawn as on the pure culture of bacteria.
5. There may be colonies of other bacteria, but the *E. coli* should all be killed by the heat.
6. The autoclaved milk should not have any bacterial growth since autoclaving sterilizes the milk. Students will probably find that the autoclaved milk changes color and smells like evaporated milk.

CHECKING YOUR HYPOTHESIS
Answers will vary. A hypothesis that heating will inhibit bacterial growth should be supported.

22-1 EXPLORATION

LAB

Observing Algae

OBJECTIVES
- Prepare wet mounts of algae for microscopic examination.
- Locate and identify chloroplasts, holdfasts, and pyrenoids of various species of algae.
- Distinguish among unicellular, filamentous, and colonial forms of organization.

PROCESS SKILLS: observing, classifying, recognizing and using spatial relationships, using numbers

TIME ALLOTMENT: 2 class periods

MATERIALS
microscopes (30)	laboratory aprons (30)
microscope slides (180)	Living specimens of
coverslips (180)	*Closterium,*
droppers (6)	*Oedogonium,*
iodine solution	*Scenedesmus, Synedra,*
paper towels (1 roll)	*Ulothrix, Volvox*

PREPARATION
- Be sure to order algae cultures to arrive prior to your lab day. Some biological supply houses sell algae sets. Check the catalog to see if the specimens required for this Exploration are more economically obtained as part of a set.
- Prepare iodine solution according to directions on page T13.

TEACHING THE LAB
- If sufficient microscopes are available, have students work individually.
- Set up algae specimens with individually labeled droppers. This will prevent contamination.

DATA AND OBSERVATIONS
Table 1.

Algal specimen	Cellular organization	Number of cells
Synedra	unicellular	1
Closterium	unicellular	1
Oedogonium	filamentous	varies
Ulothrix	filamentous	varies
Volvox	colonial	thousands
Scenedesmus	colonial	4

See the figures below.

ANALYSIS
1. *Synedra* and *Closterium*
2. Usually 4; most colonies have 2 to 6 cells.
3. The individual cells of *Volvox* are very small and they are very numerous. Two flagella on each individual cell allow the colony to move.
4. *Closterium, Oedogonium,* and *Ulothrix* contain pyrenoids.
5. Pyrenoids produce and store starch.
6. Iodine stains starch blue-black. Pyrenoids contain starch, so they are stained by the iodine.

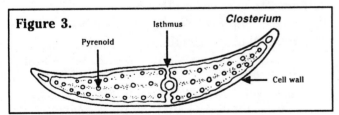

Figure 1.

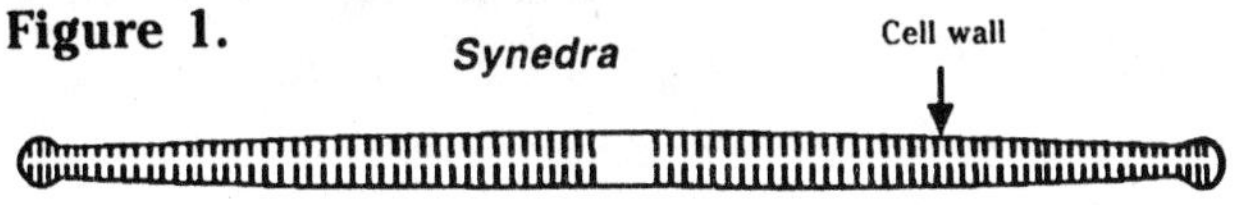

Synedra   Cell wall

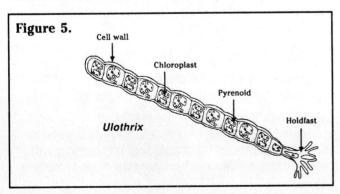

Figure 3.   *Closterium*

Pyrenoid   Isthmus   Cell wall

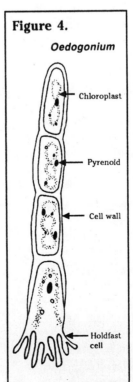

Figure 4.

Oedogonium

Chloroplast, Pyrenoid, Cell wall, Holdfast cell

Figure 5.

Cell wall, Chloroplast, Pyrenoid, Holdfast

Ulothrix

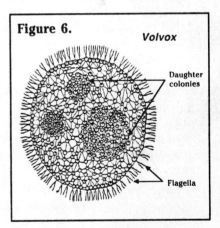

Figure 6.

Volvox

Daughter colonies, Flagella

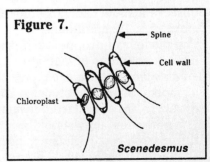

Figure 7.

Spine, Cell wall, Chloroplast

Scenedesmus

How Can Digestion Be Observed in Protozoans?

OBJECTIVES
• Prepare an indicator-stained food source for *Paramecium*.
• Observe the feeding behavior of *Paramecium*.
• Hypothesize the changes in the pH of the food vacuole as food is digested by *Paramecium*.

PROCESS SKILLS: forming hypotheses, formulating models, observing, predicting

TIME ALLOTMENT: 1 class period

MATERIALS
small beakers (5)	petroleum jelly (1 jar)
cream or whole milk (100 mL)	droppers (45)
wood splints (5)	methyl cellulose
Congo red indicator	culture of *Paramecium*
microscope slides (30)	microscopes (30)
coverslips (30)	clock or stopwatches (15)
toothpicks (30)	

PREPARATION
• Arrange for living materials to arrive a day or two before teaching the lab.
• Prepare a solution of methyl cellulose by dissolving 10 g of the compound in 90 mL of hot water. Store in the refrigerator. Plastic 15-mL dropping bottles are ideal for dispensing methyl cellulose. Commercially prepared solutions of methyl cellulose can be purchased from biological supply houses.
• Whole milk, heavy cream, or light cream can be used. Do not use skim milk or low-fat milk, since the Congo red stains fat droplets, and they are the food that will be observed in the lab.
• Congo red is available in powder and in solution. If you have Congo red solution, have students add a few drops of the solution to the milk in the beaker.

TEACHING THE LAB
• Students can work in pairs. If there are sufficient microscopes, have each student make and observe a slide. Each six students will need no more than 10 mL of milk.
• The number of droppers required can be reduced if stock solutions of methyl cellulose and *Paramecium* culture each have their own droppers. Supervise students to be sure that droppers are not mixed between solutions.
• The petroleum jelly should seal the coverslip on the slide so that liquids will not evaporate. The slide will last for several hours. If there is extra time in the lab period, you may wish to have students continue making observations of the paramecium.

HYPOTHESIS
Students will probably write that a color change from red to blue will occur.

DATA AND OBSERVATION
Results will vary, but students should see a color change from red to blue, and back to red. The milk has a pH of about 6.5, which is slightly acid. In the first stages of digestion, the food is acidified, so the indicator turns blue. Later stages of digestion require a basic pH. The food is made more basic and the indicator turns red again. The rate at which these changes take place varies.

Students should see the food vacuoles form at the gullet and move through the cytoplasm. Students may see the discharges of wastes through the anal pore.

Student drawings:

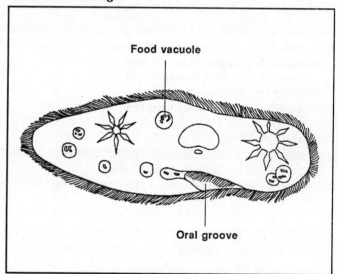

Food vacuole

Oral groove

ANALYSIS
1. Red. The milk has a pH above 5.
2. Answers will vary. Stained vacuoles should appear in about 10 minutes.
3. At first, the vacuoles are red. The food is still at a pH above 5.
4. Times will vary, but the color change will be to blue. This indicates that the pH is below 3.
5. Answers will vary. If a second change is observed, it will be back to red. This indicates a rise in pH above 5.
6. The thymol blue will show if the acid stage reaches a pH of below 1.2. It will show if the basic stage reaches a pH of above 10.

CHECKING YOUR HYPOTHESIS
Students should find that as the food is digested, the pH drops, then rises again as indicated by the color changes from red to blue and back to red again.

EXPLORATION

Identification of Common Molds

OBJECTIVES
- Observe the color and appearance of fungus colonies.
- Examine the reproductive structures of various types of fungi.
- Draw, label, and identify the four different types of fungi.

PROCESS SKILLS: observing, classifying, communicating

TIME ALLOTMENT: 1 class period

MATERIALS
cultures of fungi (4)   microscopes (30)
cellophane tape (1 roll)   microscope slides (120)

ALTERNATE MATERIALS
- You should provide cultures of fungi that represent all four divisions. *Rhizopus* is a good example of a zygomycote, a mushroom or puffball represents Division Basidiomycota, yeast or *Neurospora* represents Division Ascomycota, and *Aspergillus* or *Penicillium* represents Division Deuteromycota.

PREPARATION
- You may wish to subdivide the fungus cultures you purchase from a supply house so students are not all examining a single culture at once. Transfer some of the fungus culture to plates of nutrient medium and incubate according to supply house instructions.

TEACHING THE LAB
- Students should work individually in this Exploration unless microscopes are in short supply.
- Some students may be allergic to spores of fungi. Before beginning this Exploration, check with your students to determine if any have allergies or are medically at-risk.
- Provide reference books on fungi for students to use when they identify the divisions of their fungus cultures.
- Many students have difficulty in removing reproductive structures when using either dissecting needles or scalpels. For this reason, the use of cellophane tape has been suggested. The use of tape removes the need for coverslips that destroy the reproductive structures if improperly placed.

DATA AND OBSERVATIONS
Appearance of colonies and drawings of structures will vary with the fungus cultures examined.

ANALYSIS
1. one cell in size, round, microscopic, and usually produced in very large numbers
2. Individual spores look very much alike, while the entire reproductive structures are often different.
3. Ascomycota, Basidiomycota, Zygomycota, Deuteromycota; specific examples will vary, depending on the cultures used.

Which Foods Can Bread Mold Use for Nutrition?

OBJECTIVES
- Examine bread mold by using a hand lens or stereomicroscope.
- Hypothesize the conditions a bread mold needs for growth.
- Determine if bread mold can use a variety of food sources.
- Determine if bread mold needs moisture.

PROCESS SKILLS: observing, inferring, communicating, predicting, interpreting data, forming hypotheses, experimenting

TIME ALLOTMENT: 1 class period

MATERIALS
hand lenses or
  stereomicroscopes
  (30)
petri dishes containing
  living bread mold (15)
small jars with covers
  (90)
water

labels or wax marking
  pencils (15)
cardboard from a box
  (15 pieces)
dehydrated potato flakes
  (1 package)
raisins (1 package)
cotton swabs (90)

ALTERNATE MATERIALS
- A wet mount can be made of the bread mold for more detailed study if hand lenses or stereomicroscopes are not available.

PREPARATION
- Prepare several bread mold samples before class begins.
- Grocery store instant potatoes can be used for potato flakes.

TEACHING THE LAB
- Have students work in pairs in Part B.
- Instruct students not to chew gum or eat while carrying out the procedure. Instruct them to wash their hands immediately after finishing the Investigation.
- Instruct students to look closely at the jar of wet cardboard. Hyphae of bread mold should be visible.
- Have students dispose of their bread mold cultures according to accepted procedures in your school.
- **Troubleshooting:** Some students may be allergic to spores of fungi. Before beginning this Investigation, check with your students to determine if any have allergies or are medically at-risk.

HYPOTHESIS
Mold needs carbohydrates and water to grow.

DATA AND OBSERVATIONS
Figure 1.

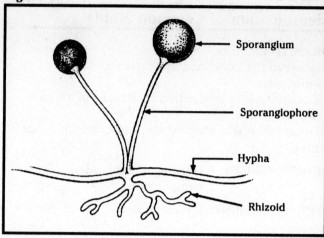

Table 1.

Results of Mold Growth		
Jar	Contents	Mold growth?
1	Dry potato flakes	No
2	Wet potato flakes	Yes
3	Dry raisins	No
4	Wet raisins	Yes
5	Dry cardboard	No
6	Wet cardboard	Yes

ANALYSIS
1. 2, 4, 6
2. The presence of moisture and carbohydrates, an energy source, are needed for growth.
3. Bread mold will not grow without moisture.

CHECKING YOUR HYPOTHESIS
Answers will vary. Students who hypothesized that bread mold would grow only in the presence of moisture and food will say that their hypotheses were supported.

Lichens

OBJECTIVES
• Determine the specific lichen type of three lichen samples.
• Diagram the macroscopic appearance of three lichen samples.
• Observe and diagram the microscopic appearance of a typical lichen.

PROCESS SKILLS: observing, classifying, communicating

TIME ALLOTMENT: 1 class period

MATERIALS
microscopes (30)
microscope slides (30)
coverslips (30)
droppers (30)
water

Cladonia (reindeer moss)
three lichen samples labeled A, B, and C (30 of each)

ALTERNATE MATERIALS
• If *Cladonia* is not available, substitute any other lichen type or prepared slides.

PREPARATION
• Do not mark samples in the order listed for lichen types in Part A.

TEACHING THE LAB
• Students may work well in small groups if lichen material is in short supply.
• If available, a microscope having a 45X objective and a 20X zoom eyepiece can dramatically show the difference in the sample lichen cells.
• Lichen can be stored dry and preserved for long periods of time.
• As the students observe the samples under a microscope, have them determine that the fungus makes up the largest part of the lichen.
• A sample of lichen can be kept after observations and diagrams have been made.The lichen can maintain its good condition by having enough light and moisture.

DATA AND OBSERVATIONS
Table 2: Answers will vary depending upon samples available.

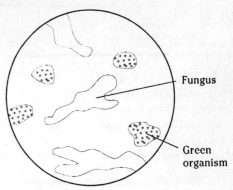

Fungus

Green organism

Cladonia **under high power**

ANALYSIS
1. **a.** an organism composed of a green alga or cyanobacterium and a fungus
   b. living together in close association
2. Two organisms, a fungus and a green organism, live together in close association.
3. The alga or cyanobacterium is small and green; the fungus is larger and colorless.

Roots and Stems

OBJECTIVES
- Identify and label root tissues.
- Describe the functions of root tissues.
- Identify and label the stems on one-, two-, and three-year-old trees.
- Explain the functions of stem tissues.

PROCESS SKILLS: observing, classifying, recognizing and using spatial relationships, interpreting data

TIME ALLOTMENT: 1 class period

MATERIALS
cross sections of parsnip root (30)	iodine stain (10 mL)
	droppers (30)
single-edged razor blades (30)	coverslips (30)
	microscopes (30)
microscope slides (30)	laboratory aprons (30)

PREPARATION
- Purchase parsnip roots in a grocery store. Give each students a slice about 3 mm thick. Do not use the narrow end of the root; the students' pieces will be too small.
- Prepare iodine stain as directed on page T13.

TEACHING THE LAB
- Students should work individually.
- Be sure students cut root sections thin enough to allow the cells to be observed.
- When students examine the slide, direct them to start at the outside edge of the parsnip and move toward the center (Area A to Area B)
- Prepared slides of the tree stems can be purchased and used to extend this lab.

Part C answers:
1. a. 1   b. 2
2. a. spring
   b. More water is available in spring, so cells are larger.
   c. cambium.
   d. pith.

Figure answers (top to bottom)

Figure 3: root hair, epidermis, cortex, endodermis, phloem, xylem

Figure 6, *left:* cork, cortex, xylem

right: cork cambium, bast fibers, phloem, vascular cambium, summer xylem, spring xylem, pith

Figure 7: second-year xylem, first-year xylem

Figure 8, *left:* bark, vascular ray, wood

right: third-year xylem, second-year xylem, first-year xylem

DATA AND OBSERVATIONS
Table 3.

Parsnip Root Section Under Low-Power Magnification	
Region A	**Region B**
	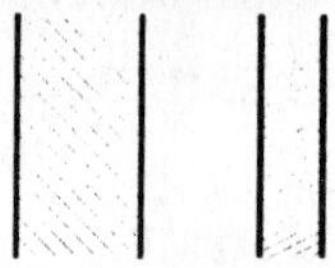

See Table 4 below.

ANALYSIS
1. The cells of the central cylinder are tube-shaped and they transport materials just as water pipes and blood vessels do.
2. Cortex cells are large thin-walled cells and are used for storage.
3.

Roots	Stems
Epidermis, endodermis	Cork
Cortex	Cortex, pith
Phloem, xylem	Phloem, xylem
Root hairs	—
—	Cork cambium, vascular cambium
—	Xylem, bast fibers

4. a. xylem, phloem.
   b. xylem-water and minerals, phloem-food
5. a. Both function as protective coverings.
   b. They are properly named—epidermis is on the outside, endodermis is on the inside.
6. Thick fibers help support the stem.
7. This thin waxy layer helps prevent water loss.
8. a. 3
   b. yes
   c. The number of bands of xylem present in a tree's stem should equal its age.
9. Answers might include water, light, minerals, disease, and temperature.

Table 4.

Root Regions and Their Functions			
Region	**Name**	**Description**	**Function**
A	Cortex	Widest area of root	Storage
B	Central cylinder	Center area of root	Transporting (Some storage cells are also observed here.)

INVESTIGATION

How Do Gymnosperm Stomata Vary?

OBJECTIVES
- Make a hypothesis that relates the number, location, and distribution of stomata on a gymnosperm leaf to the type of environment in which the species of gymnosperm lives.
- Calculate the number of stomata per millimeter of leaf surface using prepared slides of different species of gymnosperms.
- Relate the structure of the gymnosperm leaf to the survival of the plant.

PROCESS SKILLS: observing, measuring, forming hypotheses, inferring, interpreting data

TIME ALLOTMENT: 1–2 class periods

MATERIALS
Prepared single-leaf slides of various species of gymnosperm, such as:
  hemlock leaf
  redwood leaf
  yew leaf
  spruce leaf

compound microscopes (10)
calculators (10)
transparent metric rulers (10)
field guide to conifers

PREPARATION
- Prepared slides can be purchased from biological supply houses. You will probably need to order one set per team or per two teams of students.
- If you wish, photocopy field guide information about gymnosperms for students to use in completing the lab.

TEACHING THE LAB
- Have students work in teams of three.
- Review microscopic techniques.
- You may wish to demonstrate the technique of calculating the diameter of the field of view.
- Review general characteristics of gymnosperms, guard cells, and stomata.
- Discuss biomes and the types of plants associated with them.

HYPOTHESIS
Hypotheses will vary. Students may suggest that gymnosperms in temperate, moist climates with average elevation will have more stomata and those in harsh climates will have fewer stomata.

DATA AND OBSERVATIONS
Table 1.

Diameter of the field of view on low power =	1.5–1.9 mm
Magnification of the high power objective =	40
Magnification of the low power objective =	10
High power objective magnification divided by the low power of magnification =	4
Diameter of the high power field of view =	.4–.5 mm

Table 2. See below.

ANALYSIS
1. The height of the trees may vary, taller ones being exposed to winds (which cause drying out) and shorter ones being protected from the elements by the taller trees. The size and shape of the needle also may vary. Some stomata are recessed, while others are located on a flat surface. The tissue of the leaf may vary, with more tightly packed cell layers vs. more loosely packed cells.
2. Those plants carrying the genes for producing a leaf anatomy that enables the plant to survive and flourish under the given environmental conditions would be the organisms that pass on their traits.
3. Plants growing in temperate, moist environments tend to have broader leaves than those in harsh climates. Leaves from conifers in harsh environments have more tightly packed cells inside the leaf than those found in less harsh environments. There is a more loosely packed spongy layer of cells near the stomata of leaves from temperate, moist environments.

CHECKING YOUR HYPOTHESIS
Answers will vary. Any hypothesis that suggests a relationship between harsh conditions (cold, windy, dry) and fewer stomata vs. temperate conditions with more stomata will be supported.

Name	Total Stomata	Stomata per mm.	Location	Distribution	Habitat
Hemlock leaf	8	8	Lower surface	4 each side central vessel	Temperate North America. Usually 40–50 ft. high. Rocky banks of streams at elevations of 2500–4000 ft.
Redwood leaf	20	8	Lower surface	Evenly distributed to cover lower surface	Oregon southward to Monterey, Ca. Rarely found more than 20–30 miles from the coast, or beyond the influence of the ocean fogs or over 3000 ft. above sea level. Occupies sides of ravines.
Yew leaf	6	4	lower surface of leaf	Three on each side	Shrubs widely distributed in northern hemisphere Banks of mountain streams, deep gorges and damp ravines, growing usually under large coniferous trees
Spruce leaf	2	8	Sides of square leaves	One on each side of square needle	Found on well drained slopes of barren stony hills. All varieties appear to be in harsh climates or at high altitudes.

How Are Ferns Affected by Lack of Water?

OBJECTIVES
- Hypothesize how different stages of the life cycle of a fern are affected by water.
- Observe the release of spores from a fern plant.
- Observe the germination of fern spores.
- Observe the release of sperm from a fern prothallus.

PROCESS SKILLS: observing, inferring, communicating, predicting, forming hypotheses, experimenting

TIME ALLOTMENT: 1 class period, then 20 minutes 1 to 2 days later

MATERIALS
mature fern plants (1 or 2)
scalpels (15)
microscope slides (120)
coverslips (120)
droppers (2)
glycerin (10 mL)
microscopes (30)
wax marking pencils (15)
toothpicks (30)
petroleum jelly (1 jar)
fern-spore culture
mature prothalli (30)
pencils with erasers (15)
water

PREPARATION
- Order living materials to arrive a few days before the lab. Living materials are available from biological supply houses. Fern-spore cultures are available with liquid or solid media. Liquid medium is better for this lab.

TEACHING THE LAB
- Have students work alone if there is sufficient number of microscopes.
- Be sure that the sporophyte plant has been watered well. The sporangia will open best if they have ample moisture before the glycerin is added.

- Caution students to prepare the slide of the prothallus carefully. The prothallus should be right-side up. It should be moist, but the slide must not be too wet or it will flood the microscope stage when the coverslip is pressed.
- Use plastic coverslips for this Investigation, as students may break glass ones when pressing down on them.

HYPOTHESIS
Water will be needed for germination of spores and release of sperm, but not for release of spores.

DATA AND OBSERVATIONS
See the table below.

ANALYSIS
1. Glycerin simulates dry conditions, because it absorbs moisture from the plant tissues.
2. Some cells of the sporangium shrunk as moisture was removed. This caused the sporangium to open, releasing the spores inside.
3. Spores are released under dry conditions.
4. Spores began to germinate in the sealed slide. After a few days, the unsealed slide was dry, but the sealed slide had some moisture left. The germinating spores on the dry slide became dried out. The germinating spores on the sealed slide continued to develop.
5. Sperm were released when the prothallus was moist.

CHECKING YOUR HYPOTHESIS
Answers will vary. Students who said that water would be needed for germination of spores and release of sperm, but not for release of spores will say that their hypothesis was supported.

Table 1.

Observations	Wet prothallus	Dry prothallus
Release of sperm	Sperm are released.	No sperm (or very few) are released.
Movement of sperm	Sperm swim; some may be seen moving toward eggs.	Sperm do not swim.

What is the Effect of Light Intensity on Transpiration?

OBJECTIVES
- Prepare a setup that will test the effect of light intensity on a plant's transpiration rate.
- Make a hypothesis to describe the effect of light intensity on the rate of transpiration.
- Observe the effect of high and low light intensity on transpiration rate.
- Compare the rates of transpiration in a plant exposed to two different light intensities.

PROCESS SKILLS: observing, inferring, communicating, measuring, predicting, using numbers, interpreting data, forming hypotheses, separating and controlling variables, experimenting

TIME ALLOTMENT: 1 class period

MATERIALS

stop watches (8)	10 cm rubber surgical
22-cm Pasteur pipettes (8)	tubing, 5 mm bore (8 pieces)
permanent markers (8)	incandescent lamps (8)
pans, approx. 30 × 50 × 20 cm, with water approx. 10 cm deep (8)	25- and 100-watt bulbs (8 each)
pruning shears (8)	dicot branches with leaves (8)
petroleum jelly (8 jars)	thermal mitts (8)
metric rulers (8)	laboratory aprons (30)

ALTERNATE MATERIALS
- You can make pipettes by using glass tubing. Form pipette tips by melting the glass in a Bunsen burner. Then draw the tubing out into small tips.
- Young room-grown bean plants can be used instead of cutting the number of branches needed from a tree or shrub.

PREPARATION
- Try to have branches for all groups approximately the same size with the same numbers of leaves so that results will be consistent from group to group. Since the transpiration rate is different for different species of plants, use the same species for all groups.

TEACHING THE LAB
- Have the students work in groups of four.
- The best branches to use are those with smooth woody stems. The sample data are for a lilac branch with four mature leaves. You can use narrower tubing for smaller branches. You can

slow the rate of water movement in the pipette by reducing leaf area.
- With a Pasteur pipette, after the first 3.5–4.0 cm, the inside diameter increases. This difference in diameter is not easily seen, but the retreat of water will slow as the pipette widens.
- Caution students to use care with pipettes. The tips of Pasteur pipettes break easily. A pipette with a broken tip can still be used after remarking the volume to be use, but the observation time will increase with increasing diameter of the tube.
- Students will notice that the 100-watt bulb produces more heat than the 25-watt bulb. Explain that two factors are involved in this Investigation: heat and light intensity. Stomata open wider with greater light intensity and transpiration increases with an increase in temperature.
- **Troubleshooting:** Students may need help in calculating averages in step 13.

HYPOTHESIS
The greater the light intensity, the greater the rate of transpiration will be.

DATA AND OBSERVATIONS
Table 1.

	Elapsed time (in seconds)			
Light intensity	Trial 1	Trial 2	Trial 3	Average
Low (25-watt bulb)	399 s	380 s	402 s	394 s
High (100-watt bulb)	195 s	182 s	169 s	182 s

ANALYSIS
1. As water vapor is given off by the leaves, water is drawn up into the vascular system of the branch. The movement of water in the tube shows this.
2. Answers will vary. It should take about half the time for water to move 8 cm under a 100-watt bulb as under a 25-watt bulb.
3. Yes, the greater the light intensity, the greater the rate of transpiration will be.
4. Light influences the opening and closing of the stomata. Stomata do not open in the absence of light. Also, light produces heat on the leaf surfaces. The greater the light intensity, the more heat will be produced and the greater the transpiration rate.
5. A branch with twice as many leaves will move water in the tube twice as fast because the branch will have twice the surface area at which transpiration occurs.

CHECKING YOUR HYPOTHESIS
Answers will vary. Students who predicted that an increase in light intensity increases the rate of transpiration will say that their hypotheses were supported.

Do Dormant and Germinating Seeds Respire?

OBJECTIVES
- Prepare respiration chambers to measure the amount of oxygen used by dormant and germinating seeds.
- Compare the respiration rate of dormant and germinating seeds.
- Predict the volume of oxygen used by seeds in several experimental conditions.
- Hypothesize how much oxygen is used by dormant and germinating seeds.

PROCESS SKILLS: observing, inferring, communicating, measuring, predicting, using numbers, interpreting data, forming hypotheses, experimenting

TIME ALLOTMENT: 1 class period, then 15 minutes 24 hours later

MATERIALS
test tubes (30)	water, colored (350 mL)
bean seeds, soaked (50)	measuring
bean seeds, dry (50)	half-teaspoons (10)
absorbent cotton (1 roll)	masking tape (10 rolls)
metric rulers (10)	marking pens (10)
soda lime (100 grams)	rubber bands (10)
small beakers (10)	laboratory aprons (30)

PREPARATION
- Prepare all materials before class begins.

TEACHING THE LAB
- Have students work in groups of three.
- Students can push the cotton plugs into the tubes with a pencil.

HYPOTHESIS
The test tube containing the germinating seeds will show the greatest difference in water height, indicating that the most oxygen has been used.

DATA AND OBSERVATIONS
Table 1.

Height of Water in Test Tubes			
Test tube contents	Height of water at start	Height of water after 24 hours	Difference in water height
Dormant seeds	0 mm	0 mm	0 mm
Germinating seeds	0 mm	16 mm	16 mm
No seeds	0 mm	0 mm	0 mm

ANALYSIS
1. air; Air in the test tube prevents water from moving into the tube.

2. germinating seeds; water has moved into the tube; oxygen
3. cellular respiration
4. germinating; the height of the water in the tube after 24 hours is greater for germinating seeds.
5. It shows that the water rise is not due to the presence of soda lime.
6. no water level change; The amount of oxygen used is about equal to the amount of CO_2 released.
7. dormant; no water level changes; dormant; The food within seeds would last for a long time, thus allowing seeds to survive over the winter.
8.

Table 2.

Number of seeds	Expected height of water 24 hours later
20 germinating seeds	44 mm
10 germinating seeds	20 mm

Answers will vary. However, the tube with more seeds should show the higher water level.

9. The greater the number of seeds used, the greater amount of oxygen used and the higher the water rises.

10.

Table 3.

Treatment	Expected height of water 24 hours later
20 germinating seeds with soda lime	42 mm
20 germinating seeds without soda lime	1 mm

Student answers will vary.

11. Carbon dioxide replaces the oxygen used, so no change in water level occurs.

12.

Table 4.

Seed Treatment	Expected height of water 24 hours later
Boiled germinating seeds with soda lime	0 mm
Unboiled germinating seeds with soda lime	30 mm

Student answers will vary.

13. Boiled seeds are no longer living; thus no respiration occurs.

CHECKING YOUR HYPOTHESIS
Hypotheses will be supported if they predicted that the test tube with germinating seeds showed the greatest change in water level.

How Do Hormones Affect Plant Growth?

OBJECTIVES
- Record the positions and directions of growth of new roots and stems for four corn seeds over a minimum 48-hour period.
- Make a hypothesis that explains how the direction a seed faces influences the growth of the stem and roots.
- Compare the direction of growth shown by young corn roots and stems when positioned in different orientations to the force of gravity.

PROCESS SKILLS: observing, inferring, communicating, recognizing and using spatial relationships, predicting, forming hypotheses, experimenting

TIME ALLOTMENT: 1 class period to set up; one-half period 48 and 72 hours later for observations

MATERIALS
bowls or shallow dishes (15)	paper towels (1 roll)
corrugated cardboard (15 pieces)	soaked corn seeds (60)
	plastic bags (15)
staplers (8)	straight pins (60)
metric rulers (15)	masking tape (8 rolls)
	water

ALTERNATE MATERIALS
- Beans may be used instead of corn.

PREPARATION
- Soak the seeds for 24 hours before class.

TEACHING THE LAB
- Have students work in teams of two.
- It is important that students orientate their seeds as shown in Figure 2.
- If no visible signs of root and stem growth are detected after 48 hours, have students examine the seeds again after an additional one or two days.

HYPOTHESIS
No matter what the position of a seed, the roots will turn to grow toward the force of gravity (down) and the stems will grow away from the force of gravity (up).

DATA AND OBSERVATIONS
See the diagram below.

ANALYSIS
1. a. down
   b. down
   c. down
   d. down
2. a. It has no influence.
   b. They grow toward the force of gravity.
3. a. up
   b. up
   c. up
   d. up
4. a. It has no influence.
   b. Stems grow away from the force of gravity.
5. The plant hormone auxin affects how roots and stems grow. Roots and stems have more auxin on the bottom and less on the top. On roots this makes the cells curve down toward the force of gravity. The opposite is true of stem cells.
6. They respond differently. The lower surface of a root grows more slowly when auxin increases. The lower surface of a stem grows more rapidly when auxin increases.
7. No, seed position has no effect on the direction of stem and root growth.

CHECKING YOUR HYPOTHESIS
Answers will vary. Students who stated that position of the seeds does not affect the direction of growth of roots and stems will say their hypothesis is supported.

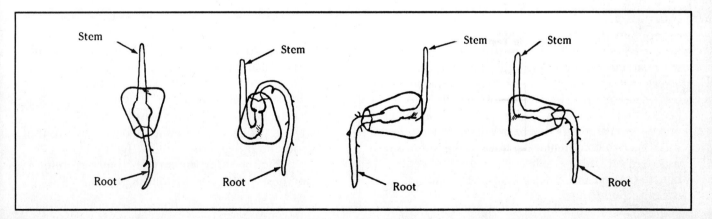

EXPLORATION

Symmetry

OBJECTIVES
• Identify the symmetry of a variety of animals.
• Relate symmetry to food capture and locomotion.

PROCESS SKILLS: observing, communicating, classifying, experimenting, inferring

TIME ALLOTMENT: 1 class period

MATERIALS

museum mount of a bird
preserved or dried
   specimens of a sand
   dollar, sea urchin,
   brittle star, jellyfish,
   butterfly or moth, sea
   anemone, and squid
live crayfish, goldfish,
   lizards, grasshoppers
living cultures of
   Daphnia
   Planaria
   Hydra
   rotifers
yeast boiled in Congo red
grass
beef liver

living flies
lettuce
guinea pig pellets
fish food
aquarium or fish bowl
terrarium
spring, pond, or aquarium
   water (2–4 L)
watchglasses (10)
compound microscopes
   (10)
pipettes (10)
hand lenses or dissecting
   microscopes (10)
depression slides (10)
coverslips (10)
glass jars, large (20)

PREPARATION
• Before the lab period, organize materials into a central lab position.
• Keep the *Hydra* culture away from direct sunlight and maintain the culture at room temperature. The *Hydra* should not be fed for 24 hours prior to the lab in order for students to observe feeding behavior.
• See pages xii–xiii in the students' manual for tips on setting up an aquarium and terrarium.
• If you wish, *Planaria* and rotifers can be collected from pond water. *Planaria* are usually found on the undersurface of stones and rotifers can be found among vegetation.

TEACHING THE LAB
• Students can work individually for Part A, but have them work in groups of three for Part B, with each group handling a different species.
• Stress that all living organisms must be treated with gentleness and proper care. Emphasize the need to keep certain organisms, such as *Hydra*, from drying out. Caution students that lizards in a dry terrarium will require a continual water source and need to be provided with adequate heat since they are cold-blooded. Also caution students that regular tap water will kill *Hydra*.

DATA AND OBSERVATIONS

Table 1.   **Table 2.**

Organism	Symmetry	Organism	Symmetry
horseshoe crab	bilateral	sand dollar	radial
sea star	radial	sea urchin	radial
turtle	bilateral	brittle star	radial
skate	bilateral	jelly fish	radial
dog	bilateral	sea anemone	radial
sponges	asymmetry	butterfly/moth	bilateral
millipede	bilateral	squid	bilateral
scorpion	bilateral		

Table 3.

Organism	Symmetry	Description of Movement and Eating Behavior
Planaria	bilateral	gliding movement, extends and contracts body in direction of food
Rotifer	bilateral	cilia surrounding mouth draw rotifer through water and sweep food into its mouth
Hydra	radial	seem to just wait for *Daphnia* to drift by; pull in prey with tentacles. Seem able to move circularly
Crayfish	bilateral	propelled itself backward when hand reached toward aquarium; raised itself on its legs and walked slowly toward pellets
Goldfish	bilateral	swims rapidly toward food and grabs it with mouth
Lizard	bilateral	remains still and then suddenly moves directly toward fly, capturing it with tongue and drawing it into mouth
Grasshopper	bilateral	jumps directly toward lettuce

ANALYSIS
1. Organisms with radial symmetry seem stationary but able to revolve in all directions in response to prey. Organisms with bilateral symmetry have a directional movement, usually toward the anterior end of the organism.
2. Food capture in organisms exhibiting radial symmetry seems random. In the case of *Hydra*, for example, whatever comes to the organism that is edible will be grasped by the tentacles and pushed into the cavity. Animals with bilateral symmetry move in a more aggressive way to obtain and eat the food.
3. Advantage: can detect and turn toward prey coming from any direction. Disadvantages: no anterior and posterior ends, no centralization of a nervous system and sense organs for aggressive movement and food capture.
4. The trend seems to be toward bilateral symmetry because it enables organisms to find food and move more efficiently. It makes it possible for an organism to be more mobile and more active in responding to its environment.

Hydra Behavior

OBJECTIVES
- Prepare a wet mount of *Hydra* for microscopic observation.
- Observe the triggering of nematocysts caused by dilute acid.
- Observe the reactions of *Hydra* to food and non-food stimuli.

PROCESS SKILLS: observing, communicating, experimenting

TIME ALLOTMENT: 1 class period

MATERIALS
droppers (45)
Hydra culture
depression slides (30)
stereomicroscopes (15)
dilute acetic acid (25 mL)
small beakers (15)
aquarium water (1 L)
dissecting probes (15)
Daphnia culture

ALTERNATE MATERIALS
- If depression slides are not available, use the bases of petri dishes or regular slides with coverslips. If coverslips are used, put brush bristles under the coverslips to keep from crushing the hydra.
- Other dilute acids could also be used.
- *Tubifex* worms could be used in addition to or in place of *Daphnia*.
- If aquarium water is not available, aged tap water can be used. Place tap water in a pail and allow to age for several days.

PREPARATION
- Before the lab period, put all materials together in a central location.
- Test the concentration of the dilute acid in advance so only one or two drops are needed to stimulate the triggering of nematocysts.

TEACHING THE LAB
- Have students work in pairs.
- Caution students to treat *Hydra* with care as they are living animals.
- Remind students to be sure the hydra do not dry out.
- Remind students to be gentle when shaking and touching hydra.
- Use hydra that have not been fed for 4 to 5 days. Hydra that are not hungry will not attack the *Daphnia*. If a hydra does not attack *Daphnia*, have the students try others.

DATA AND OBSERVATIONS
Table 1.

Reactions of *Hydra*	
Stimulus	**Response**
Shaking of slide	organism contracts
Touching with probe	organism contracts, tentacle withdraws
Presence of *Daphnia*	nematocysts are triggered, *Daphnia* are moved to mouth

ANALYSIS
1. The acid is an irritant; the hydra triggers nematocysts as a defense mechanism.
2. The reactions to shaking and touching are protective responses. The response to *Daphnia* is a feeding response.
3. When the presence of prey is detected, nematocysts are triggered, subdued prey are moved by tentacles to the mouth, prey are ingested and digested.
4. Nematocysts paralyze prey instantly, making pursuit unnecessary.
5. protection from potential predators

Earthworm Dissection

OBJECTIVES
• Dissect and identify internal and external features of the earthworm.
• Draw and label a diagram of the external anatomy of the earthworm.
• Label a diagram of the internal anatomy of the earthworm.
• Describe the major features of the earthworm phylum.

PROCESS SKILLS: observing, communicating, recognizing and using spatial relationships, defining operationally

TIME ALLOTMENT: 1 class period

MATERIALS
colored pencils (32)
dissecting pins (80–120)
dissecting pans (8)
hand lenses or stereo-microscopes (8)
single-edged razor blades or scalpels (8)
earthworms (preserved or freshly anesthetized) (8)
laboratory aprons (30)

PREPARATION
• Earthworms can be collected by the class or ordered from a supply house.
• Earthworms can be anesthetized by placing a bed of cotton in a jar with a tight lid. Pour anesthetizing agent into the cotton, moistening it. Place the worms on the cotton, close the lid, and leave the worms for 5 to 10 minutes. This can be done between classes. Any one of the following can be used as an anesthetizing agent: 0.2% chloretone, acetone, alcohol, ether, or chloroform.

TEACHING THE LAB
• Have students work in groups of four.
• Be sensitive to views of students who are opposed to dissection. Be sure that students show a respect for animals that were once living.
• Preserved earthworms can be used but they show considerable loss of color, making it difficult to distinguish internal organs. Freshly anesthetized specimens are much better.
• In fresh specimens, students can cut off one of the seminal vesicles, smear it in a drop of water on a slide, and add a coverslip. Students can observe earthworm sperm on high power.
• In fresh specimens, students can rub the anesthetized earthworm with a dry paper towel and remove small pieces of the cuticle covering the earthworm's skin.
• Give students instructions on how to dispose of their specimens.

DATA AND OBSERVATIONS
External view of an earthworm:

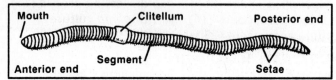

See the figure below.

ANALYSIS
1. dorsal
2. There are four pairs per segment—two pairs on the ventral surface and two pairs at the sides. They point backwards. They are used to grip the soil surface during locomotion.
3. mouth, pharynx, esophagus, crop, gizzard, intestine, anus
4. Any earthworm encountered can serve as a mate.
5. closed circulatory system, complete digestive system, true coelom, true segmentation

Figure 2.

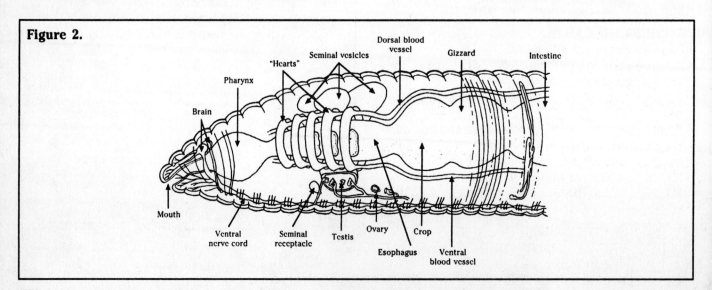

How Do Snails Respond to Stimuli?

OBJECTIVES
- Observe a snail's response to gravity, touch, and light.
- Test a snail's response to a chemical substance and to substances common in its environment.
- Make a hypothesis that describes a snail's response to an acid substance.
- Measure the rate at which a snail moves.

PROCESS SKILLS: observing, inferring, predicting, measuring, forming hypotheses, interpreting data

TIME ALLOTMENT: 1 class period

MATERIALS
white paper (30 pieces)	vinegar or lemon juice (1 bottle)
stereomicroscopes or hand lenses (6)	small plastic trays with lids, approx. 15 cm × 20 cm × 5 cm (6)
lettuce (6 leaves)	
pencils (6)	
glass jars (1 pint) or 250-mL beakers (6)	cardboard, approx. half the size of the tray (6 pieces)
paper towels (1 roll)	
live snails (30)	metric rulers (30)
camel hair brushes (6)	stopwatches (6) or clock
moist soil (1 small bag)	

ALTERNATE MATERIALS
- Aquatic snails can be used for this Investigation, but they must be kept quite moist and their reactions are slower. Isopods can also be used.

PREPARATION
- Land snails could be collected by students for this Investigation. The snails can be brought to class and kept in a terrarium until needed.

TEACHING THE LAB
- Have students work in groups of five, with one student doing Part A, a second student doing Part B, and the other three students testing the three substances in Part C. All students should do Part D. During the 15-minute observation periods, students can also make observations of those parts of the Investigation done by their teammates.
- You may wish to have students write hypotheses regarding snail behavior other than the response to vinegar or lemon juice.
- Part D can be done as a large group activity.
- **Troubleshooting:** Students may have difficulty doing the calculations in step 5 of Part D. Review finding averages and calculating rate per second, if necessary.

HYPOTHESIS
Answers will vary. Many students will say that snails will exhibit a negative response to vinegar or lemon juice.

DATA AND OBSERVATIONS
See the tables below.

ANALYSIS
1. a. away from gravity
   b. negative geotaxis
2. a. Answers will vary, but in the light some snails may withdraw into their shells while others will change their direction of movement toward the darkness.
   b. This behavior will lead them to dark, protective coverings or moist habitats so that they don't dry out.
3. Some snails will withdraw into their shells, while others will move away from the touch.
4. a. They crawled over soil and lettuce.
   b. These are not harmful substances—one is a possible food.
5. a. Answers will vary. The snails may move toward or away from a chemical substance depending upon whether it is potential food or harmful.
   b. The chemical substance may be a source of food.

CHECKING YOUR HYPOTHESIS
Answers will vary. Students who hypothesized that snails would exhibit a negative response to vinegar or lemon juice will respond that their hypotheses were supported.

Table 1.

Responses of Snails to Stimuli	
Test	**Response**
Geotaxis	Snails exhibit a strong negative geotaxis.
Touch	Snails respond to touch by withdrawal.
Phototaxis	Snails travel to the dark side, responding negatively to light.
Chemotaxis	
Lettuce	Snails respond positively to lettuce by crawling over it.
Soil	Snails respond positively to soil by crawling over it.
Vinegar or lemon juice	Snails exhibit negative response by moving away from the vinegar or lemon juice

Table 2.

A Snail's Pace	
	mm/60 s
Trial 1	60
Trial 2	45
Trial 3	57
Trial 4	55
average speed =	54.25 mm/60 s
average speed =	0.9 mm/s

EXPLORATION

Squid Dissection

OBJECTIVES
- Dissect and identify the organs and major organ systems of the squid.
- Describe the major features of the squid phylum.
- Determine the function of various squid features.

PROCESS SKILLS: observing, communicating, recognizing and using spatial relationships

TIME ALLOTMENT: 1 class period

MATERIALS

squid (fresh or thawed) (8)	dissecting pans (8)
scissors (8)	stereomicroscopes or hand lenses (8)
dissecting pins (40–80)	laboratory aprons (30)

PREPARATION
- Purchase squid. They may be ordered in 5 pound boxes from a supermarket or fish store. This size box usually will provide about 40 to 60 squid that are each about 30 cm long.

TEACHING THE LAB
- Have students work in groups of four.
- Be sensitive to students who oppose dissection of animals.

DATA AND OBSERVATIONS
See the table below.

ANALYSIS
1. A squid takes water into the mantle cavity and closes the collar tightly around the head. The mantle muscle contracts and water is squeezed out of the siphon, producing jet propulsion. Squid can also crawl using their tentacles and suckers.
2. Structures that produce jet propulsion, tentacles, suckers, and a beak are all features used for a predatory existence.
3. Without a heavy mollusk shell, the squid becomes more buoyant and capable of rapid movement.
4. A focusing lens orients light onto a retina.
5. bilateral symmetry, muscular foot (divided into tentacles in the squid), complete digestive tract, mantle, circulatory system present with a heart, separate sexes

Table 1.

Organ	Function
Mantle	Muscular body covering, used in jet-propulsion
Siphon	Muscular tube, used in jet-propulsion
Gills	Take oxygen out of the water
Suckers	Used to grasp prey as well as in crawling-type locomotion
Pen	Vestigial internal support
Ink sac	Squirted as a defense mechanism, smokescreen
Nephridium	Removes liquid waste from the blood
Nidamental gland	Secretes protective covering over eggs

How Does Temperature Affect Mealworm Metamorphosis?

OBJECTIVES
- Observe the four stages of the life cycle of the mealworm, *Tenebrio*.
- Hypothesize how an increase in temperature will affect the rate of metamorphosis.
- Conduct an experiment to test the effect of temperature on the development and emergence of an adult mealworm from a pupa.

PROCESS SKILLS: observing, inferring, communicating, predicting, using numbers, interpreting data, forming hypotheses, separating and controlling variables, experimenting

TIME ALLOTMENT: 1 class period, then 5 minutes a day until all adults have emerged

MATERIALS
samples of mealworms (egg, larva, pupa, adult) (8 of each)
mealworm pupae (of same age) (32)
wax marking pencils (8)
stereomicroscopes (8)
chart of *Tenebrio* life cycle
plastic vials (32)
foam plugs (32)
incubator (at 30°C)
thermometer

ALTERNATE MATERIALS
- Plastic vials are the same type as those used for *Drosophila* culture. Pill bottles can also be used.

PREPARATION
- At the beginning of the school year, start a large mealworm culture in a 5-gallon bucket. Fill the bucket 1/2 to 2/3 full of bran meal. Place 25 to 30 mealworms, acquired from a pet shop or biological supply house, in the bucket on top of the bran meal. Crinkle paper towels to cover the bran. Place 4 to 5 apple or potato slices on top of the paper towels. Change the slices every week or so. This culture can be used to supply this Investigation with mealworm pupae.
- To ensure that each student's pupae are the same age, culture the larvae until they pupate and collect the pupae daily. New pupae are white; they turn yellowish-brown as they mature.
- Mealworm larvae or pupae can also be purchased for the class from a biological supply house.

TEACHING THE LAB
- Have students work in groups of four.

- After adult mealworms emerge and are recorded, students can add them to the culture pail.
- Students may be confused by the different names used for mealworms. *Tenebrio* is the genus name. Mealworm and darkling beetle are common names for the same insect.

HYPOTHESIS
An increase in temperature decreases the length of time of metamorphosis in *Tenebrio*.

DATA AND OBSERVATIONS
Table 1.

Tenebrio Metamorphosis		
Temperature	Starting date	Length of time for emergence (Days)
Room temp. A		12
Room temp. B		14
30°C A		6
30°C B		8

Table 2.

Calculations			
Temperature	Total number of days for entire class	Total number of pupae for entire class	Average time for emergence
Room temp. (21°C)	206	16	12.9
30°C	115	16	7.2

ANALYSIS
1. An increase in temperature decreased the length of time for metamorphosis.

2. Metabolism and hormone action increase with increased temperature.

3. Many observations are more accurate than only a few. The pupae may not all be exactly the same age, so an average age is used.

4. egg ⟶ larva ⟶ pupa ⟶ adult ⟶

CHECKING YOUR HYPOTHESIS
Answers will vary. Students who predicted a shorter metamorphosis at 30°C will say that their hypotheses were supported by their data.

Identifying Insects

OBJECTIVES
- Become familiar with various insect anatomical characteristics.
- Become familiar with using a biological key.
- Use the characteristics of insects and a biological key to identify insects to their order.

PROCESS SKILLS: observing, classifying, recognizing and using spatial relationships, defining operationally

TIME ALLOTMENT: 1 to 2 class periods

MATERIALS
white adhesive labels
hand lenses or stereomicroscopes (several)
insect collections from Exploration 1–1

ALTERNATE MATERIALS
- Adhesive tape could be used for labels.

PREPARATION
- Obtain several copies of insect field guides for student use.

TEACHING THE LAB
- The term *tarsal* refers to the terminal portion of the leg of an arthropod.
- This key is to the common orders of insects. Many uncommon orders of insects have been left out for simplicity sake. For example, the key does not have the orders for scorpion flies (Mecoptera), book lice (Psocoptera), or bird lice (Mallophaga).
- If each student prepared his or her own insect collection for Exploration 1–1, it will probably take more than one class period for them to each identify 15 insects. If student groups made the collections, they can work together to key the 15 insects in the group's collection.
- Collections can be put on display.

DATA AND OBSERVATIONS
Each collection will contain a unique variety of insects. Students should be able to key their specimens and label them.

ANALYSIS
1. Answers will vary. The key could have been started with any feature that is not shared by all insects.
2. Yes, an order is a division of a class.
3. Answers will vary.
4. No, insects of the same species can vary greatly in color. The ladybird beetle is usually a good example of this.
5. type of wings
type of antenna
type of mouth parts
presence or absence of abdominal appendages

Comparing Arthropods

OBJECTIVES
- Examine the characteristics of a spider, a crayfish, and a grasshopper.
- Determine which characteristics are phylum characteristics.
- Determine which characteristics are class characteristics.
- Construct a dichotomous taxonomic key.

PROCESS SKILLS: observing, classifying, recognizing and using spatial relationships

TIME ALLOTMENT: 1 class period

MATERIALS
preserved spiders (8)
preserved crayfish (8)
preserved grasshoppers (8)
dissecting probes (16)
dissecting pans (8)
stereomicroscopes or hand lenses (8)

PREPARATION
- Large spiders are needed for specimens. These can be ordered from biological supply houses.

TEACHING THE LAB
- Have students work in groups of four.
- The specimens, if handled with care by the students, can be used from year to year.
- Students may need their textbooks to help them identify body structures.

DATA AND OBSERVATIONS
See the table below.

Student Key
Student keys will vary. One example of a key is given here.
1. A. Two body segments . . . . Go to 2.
   B. Three body segments . . . . . Class Insecta
2. A. Four leg pairs . . . . Class Arachnida
   B. Five leg pairs . . . . Class Crustacea

ANALYSIS
1. having segmented bodies, jointed appendages, modified appendages,exoskeleton
2. all characteristics except those listed in question 1
3. Members of a class will have more features in common than members of a phylum.
4. The characteristic that organisms share are usually the result of having similar genetic make-ups. Organisms with many characteristics in common will have similar genetic make-ups and be in the same group. Organisms with few characteristics in common will have dissimilar genetic make-ups and will be in different groups.

Table 1.

Characteristics of Arthropods			
Characteristic	Crustacea (crayfish)	Insecta (grasshopper)	Arachnida (spider)
Segmented body?	Yes	Yes	Yes
Type of skeleton?	Exoskeleton	Exoskeleton	Exoskeleton
Type of appendages?	Jointed	Jointed	Jointed
Appendages structurally different?	Yes	Yes	Yes
Functions of appendages?	Sensory Manipulation Crushing Walking Swimming	Sensory Chewing Walking Jumping Flying	Poisoning Manipulation Walking
Wings present?	No	Yes	No
Number of body regions?	2	3	2
Body regions fused?	Yes	No	Yes
Number of simple eyes?	3	3	8
Number of compound eyes?	2	2	0
Number of antenna pairs?	2	1	0
Number of leg pairs?	5 pairs	3 pairs	4 pairs

Do Starfish Respond to Gravity?

OBJECTIVES
- Observe the external structures of a living starfish.
- Observe the walking behavior of a living starfish.
- Construct a hypothesis that describes a starfish's reaction to gravity.
- Determine whether or not a starfish responds to the force of gravity.

PROCESS SKILLS: observing, inferring, communicating, recognizing and using spatial relationships, predicting, interpreting data, forming hypotheses, experimenting

TIME ALLOTMENT: 1 class period

MATERIALS
living starfish (8)
seawater
colored celluloid
   (8 sheets)
glass containers for
   starfish (8)
glass plates (8)
drawing paper (8 pieces)

PREPARATION
- See Working with Animals on page T9 for instructions on setting up a marine aquarium and making salt water for starfish.

TEACHING THE LAB
- Have students work in groups of four.
- The glass containers for the starfish should be large enough to hold the starfish and allow space for movement. Glass baking pans work well.
- Celluloid or acetate should be colored so that the movement of the starfish can be readily observed in the water.
- It is critical that equipment used for live starfish be clean of all preservatives. It is recommended that certain glassware be reserved for work with live animals only.

HYPOTHESIS
Starfish will right themselves in response to being turned over. Starfish will move downward on an inclined surface. Both of these responses are examples of positive geotaxis.

DATA AND OBSERVATIONS
1. The celluloid moved backward as the starfish moved forward.
2. It began to right itself. First one ray bent over, and then others followed. When the tube feet of several rays contacted the surface, the animal gripped the surface and turned the rest of its body over.
3. Yes.

4. Answers will vary between two and three.
5. It moved downward.
6. It reversed its direction and moved downward.

ANALYSIS
1. Yes, it moved in the opposite direction.
2. The starfish showed positive geotactic behavior. It moves in the same direction as gravity.

CHECKING YOUR HYPOTHESIS
Answers will vary. Students who predicted that starfish respond positively to the force of gravity will say that their hypotheses were supported.

Starfish Dissection

OBJECTIVES
- Dissect a preserved starfish.
- Identify and label the major external and internal structures of a starfish.
- Describe how the water vascular system enables a starfish to move.

PROCESS SKILLS: observing, communicating, recognizing and using spatial relationships

TIME ALLOTMENT: 1 class period

MATERIALS
scissors (8)	laboratory aprons (30)
preserved starfish (8)	stereomicroscope
dissecting trays (8)	demonstration slide of
dissecting probes (8)	gills and pedicellarias

PREPARATION
- Set up a demonstration slide on a stereomicroscope showing the gills and pedicellarias. Students should examine this slide in step 4 of Part A. This slide can be purchased from Carolina Biological Supply Company.

TEACHING THE LAB
- Have students work in groups of four.
- Have students dispose of their specimens according to your school's procedures.
- **Troubleshooting:** In preserved specimens, the stomach rarely remains intact. It has a tendency to stick to the central disc and get torn when the disc is removed. Point this out to students so that they do not get upset if their specimen's stomach is torn.

DATA AND OBSERVATIONS
See the figure below.

ANALYSIS
1. madreporite plate—entrance for seawater into the water vascular system
   spines—protection against predators
   gills—respiration
   digestive glands—digestion
   pedicellarias—keep upper surface of skin clear of debris
   tube feet—locomotion and respiration
   gonads—reproduction/gamete production
   radial canals—distributes seawater to the tube feet
2. The starfish alternates contraction and expansion of the ampullae. Contraction causes the tube foot to adhere to the surface while expansion releases the tube foot from the surface. The alternating release and attachment to a surface allows for movement.
3. the rays, digestive glands, gonads, and radial canals

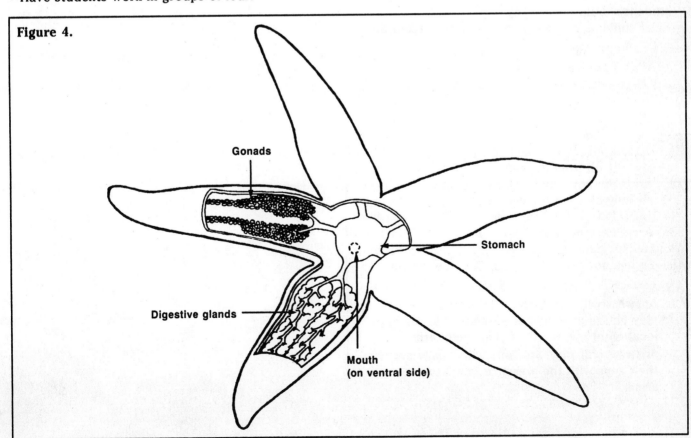

Figure 4.

Gonads

Stomach

Digestive glands

Mouth
(on ventral side)

32–3 EXPLORATION

LAB

Making an Echinoderm Key

OBJECTIVES
• Construct a key that can be used to identify various echinoderms.
• Utilize the key to identify five echinoderms.

PROCESS SKILLS: observing, classifying, communicating

TIME ALLOTMENT: 1 class period

MATERIALS
Specimens, models, Plastimounts, or photographs of the following:
starfish (8)
sea cucumber (8)
sea urchin or sand dollar (8)
brittle star (8)
sea lily (8)

PREPARATION
• Review the procedure for making a key. The Skill Handbook in the text is useful.
• Specimens or Plastimounts are preferred to models and pictures. Plastimounts are preferred for brittle stars since they break so easily.

TEACHING THE LAB
• Have students work in groups of four, but each student should make a key.
• Refer students to the keys in Investigation 20–1 and Exploration 31–2 or the Skill Handbook in the text.

DATA AND OBSERVATIONS
Many keys can be constructed. This is an example. Accept all keys that work.

1A Has arms . . . . Go to 2
1B Has no arms . . . . Go to 4
2A Arms are very long and come
    sharply off from the central disc. . Brittle star
2B Arms are not long and do not come
    sharply off from the central disc . . Go to 3
3A Arms have many branches . . . . . Sea lily
3B Arms have no branches . . . . Starfish
4A Has long spines . . . . Sea urchin
4B Spines not evident . . . . . Sea cucumber

ANALYSIS
1. Answers will vary. Some characteristics used may be the presence or absence of arms, location of the mouth, and shape of arms.

2. Answers will vary. Students will usually begin their keys with the presence or absence of arms.

33-1 EXPLORATION

LAB

Frog Dissection

OBJECTIVES
- Identify the external organs of a preserved frog.
- Perform procedures to identify the internal organs of a frog.
- Draw and label various features of a frog.
- Give a function of each major organ of a frog.

PROCESS SKILLS: observing, communicating, recognizing and using spatial relationships

TIME ALLOTMENT: 2 class periods

MATERIALS
frogs (preserved) (8)	forceps (8)
scissors (8)	microscope slides (16)
dissecting pins (about 80)	coverslips (8)
dissecting pans (8)	microscopes (8)
stereomicroscopes (8)	laboratory aprons (30)
dissecting probes (8)	

PREPARATION
- Charts of frog anatomy can be helpful to the students. These can be displayed around the room.
- Use Transparency 85, Structure of a Frog, from the Transparency Package for guided practice.

TEACHING THE LAB
- Have students work in groups of four.
- Further dissection of the three-chambered heart could be done. Students could be asked to locate the vena cava, sinus venosis, aorta, and pulmonary arteries and veins.
- If the frogs need to be stored overnight, wrap them in moistened paper towels, stack them in a dissecting pan, and store in a refrigerator.
- Caution students not to handle the frog excessively with bare hands. Formalin can be irritating to the skin.
- Use this Exploration to compare frog anatomy with human anatomy.

DATA AND OBSERVATIONS

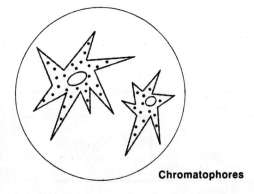

Chromatophores

ANALYSIS
1. The nictitating membrane covers the eye when the frog is underwater.
2. **a.** The eyes of the frog protrude from the side of the head.
   b. This adaptation gives the frog a wider angle of view to watch for predators and prey. The eyes also stick up above the water when the frog is submerged. This adaptation allows the frog to observe its environment while hidden.
3. The dorsal coloration of the frog is spotted and blends in with its dark, variously colored surroundings. The ventral coloration is white and is also good camouflage for when the frog is in the water. Predator or prey looking up from below cannot easily see the frog.
4. no
5. The tongue is attached at the front of the mouth.
6. The glottis leads to the trachea, which leads to the lungs.
7. The eustachian tubes lead to the tympanic membranes.
8. The lining of the stomach has large folds while the small intestine has tiny folds called villi.
9. The villi increase the surface area for absorption.
10. three
11. The pyloric sphincter regulates what enters the small intestine from the stomach.
12. Air is drawn into the mouth, through the glottis, into the lungs, back into the mouth, and out of the body through the nares.
13. Fat bodies are usually larger in females giving them a large energy reserve needed for nourishment of the eggs.
14. Frogs rely on sight and smell for food-getting and to avoid predators.

Capillary Circulation in Fish

OBJECTIVES
- Determine the diameter of a field of view.
- Locate and identify an arteriole, a capillary, and a venule in the tail of a goldfish.
- Observe and describe the direction of the blood flow in the tail.
- Measure and record data for the diameter of each blood vessel.

PROCESS SKILLS: observing, classifying, recognizing and using spatial relationships, measuring, using numbers

TIME ALLOTMENT: 1 class period

MATERIALS
absorbent cotton (60 5-cm squares)	glass slides (60)
aquarium water	microscopes (30)
petri-dish halves (30)	droppers (30)
goldfish in aquarium (30)	transparent plastic metric rulers (30)
aquarium net	

PREPARATION
- Set up and fill the aquarium with water at least two weeks before introducing goldfish into the aquarium.

TEACHING THE LAB
- Students should work individually if enough microscopes are available.
- You may wish to remove goldfish from the aquarium yourself, rather than having students do it.
- Remind students to handle the goldfish carefully, work quickly, and return the fish to the aquarium after they have made their observations.
- Students must add aquarium water to the cotton with a dropper to provide oxygen to the fish. Caution students that keeping the fish out of water for an extended period of time will harm them.

DATA AND OBSERVATION
See the table below.

ANALYSIS
1. Blood throbs in the arterioles because of the "pulse wave" from the heart.

2. Blood flows at a uniform rate in the venules because the pulse wave is lost in the capillary net. The pulse wave is not transmitted through the capillary into the venules.

3. Arterioles carry blood to the tail from the heart. However, the blood does not flow in only one direction in the arterioles. The arterioles cross the bony structure in the tail.

4. Capillaries have the smallest diameter. Blood cells pass through these vessels individually. Oxygen diffuses into the cells and carbon dioxide is absorbed.

5. Venules have the largest diameter. They return the deoxygenated blood to the heart by way of veins and have no pulse wave.

Table 1.

	Field of view	Arteriole	Capillary	Venule
Diameter in micrometers	Answers will vary, depending on the magnification. Figure 1 shows a field of view of about 1300μm.	35–85μm	20–25μm	50–100μm

EXPLORATION

Examining Bird Feet

OBJECTIVES
• Identify adaptations of birds' feet that make them suited for particular habitats and lifestyles.
• Compare and contrast the feet of various species of birds.
• Relate the characteristics of birds' feet to their functions.

PROCESS SKILLS: observing, classifying, inferring, communicating, recognizing and using spatial relationships, predicting, formulating models, defining operationally

TIME ALLOTMENT: 1 class period

MATERIALS
field guides to birds (10–15)

PREPARATION
• Review the concept of adaptation from Chapter 18.

TEACHING THE LAB
• Students should work in teams of two or three.
• Have stuffed and mounted birds available for students to examine. Have field guides and other books with illustrations of birds available.
• You may want to introduce this Exploration by drawing Figure 1 on the board and asking students why a bird can sleep on a perch and not fall off. The answer is that the feet have a built-in locking device. Muscles and tendons that run the entire length of the leg automatically curl the toes when the bird crouches. The bird must raise its entire body and straighten its legs before it can unlock its feet. The curved claws of birds of prey work in the same way. When a hawk strikes, its legs bend and its talons automatically clinch.

DATA AND OBSERVATIONS
Table 1.
Structural adaptations
A. Some toes are in front, some in back, curved claws enable scratching.
B. All toes are pointing forward, make running more efficient.
C. Webbing provides more surface area to the feet, allowing more effective paddling.
D. Stout strong toes with curved claws permit grasping and holding prey.
E. Some toes are in front and some in back enabling the bird to grip a branch.
F. Some toes are in front and some in back with curved claws, making upward movement effective while preventing downward slipping.

G. Long toes distribute weight over soft bottoms of ponds.
H. All toes pointing forward makes clinging effective.
I. Birds that spend most of their time in flight have very small feet.
J. Paddle-like feet can be used for swimming or pushing the bird on its breast and belly on icy snowfields.
K. Very long thin toes distribute weight evenly over leaves of water plants, on which the birds walk.
L. Spurs and knifelike claws are used for attack and defense.

Table 2.
Penguin   J, icy snowfields
Woodpecker   F, woodlands
Osprey   D, meadows with streams
Heron   G, ponds with soft, muddy bottoms
Ruffed grouse   A, meadows
Killdeer   B, hard, flat ground
Hummingbird   I, aerial birds
Cliff swallow   H, vertical surfaces of cliffs
Jacana   K, shorelines of ponds with water plants
Robin   E, meadows and woodlands
Mallard duck   C, lakes, ponds, streams
Pheasant   L, meadows

ANALYSIS
1. A running bird gains more distance with each step if the toes are pointed forward, so it gains more leverage pushing back with the toes.
2. It increases the surface area for paddling in the water.
3. The weight of the bird is distributed evenly over a wide surface area so the bird will not sink.
4. The claws grow long and need to be trimmed. Feet are not well exercised and get pressure sores.
5. Birds' feet will be less likely to be injured due to being on more natural surfaces.

INVESTIGATION

LAB

How Do Densities of Bird and Mammal Bones Compare?

OBJECTIVES
• Hypothesize which bones will be more dense—bird or mammal.
• Calculate the densities of various bird and mammal bones.
• Develop reasons that will explain the results of your measurements.

PROCESS SKILLS: observing, inferring, measuring, predicting, using numbers, forming hypotheses, experimenting

TIME ALLOTMENT: 1 class period

MATERIALS
clean, dry mammal bones:
  pieces of ribs (16 of beef, 16 of pork)
500-mL graduated cylinders (8)
dissecting probes (8)
pencils (8)

clean, dry bird bones:
  leg bones and wing bones (8 each of turkey and duck)
100-mL graduated cylinders (8)
balances (several)

ALTERNATE MATERIALS
• Chickens and smaller birds may be used but it is more difficult to calculate volumes of these smaller bones.

PREPARATION
• Ask a butcher to cut some beef and pork rib bones that will easily fit into graduated cylinders. When the bones are collected, boil them in water for about 20 minutes; scrape clean with a small knife, and toothbrush. Dry in a standard oven.

TEACHING THE LAB
• Students can work in groups of four.
• Make sure students understand the concept of density. Explain that volume is being measured by observing how much water the bone displaces.

• Emphasize to students to weigh their bones before measuring volumes.
• Explain to students that when they push a bone under water with the probe, they should not push it deep into the water, as the volume of the probe may then be a factor.

HYPOTHESIS
Answers will vary. Many students will say that bird bones are less dense than pig or cattle bones, and wing bones are less dense than leg bones.

DATA AND OBSERVATIONS
Answers will vary. Compile a class list on the board and take class averages for bone types.

ANALYSIS
1. Both pork and beef bones will be around 1 g/cc.
2. Both turkey and duck leg bones will be around 1 g/cc.
3. turkey = 0.8, duck = 0.6. Both will be less dense than pig and cow, and duck will be less dense than turkey.
4. The bones studied have the same function in supporting the animals' body mass when upright and moving on the ground.
5. Ducks migrate and fly on a daily basis more than turkeys do. Less dense wing bones is an adaptation to longer, more sustained flights.
6. Less density in wing bones makes them easier to move and also results in less weight to be supported in the air.
7. They float. The density of water is 1 g/cc.
8. feathers, air sacs, large sternum, no heavy jawbone, no internal development of young in the mother's body
9. bats that fly, whales and other mammals that live their lives in water; mammals that live in water do not have to support all their body weight on their skeletons as do land mammals because the water has a buoyant effect

CHECKING YOUR HYPOTHESIS
Answers will vary. A hypothesis that bird bones are less dense than mammal bones will be supported. However, the data will show that not all bird bones are less dense.

Table 1.

Bone	Mass (grams)	Volume of water in cyclinder (mL)	Volume to which water rises (mL)	Volume of bone (cc)	Density
BR1	15.1	320	335	15	1
BR2	21.7	318	335	17	1.3
PR1	10.1	75	85	10	1
PR2	11.4	71	81	10	1.1
TL	6.0	80	86	6	1
TW	7.3	303	312	9	0.8
DL	5.9	72	78	6	1
DW	2.9	69	73.5	4.5	0.6

35-1 INVESTIGATION

LAB

Mammal Teeth

OBJECTIVES
- Observe adaptations of mammals' teeth for eating particular types of food.
- Recognize the relationships of mammal feeding habits to body features other than teeth.
- Predict the types of teeth possessed by mammals adapted to specific habitats and behaviors.

PROCESS SKILLS: observing, classifying, inferring, communicating, recognizing and using spatial relationships, predicting, formulating models, defining operationally

TIME ALLOTMENT: 1 class period

MATERIALS
None

PREPARATION
- Ask students to brainstorm and make a list of things they could tell about a mammal by looking at its teeth. If you have a dog or cat skull complete with lower jaw, it could be used as an illustration. Review the concept of adaptation. Ask students to give examples of teeth adaptations in mammals.

TEACHING THE LAB
- Have students work in groups of four.

DATA AND OBSERVATIONS
See the table below.

ANALYSIS
1. 1. meat   2. insects   3. plants   4. insects   5. meat and insects   6. plants   7. meat   8. variety   9. insects   10. meat and insects   11. plants   12. insects

2. Students should follow this line of reasoning: If the animal eats food that will try to escape quickly, the animal also must be able to move quickly. If the food does not move, then the animal does not have to move quickly to get it.

3. Carnivores have large, sharp canines that kill prey; sharp, shearing premolars; large crushing molars. Herbivores have incisors adapted to gnawing wood, or cutting grasses; no canines; long molars that will not all be worn away as the animal grinds coarse plant material. Omnivores have large canines for tearing; large premolars adapted for grinding; premolars and incisors not large and sharp.

4. Large, sharp incisors and canines can make incisions. Molars and premolars are reduced in size and number. Bats that feed on blood would not chew their food.

5. Canines are large tusks that can rake the bottom of the sea floor for shellfish. They can also be used to break holes in the ice.

6. Incisors or canines would be enlarged, suited for digging in the soil. Premolars and molars would be small. Worms do not have tough, coarse fibers.

7. Incisors and canines would be flattened and wide, suited for scraping lichens from rocks and biting off short mosses. Premolars and molars would be long. Plant material is coarse and wears away teeth.

8. It would have large, sharp canines, sharp, shearing premolars, and large, crushing molars.

Table 1.

Mammal skull	Type of mammal	Characteristics
1.	Carnivore	Long canines, shearing premolars, crushing molars
2.	Anteater	Few tiny peg-like teeth
3.	Ungulate	Canines absent, premolars and molars long
4.	Insectivore	Canines and incisors peglike, premolars and molars with sharp points
5.	Mammal-like reptile	Interlocking teeth
6.	Rodent	Large incisors with chisel-like edge
7.	Carnivore	Long canines, shearing premolars, crushing molars
8.	Omnivore	Canines larger than incisors, premolars not large and sharp
9.	Insectivore	Canines and incisors peglike, premolars and molars with sharp points
10.	Early mammal	Triangular teeth
11.	Rodent	Large incisors with chisel-like edge
12.	Insect-eating bat	Sharp incisors, premolars, and molars

The above mammal skulls are identified as follows:
1. mountain lion   2. aardvark   3. Mouflon sheep   4. eastern mole   5. extinct synapsid reptile *Ophiacodon*
6. mole rat   7. dog   8. human   9. vagrant shrew   10. extinct *Deltatheridium*   11. woodrat   12. slit-faced bat

How Does Insulation Affect Thermal Homeostasis?

OBJECTIVES
- Construct hypotheses to predict how two different bodies will cool under different conditions.
- Determine the effect of insulation on the cooling of a warm body at room temperature.
- Determine the effect of insulation on the cooling of a warm body under cold conditions.
- Compare cooling rates of warm bodies at room temperature to the cooling rates of warm bodies at cold temperatures.

PROCESS SKILLS: observing, inferring, measuring, predicting, using numbers, forming hypotheses, experimenting, formulating models

TIME ALLOTMENT: 1 class period.

MATERIALS
5-L pails (3)	newspapers (160 pages)
100-mL graduated cylinders (32)	colored pencils (32)
	graph paper (30 sheets)
rubber bands (32)	hot tap water (45–55°C)
thermometers, 30 cm long (8)	ice cubes (72 cubes)
	stirring rods (8)
food storage size plastic bags (32)	masking tape (8 rolls)
	clock or stopwatches (8)

ALTERNATE MATERIALS
- Large containers such as 2-L beakers or battery jars could be substituted for the pail.

PREPARATION
- Be sure students understand the overall experimental setup before beginning the Procedure.

TEACHING THE LAB
- Have students work in groups of four. Emphasize that wrapping of the cylinder is done more easily if the fold of newspaper is at the base of the cylinder. Three groups can put their cylinders into the same pail of ice water. If the cylinders begin to float, students can hold them so their bases are at the bottom of the pail.
- **Troubleshooting:** Students may have trouble calculating the percentage decrease in temperature. Go over sample calculations in class.

HYPOTHESIS
Insulation will slow down the rate of cooling of a warm body. The greater the temperature differential between a warm body and its surrounding temperature, the faster the rate of cooling will be.

DATA AND OBSERVATIONS
Table 1.

Temperature (°C) of Water in Graduated Cylinders				
Time	Room temperature 23°C		Low temperature 5°C	
	uninsulated	insulated	uninsulated	insulated
0 minutes (start)	52	52	52	51
5 minutes	45	50	32	50
10 minutes	42	48	20	48
15 minutes	39	46	10	45
20 minutes	37	44	9	41
25 minutes	35	43	9	40
30 minutes	33	42	8	38

1. 29° 2. 47°

Table 2.

Graduated cylinder	Temperature difference	% decrease in temperature
Uninsulated at room temp.	19°C	36
Uninsulated at low temp.	44°	85
Insulated at room temp.	10°	19
Insulated at low temp.	13°	25

ANALYSIS
1. **a.** uninsulated cylinder at low temperature
   b. insulated at room temperature
   c. At room temperature, insulation allowed a cylinder to lose heat less rapidly.
   d. At low temperature, insulation allowed a cylinder to lose heat less rapidly.
2. the cylinders placed in the ice water
3. The greater the difference between the environmental temperature and body temperature, the faster the heat will be lost from the body.
4. **a.** The uninsulated cylinder lost twice as much heat as the insulated cylinder.
   b. The uninsulated cylinder lost more than three times as much heat as the insulated cylinder.
5. Insulation prevents a surface from being in direct contact with the environment. The fat in blubber and the air among feathers and in fur are good insulators.
6. Answers will vary. Adaptations include blubber or fat, the production of long fur or hairs, small ears for restricting heat loss, restricting blood supply to skin, and hibernation.

CHECKING YOUR HYPOTHESIS
Students who predicted that insulation slows down the rate of cooling and that cold surrounding temperatures speed up the rate of cooling will say that their hypotheses were supported.

How is Response Related to Nervous System Complexity?

OBJECTIVES
- Hypothesize the effect of nervous system complexity on responses to the same stimuli.
- Compare the responses of different invertebrates to the same stimuli.

PROCESS SKILLS: observing, inferring, recognizing and using spatial relationships, measuring, predicting, using numbers, forming hypotheses, experimenting

TIME ALLOTMENT: 1 class period

MATERIALS
droppers (20)	rulers (10)
flashlights (10)	stopwatches (10)
coarse salt (1 box)	hand lenses (10)
black construction paper (10 pieces)	living cultures of *Planaria*
plastic thermometer tubes with caps (30)	vinegar eels (*Turbatrix aceti*)
duct tape (1 roll)	*Daphnia*
test-tube racks (10)	filter paper (30 pieces)
wax marking pencils (10)	laboratory aprons (30)

ALTERNATE MATERIALS
- If thermometer tubes are not available, 6–8 mm glass or tygon tubing may be substituted.

PREPARATION
- Planarians are difficult to draw up into a dropper. If the dropper is inverted with the pointed end inserted into the bulb, the result is a dropper with a wide opening with which to draw up the organisms. Prepare these droppers before beginning this Investigation.

TEACHING THE LAB
- Have students work in groups of three, with each student handling one species. If students cannot see the organisms well with the hand lens, they may place the tube under a stereomicroscope.

HYPOTHESIS
Species with simple nervous systems will respond to stimuli in a predictable manner. The more complex a species' nervous system, the more variable the response will be.

DATA AND OBSERVATIONS
See the table below.

ANALYSIS
1. *Planaria* exhibited positive geotaxis; the other two species exhibited negative geotaxis. *Planaria* and *Daphnia* exhibited negative phototaxis; vinegar eels showed a weak positive or no response. All showed negative chemotaxis.
2. *Daphnia* exhibited the most diverse geotactic response, vinegar eels the most diverse response to light and *Daphnia* the most diverse response to salt.
3. Diversity of response is a sign of greater complexity of the nervous system. These organisms respond to more stimuli and so are less predictably stimulated by any one stimulus.
4. *Planaria*, a species with a simple nervous system, could react automatically without having to learn which response is desirable.
5. Response to a stimulus might not be critical to that species' survival.

CHECKING YOUR HYPOTHESIS
Answers will vary. Students who predicted that species with more complex nervous systems show more variable responses to stimuli will say that their hypotheses were supported by their data.

Table 1.

	Species' Response to Stimuli			
	Response	*Planaria*	Vinegar eel	*Daphnia*
Geotaxis	Reaction time (min)	2	4	4
	Number at top	1	20	3
	Number at bottom	5	5	1
	Response (+, −, 0)	+	−	−
Phototaxis	Reaction time (min)	1.5	2	3.5
	Number at light end	0	9	2
	Number at dark end	4	7	6
	Response (+, −, 0)	−	0	−
Chemotaxis	Reaction time (min)	2.5	5	5
	Number near salt	1	2	3
	Number away from salt	3	12	5
	Response (+, −, 0)	−	−	−

Conditioning in Guinea Pigs

OBJECTIVES
- Observe feeding behavior in guinea pigs.
- Design and carry out an experiment to produce a conditioned response in guinea pigs.
- Determine the amount of time needed to produce a conditioned response in guinea pigs.

PROCESS SKILLS: observing, communicating, experimenting, interpreting data, separating and controlling variables, defining operationally

TIME ALLOTMENT: 10 minutes per day for several weeks

MATERIALS
guinea pigs (5)
cages (5)
wood chips or other bedding material
guinea pig pellets or fresh food
bells, whistles, or other sources of sound

PREPARATION
- Arrange the guinea-pig cages in such a way that each experiment will take place as far from others as possible.

TEACHING THE LAB
- Have students work in groups of six.
- Stress the importance of careful handling of any animal. Students should handle the guinea pigs gently and carefully. Students may need to be reminded about cleaning the cages and checking that the animals have fresh water.
- If space and expenses are limited in your classroom, this experiment can be done using goldfish or guppies. Test their response to being fed with a tap on the fish tank, shining a light, or giving some other stimulus.
- Assist students in planning the animal's diet. Many of the "lab-pellet" foods contain all the nutrients the animal needs. Check the label to be sure that the food is appropriate for these animals.
- Like many domesticated animals, guinea pigs will eat foods that are not good for them. Discourage students from offering the animals cake or other inappropriate foods.
- An excellent poster-type chart giving information for the care of mice, rats, rabbits, and guinea pigs is available from Purina.
- Although guinea pigs can be fed grass and dandelion greens, you should discourage this practice unless you or the students are certain that the source of these materials has not been sprayed with fertilizer or other chemicals.

- Assist students in identifying specific behaviors. If the student notes that the animal "runs around in its cage," have the student look for specific places it runs to. If the food is always placed in the same part of the cage, an animal will often run to that spot when it expects to be fed.
- Arrange to have one student in each group take the group's animal home for the weekend so that conditioning can be continued.
- Use of a bell as the stimulus may pose a problem if bells are routinely rung in your classroom. Choose a stimulus such as a hand clap, kissing sound, or foot stamp.
- This Exploration offers an excellent opportunity for cooperative learning. Have students design and carry out the experiment as a group.

DATA AND OBSERVATIONS
Answers will vary.

ANALYSIS
1. Answers will vary. Students may choose a hand clap, whistle, or any other stimulus.
2. The new stimulus must not be one that occurs at other times because the stimulus must be associated only with being fed.
3. To be sure that only the intended stimulus is associated with being fed, other stimuli must not be present.
4. Answers will vary. Conditioning may take place in a week or it may take longer.
5. Answers will vary. Animals should show a feeding behavior response to the new stimulus.
6. Answers will vary. Animals may run to a certain area of the cage, they may make certain sounds, they may jump up and down, or so on. Each animal is different.

The Skeletal System

OBJECTIVES
- Observe a microscope slide of compact bone and identify the parts.
- Identify bones by their shapes.
- Observe the features of joints.

PROCESS SKILLS: observing, classifying, recognizing and using spatial relationships, defining operationally

TIME ALLOTMENT: 1 class period

MATERIALS
prepared slides of compact bone (30)
microscopes (30)
human skeleton or pictures of the skeleton

PREPARATION
- Be sure to order prepared slides far enough in advance of the class and to secure enough skeleton pictures and/or a skeleton.

TEACHING THE LAB
- Have students work alone in Part A and in pairs in Parts B and C.
- Remind students to use only the low-power objective. Microscope slides of bone often are too thick to fit under a high-power objective.
- If it is not possible to have a human skeleton, have a picture of a skeleton for each student or each pair of students.
- If a human skeleton is available, allow students to use it as a reference for identifying the joints shown in Figure 3.

DATA AND OBSERVATIONS
See the tables below.

ANALYSIS
1. The shape of the bone, especially grooves, projections, rounded heads and processes can be used to identify bones.
2. pivot — head or neck
3. hinge — elbow, knee
4. ball and socket — shoulder, hip
5. a. location of blood vessels and nerves
   b. carry fluids between blood vessels and osteocytes

Table 1.

Bone	Common name of bone(s)	Scientific name of bone	Location of bone
A	rib	rib	chest
B	jawbone	maxilla	head
C	shoulder blade	scapula	upper back
D	toes	phalanges	on foot
E	collar bone	clavicle	top, front of shoulder
F	spine	vertebra	down middle of back
G	knee cap	patella	in front of knee
H	skull	cranium	head
I	breastbone	sternum	middle of chest

Table 2.

Joint	Type	Location in body
A	hinge	elbow, knee
B	ball and socket	hip, shoulder
C	gliding	between vertebrae, wrist

EXPLORATION

Endocrine Gland Studies

OBJECTIVES
• Observe the structure of the thyroid and adrenal glands.
• Analyze the relationship between the structure and the function of a gland.

PROCESS SKILLS: observing, inferring, using numbers

TIME ALLOTMENT: 1 class period

MATERIALS
microscopes (30)
prepared slides of thyroid gland (30)
prepared slides of adrenal gland (30)

PREPARATION
• Obtain prepared slides from a biological supply house.

TEACHING THE LAB
• If enough microscopes are not available, students could work together in teams.
• Any human anatomy and physiology textbook would be a good reference book for this lab.
• Most gland slides are stained with HE (hematoxylin and eosin). This stain makes the nuclei dark purple and the cytoplasm pink. The colloid material will also be pink.
• When students look at the adrenal gland, be sure they do not confuse the fibrous connective-tissue covering with the outer zone of the gland.

DATA AND OBSERVATIONS
The average diameter of follicles is approximately 0.03–0.15 mm.

Table 1.

Follicle Diameter	
Follicle	**Diameter (mm)**
1	0.06
2	0.11
3	0.17
4	0.09
5	0.02
Average	0.09

ANALYSIS
1. Thyroid hormones increase metabolic activity of most body tissues.
2. Calcitonin regulates the calcium level in the blood.
3. The colloid material will shrink. The follicle cells will enlarge.
4. Part A.
5. Part B.

Figure 3.

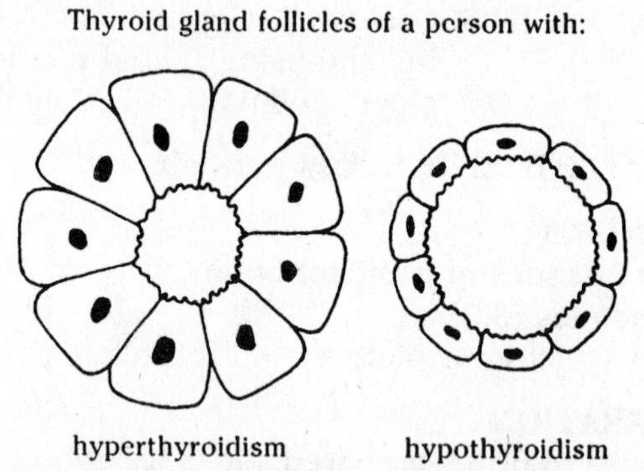

Thyroid gland follicles of a person with:

hyperthyroidism   hypothyroidism

6. a. Epinephrine increases heart rate and force, and constricts blood vessels, except those going to the muscles.
   b. Aldosterone increases sodium and water retention by the kidneys.
   c. Cortisol increases carbohydrate, protein, and lipid metabolism.
7. Epinephrine can slow bleeding from an injury by constricting the blood vessels.
8. Epinephrine causes increased blood flow to the muscles, stimulates the heart to pump blood more quickly to the muscles, and provides additional energy. The animal can run from danger.

Caloric Content of a Meal

OBJECTIVES
- Calculate the number of Calories and grams of carbohydrate, fat, and protein in two meals.
- Appraise the nutritional value of each meal.
- Plan a balanced, nutritional meal.

PROCESS SKILLS: classifying, communicating, predicting, using numbers

TIME ALLOTMENT: one class period

MATERIALS
food tables (with Calories and nutrients listed) (15)

PREPARATION
- Inexpensive paperback Calorie guides or food tables can be purchased in most grocery stores.

TEACHING THE LAB
- Have students work in pairs.
- Be sure to review the difference between a food Calorie and a scientific calorie. A food Calorie is actually a kilocalorie and is thus spelled with a capital C.
- Calculators can be used to speed up the calculations.

- **Troubleshooting:** Calorie tables available in the United States list the amounts of foods in English units such as cups or ounces, and the amounts of nutrients in metric units such as grams or milligrams. This may be confusing to students.

DATA AND OBSERVATIONS
Table 1.

Recommended Food Intake		
Nutrients	Amount per day	Amount per meal
Calories	2000–2500	666–833
Carbohydrate (grams)	120	40
Fat (grams)	60	20
Protein (grams)	60	20

See Table 2 below.

ANALYSIS
1. Calories – Meal 2
   fat – Meal 2
   carbohydrate – Meal 2
   protein – Meal 2
2. Meal 1. Meal 1 is closer to the recommendation for Calories and nutrient levels per meal.
3. protein; meats, fish, dairy products, rice, beans, other grains
4. Your weight will increase.
5. Your weight will decrease.

Table 2.

Calories and Nutrients of Two Sample Meals					
Food	Serving size	Calories	Carbohydrate (grams)	Fat (grams)	Protein (grams)
Meal 1					
Spaghetti with meat sauce	6 oz	350	70	4.5	7
Green beans	4 oz	30	6	Trace	2
Garlice bread	2 slices	130	12	Trace	2
Butter	1 Tbsp	100	Trace	11	Trace
Gelatin dessert	4 oz	75	17	0	2
Total		685	105	15.5	13
Meal 2					
Hamburger bun	1	120	21	2	3
Ground beef	4 oz	320	0	23	27
Cheese (American)	1 oz	100	Trace	9	6
Catsup	2 Tbsp	60	16	Trace	Trace
French fries	24	400	48	20	4
Cola-type beverage	10 oz	130	30	0	0
Total		1130	115	54	40

How Much Vitamin C Are You Getting?

OBJECTIVES
- Measure the volume of iodine solution needed to react with a standard concentration of vitamin C.
- Hypothesize which citrus juice sample contains the greatest amount of vitamin C per serving.
- Measure the volume of iodine solution needed to react with unknown concentrations of vitamin C.
- Calculate the concentration of vitamin C in the citrus juice samples.

PROCESS SKILLS: forming hypotheses, measuring, preparing standards, experimenting, quantifying, interpreting data

TIME ALLOTMENT: 1 class period or a double period for multiple determinations

MATERIALS

standard vitamin C solution (0.001 g/mL of water)
1% starch solution
distilled water
iodine solution
citrus juice samples (including fresh orange juice)
paper towels (1 roll)
plastic pipettes or microtip pipettes (45)

toothpicks (30)
microplates, 96-well (15)
microplates, 24-well (15)
scissors or scalpels (15)
tape (1 roll)
plastic straws (30)
wax marking pencils (15)
laboratory aprons (30)
safety goggles (30)

ALTERNATE MATERIALS
- A single, combination microplate can be used in place of the two plates. The combination plate has both large and small wells.

PREPARATION
- The most convenient way to prepare standard vitamin C solution is to dissolve one 1000-mg tablet of vitamin C in 1 L of water and stir. (The white precipitate on the bottom of the container is starch, not vitamin C.) Refrigerate the solution if it is to be used the next day.

- See page T14 to prepare an iodine solution.
- Place 2.5 g of starch in a beaker and add 100 mL of distilled water. Swirl the mixture and add boiling water to a total volume of 150 mL. Cool before use.
- See page T15 about stretching pipettes and about other tools and techniques of microchemistry.

TEACHING THE LAB
- Have students work with a partner. To complete the lab in one period, divide the class into the same number of groups as citrus juice samples being tested.
- Students must count the number of drops of iodine solution carefully. This is especially true for the standard vitamin C determination, since all other calculations will be based on the standard.
- Students should pool their data for the best results.

HYPOTHESIS
Answers will vary. Many students will probably say that fresh orange juice contains the greatest amount of vitamin C per serving.

DATA AND OBSERVATIONS
Student results will vary, depending on the juice samples used. Results for the standard should be fairly consistent. See Table 1 below.

ANALYSIS
1. 0.32
2. .00032
3. 0.048
4. 48. Fresh juices are usually higher in vitamin C than canned or frozen products. The highest vitamin C content is usually found in freshly squeezed fruit juice. Vitamin C is easily destroyed by heat and oxygen.
5. If the student forgot to add starch indicator, the solution would never turn blue-black. The starch reacts with excess iodine to give a visual cue that all the vitamin C has been reacted.
6. Answers will vary. Students may suggest calibrating the pipettes, or using precalibrated instruments to measure the volume in cubic centimeters or milliliters.

CHECKING YOUR HYPOTHESIS
Answers will vary. Students who said that fresh orange juice contains the greatest concentration of vitamin C per serving will say that their hypothesis was supported.

	Trial 1	Trial 2	Trial 3	Average
Drops Iodine Solution (Std. Vit. C)	12	15	11	12.6
Drops Iodine Solution (Juice A)	4	5	3	4
Drops Iodine Solution (Juice B)				
Drops Iodine Solution (Juice C)				
Drops Iodine Solution (Juice D)				

INVESTIGATION

How Does Exercise Affect Heart Rate?

OBJECTIVES
- Determine your heart rate by taking your pulse.

- Hypothesize the effect of exercise on heart rate.

- Compare your heart rate during rest and during exercise.

PROCESS SKILLS: observing, communicating, forming hypotheses, measuring, predicting, using numbers, experimenting

TIME ALLOTMENT: 1 class period

MATERIALS
stopwatches (15) (not necessary if a clock with second hand is visible to all students)

PREPARATION
- Before planning this activity, check with the school nurse to see if any students have limited activities due to medical conditions. Also be sure that there is ample room for exercise in your classroom. If there is insufficient space, try to arrange the use of a gymnasium or other large space.
- Give students advance notice that they will be exercising so they can wear appropriate clothing.

TEACHING THE LAB
- Demonstrate for students the proper technique for taking the pulse at the carotid artery.
- Determine the kind of exercise most appropriate for your students. Running in place or doing jumping jacks will provide good results. Whatever exercise you assign should be done at a consistent rate throughout the ten-minute exercise period.
- The entire class can do this activity at the same time if you act as timekeeper.
- Go over the timing for Parts B and C. You may wish to "walk students through" one minute of each cycle. Figure 2 looks complicated, but it should reassure students that 30 seconds provides plenty of time to get a 15-second pulse count and record the data.
- If you have an extra long lab period, allow the time keeper and experimental subject to reverse roles and repeat the procedure.

HYPOTHESIS
Hypotheses will vary. Many students will probably say that exercise increases the heart rate.

DATA AND OBSERVATIONS
Answers will vary, depending on the physical condition of the experimental subject. For an average student, the resting heart rate should be about 70 beats per minute. This rate should rise to about 180 beats per minute during exercise, and then fall back to the resting rate with recovery.

ANALYSIS
1. Answers will vary.
2. Answers will vary. The highest heart rate should occur during the exercise period.
3. Answers will vary. Most students should return to resting rate.
4. They are more physically fit.
5. The curve should be bell shaped.
6. Drugs such as caffeine and nicotine, sex (females have faster heart rates than males), age (children have high heart rates), and emotions (nervousness, embarrassment, and fear) will all increase the heart rate.

CHECKING YOUR HYPOTHESIS
Answers will vary. A hypothesis that heart rate will increase during an exercise period should be supported.

What Are the Locations of Taste and Smell Receptors?

OBJECTIVES
- Hypothesize about the locations of different taste receptors on the tongue.
- Diagram the distribution of chemoreceptors on the tongue.
- Use correct lab techniques to dilute solutions.
- Determine the effect of the sense of smell on the ability to identify tastes.

PROCESS SKILLS: observing, classifying, inferring, communicating, recognizing and using spatial relationships, predicting, interpreting data, forming hypotheses, experimenting

TIME ALLOTMENT: 2 class periods

MATERIALS
paper cups (210)	paper towels (15)
permanent marking pens (15)	paper bags (15)
	25-mL graduated
sweet solution (5.0% sucrose solution)	cylinders (15)
	stirring rods (15)
sour solution (1.0% acetic acid solution)	toothpicks (270)
	potato, cut 0.5 cm thick
bitter solution (0.1% quinine sulfate solution)	(90 pieces)
	apple, cut 0.5 cm thick
salty solution (10.0% sodium chloride solution)	(90 pieces)
	onion, cut 0.5 cm thick
	(90 pieces)
cotton swabs (630)	plastic wrap (1 roll)

PREPARATION
- Use the smallest paper cups available.
- Prepare one liter each of the salt, sour, and bitter solutions, and 2L of the sweet solution.

TEACHING THE LAB
- For sanitary reasons, students must keep the cotton swabs clean. Swabs should be kept on a clean paper towel, not on the surface of the desk. Swabs must not be reused. Impress upon students the importance of discarding swabs immediately.
- The investigator should mix the tastes as much as possible. It is a good idea to repeat a test, or to do a test with water, if the investigator thinks the subject is guessing at answers.
- **Troubleshooting:** Before doing this Investigation, check to be sure that no students have food allergies, or other problems that might be aggravated by the testing procedures.

HYPOTHESIS
Sweet taste receptors are located at the tip and front of the tongue, salt receptors at the sides near the front, sour receptors at the sides near the back, and bitter receptors at the back.

DATA AND OBSERVATIONS
Table 1.
Answers will vary, but the strongest taste for sweet will be at the tip and front, the strongest taste for salt will be at the sides near the front, the strongest taste for sour will be at the sides behind the salt, and the strongest taste for bitter will be at the back of the tongue.

Table 2.
Answers will vary.

Table 3.
Answers will vary, but the best results should occur when the senses of smell and taste work together.

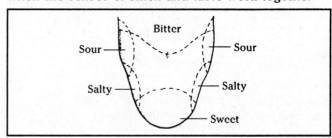

ANALYSIS
1. Answers will vary, but the strongest taste for sweet will be at the tip and front, the strongest taste for salt will be at the sides near the front, the strongest taste for sour will be at the sides behind the salt, and the strongest taste for bitter will be at the back of the tongue.
2. near the front of the tongue
3. **a.** near the back of the tongue
   b. As bitter food is swallowed, it passes over the strongest center for the reception of bitter taste.
4. Thresholds will vary from person to person.
5. The test was done with the sweet solution, and the strongest reception of that taste occurs near the front of the tongue.
6. by making intermediate dilutions between the weakest concentration that could be tasted and the strongest concentration that could not be tasted
7. Answers will vary, but the most accurate identifications will probably occur when tasting and not holding the nose.
8. When the nasal passages are blocked, only the tongue is able to participate in the sensation that is recognized as taste. The nose does not take part in any chemoreception.

CHECKING YOUR HYPOTHESIS
Answers will vary. Students who predicted that the sweet receptors are at the front and tip, salt receptors at the front sides, sour receptors at the rear sides, and bitter receptors at the back will say that their hypotheses were supported.

INVESTIGATION

When Does a Chicken Embryo Grow the Fastest?

OBJECTIVES
Hypothesize the rate of growth of an embryo.
Observe the development of a chick embryo over a period of 72 hours.
Measure the length of a chick embryo at various stages of development.
Prepare graphs to show the rate of growth and development of a chick embryo.

PROCESS SKILLS: observing, communicating, measuring, forming hypotheses

TIME ALLOTMENT: 2 class periods

MATERIALS
fresh fertilized chicken eggs
  incubated for 0 hours, 24
  hours, 48 hours, and 72
  hours (15 each)
incubator (38°C)
dissecting probes (15)
small dishes or bowls (15)
scissors (15)

forceps (15)
stereomicroscopes
  (15)
metric rulers (15)
graph paper (60
  sheets)
absorbent cotton
  (1 roll)

PREPARATION
Order living materials in advance. Plan to have eggs arrive 3–4 days before you need them. Fertilized chick embryos can be incubated to the desired stage and refrigerated for a few days to keep them at that stage. Label each egg with its hours of incubation.
Chick eggs can be incubated in a regular bacteriological incubator at 38°C with a large pan of water in the incubator. This incubator can be used for this early development but should not be used if you plan to hatch any chicks.
Incubated eggs must be turned twice a day.

• Inexpensive foam chick incubators can be purchased from a biological supply house. These incubators will work well if you wish to hatch any chicks.

TEACHING THE LAB
• Demonstrate the technique for cutting a "window" in an egg.
• The contents of an egg can be studied if it is gently poured into a shallow dish.
• **Troubleshooting:** The yolk floats in the albumin with the embryo on the upper side. If the egg is allowed to rest on its side for a few minutes before it is opened, the embryo is more likely to be in position to be seen through the "window."

HYPOTHESIS
Answers will vary. Many students will say that a chicken embryo will grow fastest during the 48 to 72 hour stage of development.

DATA AND OBSERVATIONS
Answers will vary. See the table below.

ANALYSIS
1. Food for the chick embryo.
2. In the 72 hour embryo, the limb buds are visible. Other body systems, such as digestive and skeletal systems, are much further developed when compared with the 24-hour embryo.
3. Answers may vary. The embryo normally grows fastest during the 48 to 72 hour period.

CHECKING YOUR HYPOTHESIS
Numbers may vary but the graphs should all be similar. Data should support a hypothesis of most rapid development in the 48 to 72 hour period.

Table 1.

Hours of incubation	Length of embryo in mm	Number of pairs of somites	Number of pairs of somites observed
0	Less than 1 mm	0	0
24	4	8	8
48	9	21	21
72	110	36	36

41-2 EXPLORATION

LAB

Human Fetal Growth

OBJECTIVES
- Calculate the length of a human fetus at various stages of development.
- Graph the length of a developing human fetus.
- Graph the mass of a developing human fetus.
- Determine the period of development during which the greatest changes in mass and in length occur.

PROCESS SKILLS: measuring, using numbers, formulating models, defining operationally

TIME ALLOTMENT: 1 class period

MATERIALS
metric rulers (15)

PREPARATION
- Review measuring and graphing techniques before beginning this Exploration. Refer to the *Skill Handbook* in the Student Edition of the text, page 1146.

TEACHING THE LAB
- Students can work in pairs to record measurements. Each student should graph the data.
- Review Section 41.2 in Chapter 41 of the text with students before beginning the Exploration.

DATA AND OBSERVATIONS
See the table and figures below.

ANALYSIS
1. 52.5 mm
2. 1 gram
3. **a.** yes
   b. no
4. The diagrams are shown at 40% of their natural size.
5. The skeletal system is developing.
6. **a.** week 19
   b. week 31
7. **a.** 35 weeks
   b. 32 weeks (Students may say 33 weeks.)

Table 2.

	Length of a Developing Fetus				
Age of fetus in weeks	Body length (mm)	Thigh length (mm)	Leg length (mm)	Total length (mm)	Actual length (mm)
2	—	—	—	—	2 mm
9	13	4	4	21	52.5
16	28	12	10	50	125
20	48	19	16	83	207.5
24	57	22	22	101	252.5
32	65	28	25	118	295
38	78	37	33	148	370

Figure 2.   Length of a developing fetus

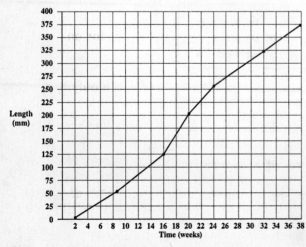

Figure 3.   Mass of a developing fetus

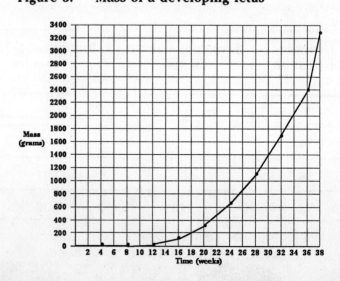

T102

INVESTIGATION

How Do Bactericides Affect the Growth of Bacteria?

OBJECTIVES
- Compare the effects of different bactericides on the growth of milk and intestinal bacteria.
- Hypothesize how bactericides affect the growth of intestinal and milk bacteria.
- Use sterile techniques for handling bacterial cultures and inoculating agar plates.

PROCESS SKILLS: observing, communicating, predicting, forming hypotheses, experimenting, measuring

TIME ALLOTMENT: 1 class period, then 30 minutes after 48 hours

MATERIALS
laboratory aprons (30)	*Escherichia coli* cultures (8)
disinfectant solution	isopropyl alcohol (100 mL)
paper towels (1 roll)	
sterile cotton swabs (32)	household bleach (100 mL)
forceps (8)	
Bunsen burners (8)	tincture of iodine (100 mL)
strikers (8)	
sterile filter paper disks (112)	3% hydrogen peroxide (100 mL)
test tubes (48)	antibiotic disks (16)
sterile petri dishes containing nutrient agar (16)	mouthwash (100 mL)
	disinfectant (100 mL)
	test-tube racks (8)
sterile petri dishes containing lactose agar (16)	masking tape (8 rolls)
	metric rulers (8)
sour milk (200 mL)	wax marking pencils (8)

ALTERNATE MATERIALS
- Stock cultures of *Bacillus subtilis* or *Proteus vulgaris* can be used instead of *E. coli*.

PREPARATION
- Have all the bactericides in capped test tubes.
- Prepare agar plates or purchase ready-made plates from a biological supply house.
- Obtain individually-wrapped sterile cotton swabs from a biological supply house or hospital supply store, or sterilize them yourself.
- Order a non-pathogenic strain of *E. coli*.
- Provide each student group with a petri dish containing 14 sterile filter paper disks. Prepare them by punching disks from filter paper with a hole punch and autoclaving in petri dishes.
- Have sufficient quantities of disinfectant solution to clean the work surfaces. A solution of 250 mL of household bleach to 4 L water can be used.
- Prepare sour milk by allowing milk to stand at room temperature for 48–72 hours.
- Antibiotic disks can be purchased from a hospital or supply house.

TEACHING THE LAB
- Have students work in groups of four.
- Review sterile procedures.
- Have students read the entire Procedure before they begin working.
- Review and have the students practice the uncapping and recapping of a test tube while holding the swab.
- Follow your school's procedures for disposing of the agar plates and cotton swabs. The cotton end of the swab can be flamed or they can be dropped in a beaker of disinfectant solution and disposed of later.

HYPOTHESIS
Students will probably hypothesize that the bactericides all kill or control bacteria.

DATA AND OBSERVATIONS
Answers will vary. The antibiotic, bleach, and disinfectant will probably show the largest zones of inhibition. The mouthwash will probably have no zone of inhibition.

ANALYSIS
1. Milk bacteria need lactose to grow; intestinal bacteria need other kinds of nutrients to grow.
2. to prevent other bacteria and mold spores from contaminating the cultures
3. It serves as a control.
4. Answers will vary. The antibiotic will probably be the most effective.
5. Answers will vary. The disinfectant will probably be the most effective.
6. Different bacteria are affected in different ways by the same bactericides. Different bactericides have different strengths and have different effects on the bacteria.
7. They inhibit growth of several kinds of bacteria.
8. Answers may vary; an antibiotic is not equally effective against all bacteria.
9. No, antibiotics may be effective for only a specific type of bacteria.
10. No, the bacteria used were non-disease-causing bacteria; observations made can only be used to draw conclusions about bacteria used.
11. No, some bactericides such as bleach are toxic.

CHECKING YOUR HYPOTHESIS
Students should find that the antibiotic, bleach, and disinfectant kill the bacteria. The alcohol, hydrogen peroxide, and iodine inhibit the growth of bacteria. The mouthwash will probably have little effect and the plain disk will not affect growth.

Testing Water Quality

OBJECTIVES
- Treat water samples to determine the presence of oxygen and carbon dioxide in the water.
- Determine the parts per million (ppm) of dissolved oxygen and carbon dioxide in water samples.
- Evaluate whether water samples may be clean or polluted.
- Analyze how the presence of oxygen and carbon dioxide in bodies of water affects life in that water.

PROCESS SKILLS: observing, measuring, using numbers, interpreting data, experimenting

TIME ALLOTMENT: 1 class period

MATERIALS
water samples from 2 sources (3L of each)
small flasks or beakers (30)
droppers (105)
masking tape (15 rolls)
solution A
solution B
solution C
solution D
solution E
safety goggles (30)
laboratory aprons (30)
phenolphthalein solution
sodium hydroxide solution

PREPARATION
- You may wish to conduct a field trip to collect water samples. Some sources include: tap water, stream, pond, lake, and fish tank with aerator.
- To prepare water sample with a high CO_2 level, add 0.1 mL concentrated HCl to 2000 mL water.
- Have all water samples at room temperatures.
- Prepare all solutions, following directions on page T14.

TEACHING THE LAB
- Have students work in groups of two.
- Caution the students about using the chemicals.
- Be sure students hold the dropper close to the water surface to avoid splashing.

- For more accurate data, have students use a graduated burette filled with sodium thiosulfate. The titrated number of mL for the solution to become colorless is equal to the ppm of oxygen.
- You may wish to do only one part of this lab. Part A takes about 20 minutes if all equipment is set up and available. Part B takes about 10 minutes.

DATA AND OBSERVATIONS
See the table below.

ANALYSIS
1. Answers will vary, but should match those in Table 1.
2. a. Answers should indicate samples with high oxygen content.
   b. samples with low oxygen content
   c. Samples in 2a have a high enough oxygen content to support fish; samples in 2b do not.
3. Oxygen is needed for respiration.
4. a. producer
   b. photosynthesis
5. decreases
6. Photosynthesis cannot occur in deep water because there would not be enough light.
7. a, b. Answers will vary with the samples.
8. a, b. Answers will vary with the samples.
   c. Samples with less than 25 ppm CO_2 will support most bivalves.
9. a. higher at night
   b. At night, plants take in oxygen and give off carbon dioxide during the process of cellular respiration. Photosynthesis cannot take place.
10. a. Answers might include error in counting drops, uncertainty of when flask returns to colorless, error in adding proper reagents, and allowing extra oxygen to mix with sample.
    b. Students might suggest repeating a number of trials on a sample.
    c. Answers might include error in counting drops, uncertainty of when flask is pink, and math errors.
    d. Students might suggest repeating a number of trials on a sample.

Table 1.

	Results of Oxygen and Carbon Dioxide Tests						
	Oxygen results			Carbon dioxide results			
	Water source	Drops of solution E used	Amount of O_2 (ppm)	Water source	Carbon dioxide present?	Drops of sodium hydroxide used	Amount of CO_2 (ppm)
Sample 1	Student water		Date will vary with source used.				
Sample 2	Lake	180	$\frac{180}{20} = 9$	Lake	yes	3	$3 \times 5 = 15$

LABORATORY MANUAL

BIOLOGY
THE DYNAMICS OF LIFE

AUTHORS

Chris Kapicka
Biology Department
University of Nevada-Reno
Reno, NV

Alton Biggs
Allen High School
Allen, TX

Linda Lundgren
Bear Creek High School
Denver, CO

Albert Kaskel
Evanston Township High School
Evanston, IL

Contributing Authors

Diane Bynum
Belaire High School
Baton Rouge, LA

Rachel Hays
Heath Junior High School
Greeley, CO

Juliana Texley
Richmond Schools
Richmond, MI

Donald Emmeluth
Fulton-Montgomery
Community College
Johnstown, NY

Priscilla Lee
Venice High School
Los Angeles, CA

Maureen Wahl
Notre Dame-Cathedral
Latin School
Chardon, OH

Tom Russo
Chemistry Education
Consultant
Kemtec Educational Corp.
West Chester, OH

Consultants

Lucy Daniel
Rutherfordton County Schools
Spindale, NC

Lorena Farrar
Westwood Junior High School
Richardson, TX

Albert Kaskel
Evanston Township High School
Evanston, IL

Ouida Thomas
Rosenberg High School
Rosenberg, TX

GLENCOE
Macmillan/McGraw-Hill

New York, New York   Columbus, Ohio   Mission Hills, California   Peoria, Illinios

A
GLENCOE
PROGRAM

Biology
The Dynamics of Life

Student Edition
Teacher Wraparound Edition
Study Guide, SE and TE
Laboratory Manual, SE and TE
Biolab and Minilab Worksheets
Chapter Assessment
Videodisc Correlations
Science and Technology Videodisc Series Teacher Guide
Transparency Package
Concept Mapping
Biology Projects
Exploring Applications of Biology
Critical Thinking/Problem Solving
Great Developments in Biology
Spanish Resources
Lesson Plans
Computer Test Bank IBM/APPLE/MACINTOSH
English/Spanish Audiocassettes

Send all inquiries to:

GLENCOE DIVISION
Macmillan/McGraw-Hill
936 Eastwind Drive
Westerville, Ohio 43081

ISBN 0-02-826666-8

Printed in the United States of America.

3 4 5 6 7 8 9 10   POH   00 99 98 97 96

TABLE of CONTENTS

HOW TO USE THIS LABORATORY MANUAL

Working in the laboratory throughout the course of the year can be an enjoyable part of your biology experience. **Biology: The Dynamics of Life,** *Laboratory Manual* is a tool for making your laboratory work both worthwhile and fun. The laboratory activities are designed to fulfill the following purposes:

- to stimulate your interest in science in general and especially in biology;
- to reinforce important concepts studied in your textbook;
- to allow you to verify some of the scientific information learned during your biology course;
- to allow you to discover for yourself biological concepts and ideas not necessarily covered in class or in the textbook readings; and
- to acquaint you with a variety of modern tools and techniques used by today's biological scientists.

Most importantly, the laboratory activities will give you first-hand experience in how a scientist works.

The activities in this manual are of two types: Investigation or Exploration. In an Investigation activity, you will be presented with a problem. Then, through use of scientific methods, you will seek answers. Your conclusions will be based on your observations alone or on those made by the entire class, recorded experimental data, and your interpretation of what the data and observations mean. In the Exploration, you will make observations but will not use scientific methods to reach conclusions.

Each activity in **Biology: The Dynamics of Life,** *Laboratory Manual* has the nine parts listed below. Understanding the purpose of each of these parts will help make your laboratory experiences easier.

1. **Introduction** — The introductory paragraphs give you background information needed to understand the activity.
2. **Objectives** — The list of objectives is a guide to what will be done in the activity and what will be expected of you.
3. **Materials** — The materials section lists the supplies you will need to complete the activity.
4. **Procedure** — The procedure gives you step-by-step instructions for carrying out the activity. Unless told to do otherwise, you are expected to complete all parts of each assigned activity. Important information needed for the procedure but that is not an actual procedural step is also found in this section. Many steps have safety precautions. Be sure to read these statements and obey them for your own and your classmates' protection.
5. **Hypothesis** — In Investigations, you will write a hypothesis statement to express your expectations of the results and as an answer to the problem statement. Explorations do not require you to make a hypothesis.
6. **Data and Observations** — This section includes tables and space for you to record your data and observations.
7. **Analysis** — In this section you draw conclusions about the activity just completed. Rereading the introduction before answering the questions is most helpful at this time.
8. **Checking Your Hypothesis** — You will determine whether your hypothesis is valid, based on your data.
9. **Further Investigations/Further Explorations** — This section gives ideas for further activities that you may do on your own. They may be either laboratory or library research.

In addition to the activities, this laboratory manual has several other features—a glossary, a description of how to write a lab report, a section on the care of living things, diagrams of laboratory equipment, and information on safety that includes first aid and a safety contract. The glossary, included in the back of the manual, defines terms used throughout the manual. A pronunciation key has also been included to help you with the more difficult words. Read the section on safety now. Safety in the laboratory is your responsibility. Working in the laboratory can be a safe and fun learning experience. By using **Biology: The Dynamics of Life,** *Laboratory Manual,* you will find biology both understandable and exciting. Have a good year!

WRITING A LABORATORY REPORT

When scientists perform experiments, they make observations, collect and analyze data, and formulate generalizations about the data. When you work in the laboratory, you should record all your data in a laboratory report. An analysis of data is easier if all data are recorded in an organized, logical manner. Tables and graphs are often used for this purpose.

A written laboratory report should include all of the following elements.

TITLE: The title should clearly describe the topic of the report.

HYPOTHESIS: Write a statement to express your expectations of the results and as an answer to the problem statement.

MATERIALS: List all laboratory equipment and other materials needed to perform the experiment.

PROCEDURE: Describe each step of the procedure so that someone else could perform the experiment following your directions.

RESULTS: Include in your report all data, tables, graphs, and sketches used to arrive at your conclusions.

CONCLUSIONS: Record your conclusions in a paragraph at the end of your report. Your conclusions should be an analysis of your collected data.

Read the following description of an experiment. Then answer the questions.

All plants need water, minerals, carbon dioxide, sunlight, and living space. If these needs are not met, plants cannot grow properly. A biologist thought that plants would not grow well if too many were planted in a limited area. To test this idea, the biologist set up an experiment. Three containers were filled with equal amounts of potting soil. One bean seed was planted in Container 1, five seeds in Container 2, and ten seeds in Container 3. All three containers were placed in a well-lit room. Each container received the same amount of water every day for two weeks. The biologist measured the average height of the growing plants in each container every day, recorded these measurements in a table, and then plotted the data on a graph.

1. What was the purpose of this experiment? _____

_____

2. What materials were needed for this experiment? _____

_____

3. Write a step-by-step procedure for this experiment. _____

_____

_____

_____

_____

_____

_____

_____

4. Table 1 shows the data collected in this experiment. Based on these data, state a conclusion for this experiment. _____

_____

_____

Table 1.

Average Height of Growing Plants (in Millimeters)										
	Day									
Container	1	2	3	4	5	6	7	8	9	10
1	20	50	58	60	75	80	85	90	110	120
2	16	30	41	50	58	70	75	80	100	108
3	10	12	20	24	30	35	42	50	58	60

5. Plot the data in Table 1 on a graph. Show average height on the vertical axis and the days on the horizontal axis. Use a different color pencil for the graph of each container.

CARE OF LIVING THINGS

Caring for living things in a biology laboratory can be interesting and fun, and it can help develop the respect for all life that comes only from first-hand experience. In a room with an aquarium, terrarium, healthy animals, or growing plants, there is always some observable interaction between organisms and their environment. There are many species of plants and animals that are suitable for a classroom, but having them should be considered only if proper care will be taken so that the organisms not only survive, but thrive. Before growing plants or bringing animals into a classroom, find out if there are any health or safety regulations restricting their use or if there are any applicable state or local laws governing live plants and animals. Also be sure not to consider cultivating any endangered or poisonous species. A biological supply house or local pet store will provide growing tips for plants or literature on animal care when these organisms are purchased.

Evaluating Resources

Before bringing any live specimens into a new environment, check with your teacher to see if the basic needs will be met in their new location. Plants need either sunlight or grow lights. Animals must be placed in well-ventilated areas out of direct sunlight and away from the draft of open windows, radiators, or air conditioners. For both animals and plants, a source of fresh water is essential. Consider what the fluctuation in temperature is over weekends and holidays and who will care for the plants or animals during those times.

Setting Up an Aquarium

A closed system such as an aquarium provides a variety of animals and plants and can be maintained easily if set up correctly. A 10 or 20 gallon tank can be a suitable home for about 5 to 10 tropical fish or more of the more temperate goldfish. An air pump, filter, heater, thermometer, and aquarium light (optional) need to be in working order. First fill an aquarium with a layer of gravel, then fill with water. If using tap water, let the water stand a day before putting any fish in the tank. During cooler months, adjust the thermostat of the heater to bring the water to the desired temperature before adding fish. Most fish require temperatures of 20° to 25°C. An inexpensive pH kit purchased from a pet store will test the acidity of the water and guide the maintenance of a healthy pH.

Choose fish that are compatible with one another. A helpful pet store clerk can help in the selection. It is worth purchasing a scavenger fish like a catfish or an algae eater that will help keep the tank clean of algae. Snails are also helpful for this purpose. After purchasing, keep fish in the plastic bag containing water that they came in. Float it in the aquarium until the water reaches the same temperature, then slowly let the fish swim out of the bag. Some fish, such as guppies, eat their young. A smaller brood tank can be place inside the aquarium to keep the mother separated from the young.

One person should be responsible for feeding the fish. Feed fish sparingly. Overfeeding is not healthy for the fish; also, it clouds the tank and causes unnecessary decay. Weekend or vacation food should also be available. These are slow-dissolving tablets that can feed the fish over vacations.

Plants can be added to an aquarium as well. *Elodea, Anacharia, Sagittaria, Cabomba* or *Vallianeria* grown in a fishtank are also useful for many biology lab activities. Monitor their growth carefully and trim plants if growth is excessive. Some fish and snails may nibble on the plants causing them to break apart and decay. Decay introduces bacteria populations that may endanger the fish, so be sure to remove any decaying plant matter.

Variations on an aquarium include setting up a "balanced aquarium" with fish, plants, and scavengers in balance so that no pump or filter is necessary. This usually takes more planning and maintenance than a filtered tank. More maintenance is needed also for a marine aquarium because of the corrosive nature of saltwater. However, if specimens of marine organisms are readily available, creating such a mini-habitat is well worth the effort.

Setting Up a Terrarium

A terrarium is a mini-ecosystem that makes a suitable habitat for small plants and cold-blooded animals like amphibians and reptiles. Start with a layer of gravel in the bottom of a 10 gallon tank. Add a layer of sand, then some topsoil. Plant small mosses and ferns in the topsoil and water moderately. At one end of the tank, bury a dish of water in the sand. This can be a water source for salamanders, frogs, or toads. Cover the terrarium. Check the humidity level regularly. If water droplets form inside the glass, or if water collects in the gravel, take the cover off for a while. Too much moisture will foster growth of mold. A specialized terrarium, such as a desert or bog, can be made by planting suitable plants and maintaining water conditions similar to those environments.

Animals such as newts and turtles require a terrarium with more water. For this type of animal, fill an aquarium one-third of the way with water and place rocks in it so that the animals can get out of the water. Other animals such as lizards, anoles, or chameleons can do well in a dry terrarium with rocks as long as they have a water bowl. Care should be taken to give cold-blooded animals enough heat during the cooler months. This can be done with a lamp. Feed cold-blooded animals according to their needs. Many enjoy live worms or insects such as grasshoppers or crickets. However, freeze-dried worms available from a pet store are a good source of protein and require less maintenance than live food.

Keeping Mammals In a Classroom

Keeping mammals takes more consideration and commitment of time and expense. A small mammal like a gerbil, guinea pig, hamster, or rabbit may be kept in a classroom, at least for a short time. Explore the possibility of dwarf breeds that are more at home in a small space. However, many mammals are sensitive, social mammals that form bonds and attachments to people. Life in a small cage alone most of the time is not suitable for a long and healthy life. For short periods of time, however, small animals may be kept in a cage, provided it is large and clean enough. Find out the exact nutritional needs of the animals; feed them on a regular schedule and provide fresh water daily. Some animals require dry food supplemented with fresh foods, such as greens. However, these foods spoil more rapidly and thus uneaten portions must be removed. Provide a large enough cage for the animal as well as materials for bedding, nesting, and gnawing. Clean the cages frequently. Letting urine and feces collect in a cage fosters the growth of harmful bacteria. Animals in a cage also require an exercise wheel. Lack of space combined with overeating can make an animal overweight and lethargic. Handle animals gently. Under no circumstances should animals be exposed to harmful radiation, drugs, toxic chemicals, or surgical procedures.

Many times students wish to bring a pet or even a wild animal that they have found into the classroom for observation. Do so only with discretion and if a proper cage is available. Protective gloves and glasses should be worn while handling any animals with the potential to bite. Be sure to check with local park rangers or wildlife specialists for any wildlife restrictions that may apply. Return any wild animals to their environment as soon as possible after observations.

Growing Plants in the Classroom

To successfully grow plants in a classroom, have on hand commercial potting soil, suitable containers such as clay or plastic pots, plant fertilizer, a watering can, and a spray bottle for misting. Always put a plant in the correct size container. One that is too large will encourage root growth at the expense of the stem and leaves. Place bits of broken clay or gravel in the bottom of the pot for drainage, then add potting soil and the plant. Place in a warm, well-lighted area and supply water. Fertilize occasionally. Give careful attention to a new plant to assess its adaptation to its new environment. Pale leaves indicate insufficient light, yellowing leaves indicate overwatering, and dropping leaves usually indicate insufficient humidity.

With very little special attention, such plants as geraniums, begonias, and coleus can be easily and inexpensively grown in a classroom. These plants are hardy and can withstand fluctuations in light and temperature. From one hardy plant, many cuttings can be made to demonstrate vegetative propagation. A cutting of only a few leaves on a stem will develop roots in 2 to 4 weeks if it is placed in water or given root-growth hormone powder.

These plants not only add color to a classroom, but are useful in biology experiments as well. The dense green leaves of geraniums are especially useful for extracting chlorophyll or showing the effects of light deprivation. The white portions of variegated coleus leaves are good for showing the absence of photosynthesis with a negative starch-iodine reaction. Pinch back the flower buds as they begin to form to encourage fuller leaf growth.

Larger plants such as a fig (*Ficus*), dumbcane (*Dieffenbachia*), cornplant (*Dracaena*), Norfolk Island pine (*Araucaria*), umbrella plant (*Schefflera*), or various philodendrons adapt well to low-light conditions and so do not need frequent watering. However, make sure humidity is suitable to avoid dropping leaves. More exotic plants might be best suited to a small-dish garden but will need special care because there is less soil to hold moisture.

During winter months, a dish garden of forced bulbs, such as paperwhite narcissus, can be easily grown by placing the bulbs in a container of water left in a cool, dark place. Blooms will appear in 3 to 4 weeks. In the early spring, shoots of early flowering shrubs, such as forsythia and pussy willow, may be forced. Cut off some healthy shoots when buds appear, wrap in wet newspaper, then bring indoors and immerse cut ends in a tall vase or jar. Also buds of fruit trees, like apple, plum, or peach, will produce leaves and flowers in this way. Be sure to maintain shoots by changing water when necessary.

LABORATORY EQUIPMENT

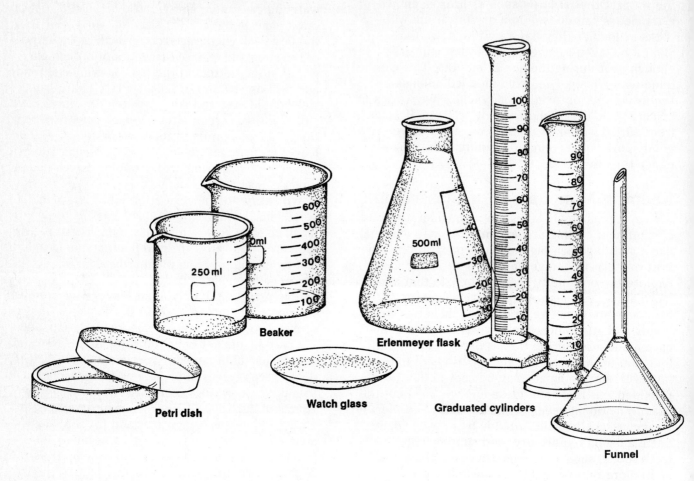

Beaker

250 ml

Erlenmeyer flask

500ml

Petri dish

Watch glass

Graduated cylinders

Funnel

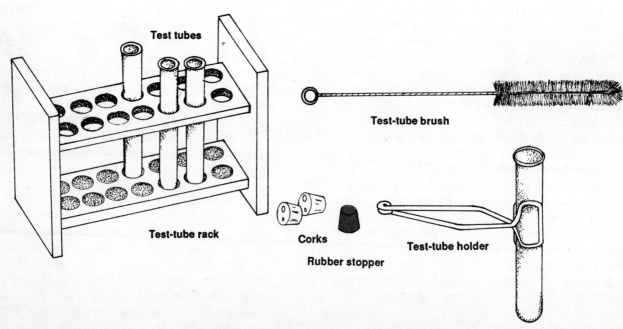

Test tubes

Test-tube brush

Test-tube rack

Corks

Rubber stopper

Test-tube holder

LABORATORY EQUIPMENT

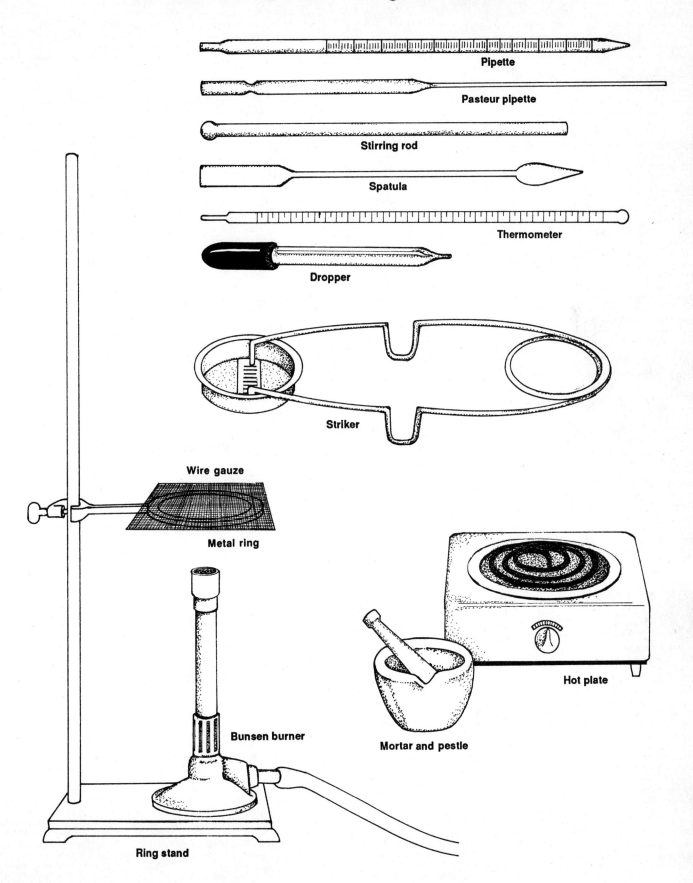

Pipette

Pasteur pipette

Stirring rod

Spatula

Thermometer

Dropper

Striker

Wire gauze

Metal ring

Hot plate

Bunsen burner

Mortar and pestle

Ring stand

LABORATORY EQUIPMENT

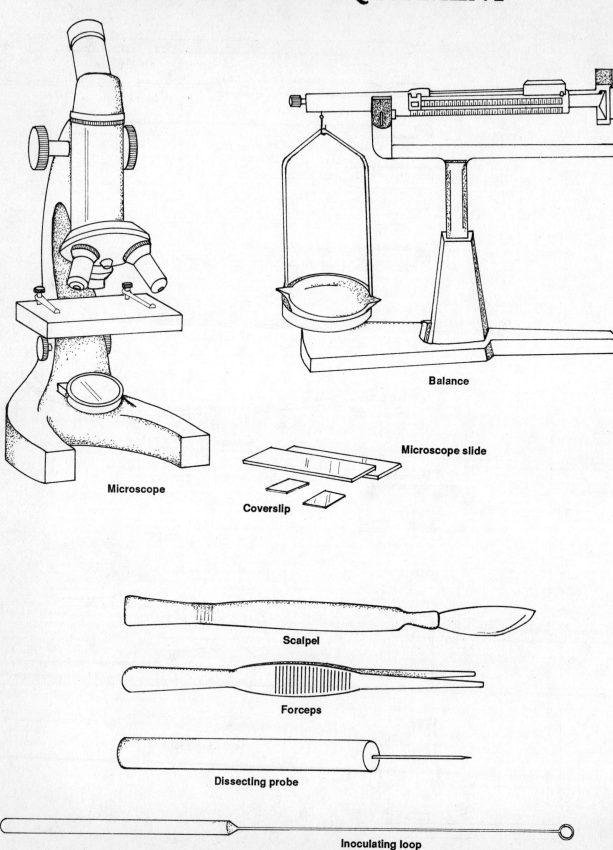

Balance

Microscope

Microscope slide

Coverslip

Scalpel

Forceps

Dissecting probe

Inoculating loop

SAFETY RULES

1. Always obtain your teacher's permission before beginning an activity.
2. Study the procedure. If you have questions, ask your teacher. Be sure you understand any safety symbols shown on the page.
3. Use the safety equipment provided for you. Goggles and a laboratory apron should be worn when any investigation calls for using chemicals.
4. Always slant test tubes away from yourself and others when heating them.
5. Never eat or drink in the lab, and never use lab glassware as food or drink containers. Never inhale chemicals. Do not taste any substance or draw any material into a tube with your mouth.
6. If you spill any chemical, wash it off immediately with water. Report the spill immediately to your teacher.
7. Know the location and proper use of the fire extinguisher, safety shower, fire blanket, first aid kit, and fire alarm.
8. Keep all materials away from open flames. Tie back long hair and loose clothing.
9. If a fire should break out in the classroom, or if your clothing should catch fire, smother it with the fire blanket or a coat, or get under a safety shower. **NEVER RUN.**
10. Report any accident or injury, no matter how small, to your teacher.

Follow these procedures as you clean up your work area.

1. Turn off the water and gas. Disconnect electrical devices.
2. Return all materials to their proper places.
3. Dispose of chemicals and other materials as directed by your teacher. Place broken glass and solid substances in the proper containers. Never discard materials in the sink.
4. Clean your work area.
5. Wash your hands thoroughly after working in the laboratory.

FIRST AID	
Injury	**Safe response**
Burns	Apply cold water. Call your teacher immediately.
Cuts and bruises	Stop any bleeding by applying direct pressure. Cover cuts with a clean dressing. Apply cold compresses to bruises. Call your teacher immediately.
Fainting	Leave the person lying down. Loosen any tight clothing and keep crowds away. Call your teacher immediately.
Foreign matter in eye	Flush with plenty of water. Use eyewash bottle or fountain.
Poisoning	Note the suspected poisoning agent and call your teacher immediately.
Any spills on skin	Flush with large amounts of water or use safety shower. Call your teacher immediately.

SAFETY CONTRACT

I, _____ , have read and understand the safety rules and first aid information listed above. I recognize my responsibility and pledge to observe all safety rules in the science classroom at all times.

_____   _____
signature   *date*

SAFETY SYMBOLS

The *Biology: The Dynamics of Life* Program uses several safety symbols to alert you to possible laboratory dangers. These safety symbols are explained below. Be sure that you understand each symbol before you begin a lab activity.

DISPOSAL ALERT
This symbol appears when care must be taken to dispose of materials properly.

ANIMAL SAFETY
This symbol appears whenever live animals are studied and the safety of the animals and the student must be ensured.

BIOLOGICAL SAFETY
This symbol appears when there is danger involving bacteria, fungi, or protists.

RADIOACTIVE SAFETY
This symbol appears when radioactive materials are used.

OPEN FLAME ALERT
This symbol appears when use of an open flame could cause a fire or an explosion.

CLOTHING PROTECTION SAFETY
This symbol appears when substances used could stain or burn clothing. A laboratory apron should be worn when this symbol appears.

THERMAL SAFETY
This symbol appears as a reminder to use caution when handling hot objects.

FIRE SAFETY
This symbol appears when care should be taken around open flames.

SHARP OBJECT SAFETY
This symbol appears when a danger of cuts or puncture caused by the use of sharp objects exists.

EXPLOSION SAFETY
This symbol appears when the misuse of chemicals could cause an explosion.

FUME SAFETY
This symbol appears when chemicals or chemical reactions could cause dangerous fumes.

EYE SAFETY
This symbol appears when a danger to the eyes exists. Safety goggles should be worn when this symbol appears.

ELECTRICAL SAFETY
This symbol appears when care should be taken when using electrical equipment.

POISON SAFETY
This symbol appears when poisonous substances are used.

PLANT SAFETY
This symbol appears when poisonous plants or plants with thorns are handled.

CHEMICAL SAFETY
This symbol appears when chemicals used can cause burns or are poisonous if absorbed through the skin.

Preparing an Insect Collection

EXPLORATION

1–1

LAB

▶ Insects are a class of animals in the Phylum Arthropoda. Some scientists estimate that there might be more than 10 000 000 species of insects on Earth although only 750 000 have been classified. Insects are important not just because there are so many of them, but also because they are responsible for pollination of flowers and for production of honey, wax, and silk. They also consume 13 percent of our food crops and are responsible for transmitting organisms that cause many human diseases. Because they are so important to our lives and they are so abundant, insects are excellent organisms for study.

OBJECTIVES

• Collect and prepare an insect collection.

• Recognize and become familiar with the features used to identify insects.

MATERIALS

scissors	white index cards	plastic film cannister	clean glass jars (approx.
white glue	(unlined) (2)	moth balls (crystals)	1-L size) with lids (2)
white paper	facial tissue (4)	95% ethanol	polystyrene foam pieces
#2 insect pins (20–30)	insect net (optional)	10x hand lens	(25 cm × 25 cm × 3 cm) (2)
fingernail polish remover	masking tape	small vials (50 mL)	wax marking pencil
(ethyl acetate)	cardboard display box	with stoppers	small cardboard boxes
			(approx. 6)

PROCEDURE

Part A. Collecting Insects

1. Reinforce the two jars with masking tape as shown in Figure 1.

2. Label both jars with a wax marking pencil "POISON." On one write "BUTTERFLIES AND MOTHS" and on the other, write "GENERAL." The jar for butterflies and moths will prevent these delicate insects from being damaged by other, tougher insects.

3. Place two pieces of crumpled tissue in each jar. These jars will serve as chambers for killing the insects.

4. Place one capful, approximately 5–10 mL, of fingernail polish remover on the tissues in the jars when ready to collect insects. The liquid will be effective for a few hours of collecting. Keep lids tightly on the jars except when placing insects in them.

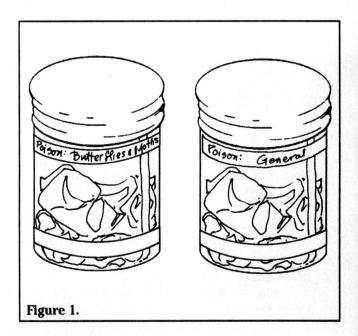

Figure 1.

1

5. Collect 15 different types of insects from several different habitats. Look under stones, boards, loose bark, shrubs and leaves, inside flowers, and around windows. Flying insects can be collected using an insect net around vegetation during the day or around lights at night. As you collect the insects, write a brief description of the insect, habitat, date, and time of day. This information will help you prepare labels when you return to the lab.

6. Place each insect in the appropriate jar for at least an hour.

7. Place the insects, once they are dead, in small cardboard boxes to hold until pinning. Label the boxes with your name. *DO NOT place insects in plastic boxes or they will mold before being pinned.*

Part B. Preparing Insects

Pin insects within 48 hours of their collection.

1. Prepare a label for each insect as shown in Figure 2. Include the following information on each label:

   ☐ State, County
   ☐ City, Location
   ☐ Date, Time of Day Collected
   ☐ Your Name

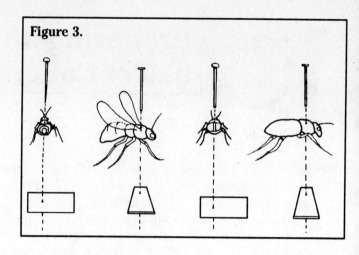

Figure 3.

Small Insects

Very small insects (gnats, ants, lice) can be tabbed.

5. Cut a very small triangle from an index card.

6. Place a small dab of glue on the point of the tab. Approach the right side of the insect's thorax with the tab, touching the insect so that it attaches to the tab. After the glue dries, run an insect pin through the tab as shown in Figure 4.

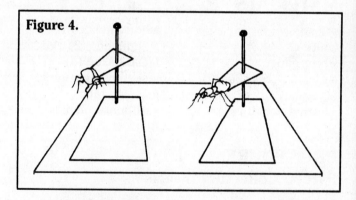

Figure 4.

Figure 2.

State: New York   St. Lawrence County
City: Cranberry Lake   bog habitat
Date: June 1, 1990, afternoon
Collector: J.W. Cook

Moths and Butterflies

Moths and butterflies need special care. After they have been pinned, their wings must be spread and allowed to dry in place.

7. With the point of the scissors, dig a small depression or "ditch" about 3 cm long and 1 cm wide in a piece of polystyrene foam. The size of the body of the insect will determine the size of the depression.

8. Pin the insect body according to steps 1–4 and place it in the depression so that the wings are even with the flat surface of the polystyrene foam. Examine Figure 5.

9. Cut two small strips of the index card and pin one end of each strip over the wings as shown in Figure 6.

Standard Insects

2. Place an insect pin through the thorax so that it runs from the dorsal side to the ventral side as in Figure 3. Place the insect on the upper one-third of the pin. About one-fourth of the top of the pin should be exposed so that all insects are at the same level in the box. Make sure the insect is level on its pin. It should not be tilted in any direction.

3. Pin each insect's label under the insect as in Figure 3.

4. Place the pinned insect onto a piece of polystyrene foam.

10. Position the wings on the right side as shown and pin down the other end of the strip with a second pin. Repeat this procedure with the wings on the left side.

11. Allow the moth or butterfly to dry for at least two days.

12. After the insect is dry, carefully remove the paper strips. Handle these fragile insects gently, touching only their pins.

Aquatic Insects

13. Place each aquatic insect directly into a small vial half filled with ethanol. Use pencil to fill out the label because many inks will wash away in the alcohol. Place the insect label into the vial with the insect. Seal the vials with liquid-tight lids.

Part C. Preparing a Display Box

1. Cut a piece of polystyrene foam so that it fits the bottom of your display box tightly. Place it in the box.

2. Pin the insects in the foam. Insects in vials can be placed in holes in the foam or held in place by tape.

3. Place an uncapped plastic film cannister in the corner of the box and fill it with moth crystals.

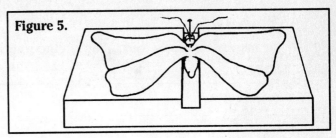

Figure 5.

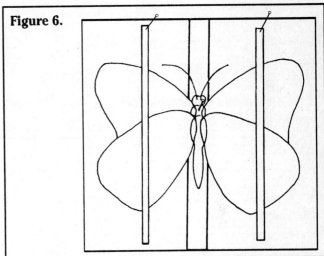

Figure 6.

4. Using a hand lens, study the external features of your insects. Answer the questions in Data and Observations.

5. Complete this Exploration and store the collection for later use in Exploration 31–2.

DATA AND OBSERVATIONS

1. List at least three observable features that all of your insects share.

_____

_____

_____

2. List the different types of insect mouthparts that occur in your collection. Use Figure 1 in Exploration 31–2 as a guide.

_____   _____

_____   _____

_____   _____

3. List the different types of appendages found at the end of your insects' abdomens.

_____   _____

_____   _____

_____   _____

ANALYSIS

1. Of the insects that have wings, how do the wings differ? _____

_____

_____

2. Do all of your insects have antennae? _____ How do your insects' antennae differ? _____

_____

_____

3. Which of the observable features of your insects could be used to separate them into different groups?

_____

_____

4. Would it be a good idea to classify insects based upon an insect's habitat? Why or why not? _____

_____

_____

FURTHER EXPLORATIONS

1. Make another insect collection for your own use and enjoyment.
You may want to specialize and collect only beetles or butterflies.
Borrow a field guide to insects from a library. This book will give
you many ideas.
2. Make a collection of other organisms such as sea shells, leaves, or
wildflowers. Use a field guide to try to classify and identify these
organisms.

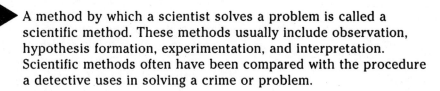

Can Scientific Methods Be Used to Solve a Problem?

2–1
LAB

INVESTIGATION

▶ A method by which a scientist solves a problem is called a scientific method. These methods usually include observation, hypothesis formation, experimentation, and interpretation. Scientific methods often have been compared with the procedure a detective uses in solving a crime or problem.

OBJECTIVES

- Use scientific methods to decide whether two liquids are similar or different.
- Make careful observations.

- Hypothesize whether the two liquids are the same or different.
- Record accurate experimental results.

MATERIALS

Erlenmeyer flasks containing liquids, labeled A and B
clock or watch with second hand

stoppers to fit flasks (2)
beaker

laboratory apron
goggles

PROCEDURE

Part A. Observation

Accurate observations are a necessary part of scientific methods.

1. Examine the two flasks. DO NOT remove the stoppers and DO NOT shake the contents.
2. Record in Table 1 two or three similarities and differences between the contents of the two flasks.

   a. Do you think both flasks contain the same

      liquid? _____

      Explain. _____

   b. Is your answer to question 2a based on

      experimentation or guessing? _____

   c. Would scientists guess at answers to questions or would they experiment first?

      _____

   d. Do both flasks contain the same volumes of

      liquid? _____

e. What gas might be in the upper half of flask

   A that is not in flask B? _____

f. Is there any direct evidence for your answer

   to question 2e? _____

3. Make a **hypothesis** about the contents of the two flasks. Are they the same or different? Write your hypothesis in the space provided.

Part B. Experimentation

In determining if the two liquids are the same or not, a scientist would carry out some experiments. Experimentation is another part of the scientific method.

Experiment 1. What happens if you shake the liquids?

1. Give each flask *one hard shake using an up-and-down motion of your hand.* Make sure your thumb covers the stopper as you shake. See Figure 1. Observe each flask carefully.
2. Record your observations in Table 2. Again, look for similarities and differences in the contents of the two flasks.

a. After shaking the flasks, do you think they

contain different liquids? _____

b. What was present in flask A that might have been responsible for the change in the

liquid? _____

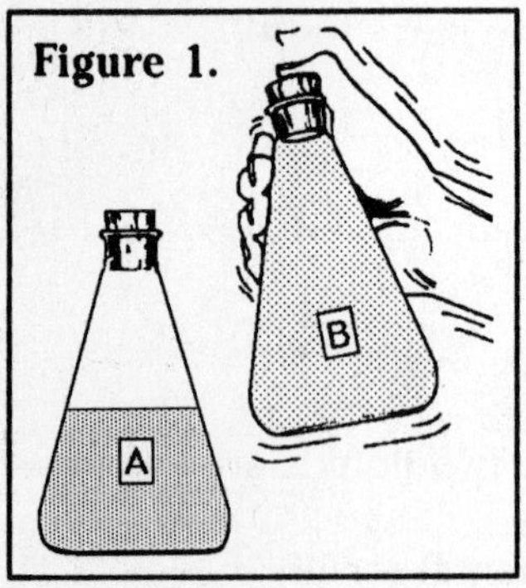

Figure 1. 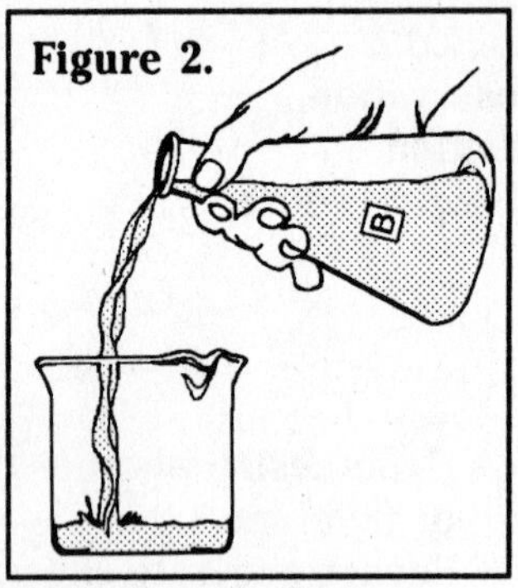   Figure 2.

Experiment 2. What happens if you remove some of the liquid in flask B so it contains the same volume of liquid as flask A?

1. Remove the stopper from flask B and pour out half of the contents into a beaker. See Figure 2. Make sure that the volume of liquid in flask B is equal to the volume of liquid in flask A. Replace the stopper.
2. Give both flasks *one hard shake using an up-and-down motion of your hand.* Hold stopper in place while shaking.
3. Observe each flask carefully.
4. In Table 3, record any similarities or differences observed.
   a. Do both flasks now appear to contain the

   same liquid? _____

   b. What might have been added to flask B that

   was not present before? _____

Experiment 3. What happens if you shake the flasks more than once?

1. Shake each flask *hard once with an up-and-down motion.*
2. Note the number of seconds it takes for each liquid to return to its original condition after shaking. Record the time under Trial 1 in Table 4.
3. Shake each flask *hard twice with an up-and-down motion.*
4. Time how long it takes for the liquids to return to their original conditions. Record your data under Trial 1 in Table 4.

5. Shake both flasks *hard three times with an up-and-down motion.*
6. Note and record in Table 4 the time it takes for the liquids to return to their original conditions.
   a. Is the time needed for the liquids in flasks A and B to return to their original conditions

   after one shake about the same? _____

   b. Is the time needed for the liquids in flasks A and B to return to their original conditions

   after two or three shakes the same? _____

   c. Look at your data in Table 4. Does flask A show an increase or decrease in time needed to return to its original condition as the number of shakes increases from one to

   three? _____

   Does flask B show a similar change? _____

7. In any experiment, running several trials reduces the probability of making errors. Run two more trials for each part of Experiment 3. Be sure to keep track of the amount of time needed for the liquids to return to their original conditions.
8. Record the results of Trials 2 and 3 in Table 4.
   a. Do three trials give better evidence that the liquid in flask A is "behaving" in a way similar to the liquid in flask B after shaking each flask

   once? _____

   twice? _____

   three times? _____

   b. Do three trials give better evidence that an increase in time is needed for the liquid to return to its original condition as the number of shakes increases from one to three

   for flask A? _____

   for flask B? _____

HYPOTHESIS

_____

_____

DATA AND OBSERVATIONS

Table 1.

First Observations	
Similarities	**Differences**

Table 2.

Results of Experiment 1	
Similarities	**Differences**

Table 3.

Results of Experiment 2	
Similarities	**Differences**

Table 4.

Three Trials of Experiment 3									
	Time in seconds to return to original condition								
	1 shake			2 shakes			3 shakes		
Trial	1	2	3	1	2	3	1	2	3
Flask A									
Flask B									

ANALYSIS

Questions 1–4 should help you to make some interpretations of what you have observed. Interpretations are reasonings based on observations and experiments. They usually are the next step in a scientific method.

1. On the basis of your first observations in Part A, could you decide if both flasks contained the same liquid? _____

2. After performing Experiment 1, could you decide if both flasks contained the same liquid? _____

3. Which experiment or experiments might have helped you to decide that the liquids in flasks A and B were similar or different? _____

   Explain. _____

4. Besides the liquid itself, what else seems to be needed for the liquid to change color? _____

   _____

Questions 5–7 should help you to form a conclusion.

5. Explain why flask B did not change color when shaken in Experiment 1. _____

6. Why must the liquids in the half-filled flasks be shaken to produce a color change? _____

   _____

7. Why did more shaking increase the amount of time needed for the liquids in flasks A and B to change back to their original color? _____

   _____

8. a. Could you have answered the first question in Part A by guessing? _____

   b. Why is experimenting a better method of problem solving than guessing? _____

   _____

   c. What is meant by the phrase "solving a problem by using scientific methods"? _____

   _____

   _____

CHECKING YOUR HYPOTHESIS

Was your **hypothesis** supported by your data? Why or why not? _____

_____

FURTHER INVESTIGATIONS

1. Design an experiment to test the effect of a new fertilizer on plant growth.
2. Design and carry out an experiment to test the effect of different colors of light on plant growth.

Using SI Units

2-2

LAB

EXPLORATION

Scientists all over the world use SI units for measuring. SI stands for the International System of Measurement. It allows scientists from different countries to communicate easily.

The SI system is more convenient than the English system of inches, feet, ounces, pounds, and so on, because all SI units use a base with standard prefixes and units are multiples of 10. For example, there are 100 centimeters in a meter and 1000 meters in a kilometer.

Converting from one unit to another involves multiplying or dividing by 10 or a multiple of 10. By memorizing a few standard prefixes and knowing when to multiply or divide, unit conversion is easy. When converting from a small unit to a larger unit, divide. When converting from a large unit to a smaller unit, multiply. What number you use to divide or multiply with depends on the units involved. For example, in converting from millimeters to centimeters, divide by 10 because there are 10 millimeters in one centimeter. SI units will be used in this Exploration to measure length, mass, and volume as you work with the growth of seeds.

OBJECTIVES

• Measure the mass of bean seeds.

• Measure, record, and graph the growth of bean seeds.

MATERIALS

bean seeds (12)
masking tape
balance
10-cm flower pots (4)
vermiculite

water
fertilizer solutions: full-strength,
    double-strength, half-strength
beakers (4)
colored pencils (4)

metric ruler
newspaper
wax marking pencil
50-mL graduated cylinder

PROCEDURE

Part A. Seed Mass

1. Separate the bean seeds into four groups of three seeds each. Label the groups A, B, C, and D.

A gram (g) is a common SI unit of mass. Other common units of mass are the milligram (mg) and the kilogram (kg). A milligram is 1/1000 of a gram. A kilogram is 1000 grams.

2. With a balance, determine the average mass in grams of the seeds in each group. To do this, find the mass of the entire group and divide by the number of seeds. Record the total mass and the average mass in Table 1.

3. Use the information in the introduction to convert the average mass of the seeds in each group to milligrams and kilograms. Record these units in Table 1.

Part B. Plant Height

1. Label four flower pots A, B, C, and D. Plant each group of three bean seeds in its labeled pot containing about 8 cm vermiculite. Plant the seeds 2 cm deep. Be sure to put Group A seeds into Pot A, Group B seeds into Pot B, and so on.

2. Label four beakers as follows: A—tap water; B—plant fertilizer; C—double-strength plant fertilizer; and D—half-strength plant fertilizer. Obtain some of each solution.

A liter (L) is a common SI unit of volume. Another common unit of volume is the milliliter (mL). A milliliter is 1/1000 of a liter.

3. Using a graduated cylinder, measure 50 mL of each solution and carefully water each pot with the corresponding solution. Wash the cylinder between solutions.

4. Continue to water all four pots with 50 mL of solution each time you water.

5. Observe the pots of seeds daily. On the pot's label, record the date when the first seedling in each pot emerges.

A centimeter (cm) is a common SI unit of length. Other units of length include the millimeter (mm), meter (m), and kilometer (km). The meter is the standard SI unit of length. The other units are derivatives of the meter. For example a centimeter is 1/100 of a meter. A millimeter is 1/1000 of a meter. There are 1000 millimeters in a meter. A meter is 1/1000 of a kilometer. There are 1000 meters in a kilometer.

6. On the third day after the seedling emerges, measure the height in centimeters of each plant in each pot. Determine plant height by measuring from the top of the vermiculite to the tip of the main plant stem, as in Figure 1. If a plant droops, gently straighten it to get its actual height. Record the plant heights in Table 2.

7. Determine the average height of the plants in each pot. Record these average heights in Table 2.

8. On the sixth day, measure, average, and record in Table 3 the heights of the plants in each pot.

9. On the ninth day, measure, average, and record in Table 4 the heights of the plants in each pot.

10. In Tables 2, 3, and 4, convert the average heights of the plants in each of the pots to millimeters, meters, and kilometers. Be sure to multiply or divide by the correct multiple of 10 to get the correct numbers.

11. Using a different color pencil for each pot, graph the average growth of the plants in pots A–D in Figure 2.

Part C. Plant Mass

1. After the ninth day, spread newspapers over your work area and carefully remove the plants and vermiculite from Pot A. Being careful not to break the roots, gently remove the vermiculite

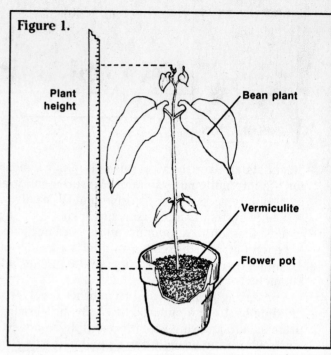

Figure 1.

Plant height

Bean plant

Vermiculite

Flower pot

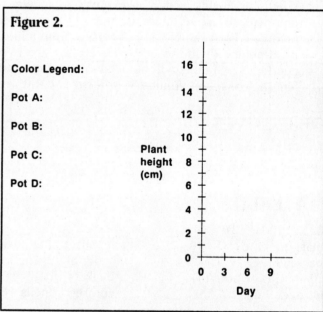

Figure 2.

Color Legend:

Pot A:

Pot B:

Pot C:

Pot D:

Plant height (cm)

Day

from around the roots. Remove any clinging material by rinsing the roots gently in tap water.

2. Using a balance, determine and record in Table 5 the combined mass of the three plants in pot A. Then calculate the average mass of a plant in Pot A and record it in Table 5.

3. Repeat steps 1 and 2 for pots B, C and D.

4. Record average seed masses from Table 1 in the proper column of Table 5.

5. Record the increase of mass in Table 5.

$$\text{increase of mass} = \text{average plant mass} - \text{average seed mass}$$

6. Determine the percentage increase in mass for the plants from each pot. Record the percentage increases in Table 5.

$$\text{percentage increase in mass} = \frac{\text{increase of mass}}{\text{average seed mass}} \times 100\%$$

DATA AND OBSERVATIONS

Table 1.

	Average Mass of Seeds			
Group	Total mass of seeds	Average mass (g)	Milligrams (mg)	Kilograms (kg)
A				
B				
C				
D				

Table 2.

	Average Height of Plants 3 Days after Germination				
Pot	Height of each plant	Average height (cm)	Millimeters (mm)	Meters (m)	Kilometers km)
A					
B					
C					
D					

Table 3.

	Average Height of Plants 6 Days after Germination				
Pot	Height of each plant	Average height (cm)	Millimeters (mm)	Meters (m)	Kilometers (km)
A					
B					
C					
D					

Table 4.

	Average Height of Plants 9 Days after Germination				
Pot	Height of each plant	Average height (cm)	Millimeters (mm)	Meters (m)	Kilometers (km)
A					
B					
C					
D					

Table 5.

Pot	Combined mass of plants	Average plant mass (g)	Average seed mass (g)	Increase of mass	Percentage increase
Plant Mass and Mass Increase					
A					
B					
C					
D					

ANALYSIS

1. What are the common SI units of mass? _____

   of length? _____

2. How do you convert: **a.** grams to kilograms? _____ **b.** grams to milligrams? _____

   c. centimeters to millimeters? _____ **d.** centimeters to meters? _____

3. The standard SI unit of volume is the liter. Other units of volume include the milliliter and kiloliter. How

   many milliliters are in one liter? _____

4. How do you convert liters to milliliters? _____

5. Why was it necessary to water each pot at the same time with the same amount of solution? _____

   _____

6. Rank the pots in order of growth from least average height to greatest average height. Explain your

   results. _____

   _____

7. Rank the pots in order of percentage increase in plant mass from least increase to greatest increase.

   _____

8. How does this ranking compare with your ranking for plant growth? _____

   _____

9. What are possible explanations for differences in your rankings? _____

   _____

FURTHER EXPLORATIONS

1. Design and conduct an experiment to determine if bean seed size
   has any effect on the size and growth of the bean plants.
2. Design and conduct an experiment to compare the relative
   effectiveness of several different types of commercial plant
   fertilizers.

Name _____ Date _____

Physical Factors of Soil

3-1
LAB

EXPLORATION

▶ Soil is a major factor influencing the survival of many living things. Many organisms live in the soil. Others are anchored in soil and obtain water and minerals from it. Still other organisms depend on these soil-dependent organisms for food. The physical properties of a particular kind of soil determine the kinds of plants that grow in the soil and the kinds of animals that live in or on it.

OBJECTIVES
- Determine the amounts of various particle types in three soil samples.
- Calculate the water contents and water-holding capacities of three soil samples.

MATERIALS

soil samples (3)
20-cm cloth squares (3)
balance

specimen jars with lids (3)
beakers (3)
scoop

water
metric ruler

pins (3)
masking tape

PROCEDURE
Part A. Particle Size
1. Label three specimen jars with the locations of the soil samples. Fill each jar halfway with soil. Add water, allowing it to soak into the soils, until the jars are full.
2. Cover the jars with lids and shake until the large soil particles break apart. Set the jars aside and let the particles settle overnight.
3. Using a ruler, measure the depth of each particle type in each jar.
4. Record in Table 1 the amounts of gravel, coarse sand, fine sand, silt, and clay in the settled soil samples. See Figure 1.

Part B. Water Content
1. Soak the cloth squares in water. Attach labels identifying the samples with pins.
2. Place a scoop of soil in each cloth. Wrap the soil samples in the wet cloths. Determine and record their masses in Table 2. Place the wrapped samples where they will dry completely, then redetermine and record their masses. Calculate the water content of each sample as a percentage of the dry mass of the soil.

$$\text{percentage water content} = \frac{\text{mass of soil and cloth} - \text{mass of dried soil and cloth}}{\text{mass of dried soil and cloth}} \times 100$$

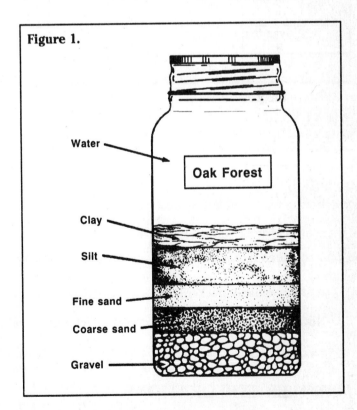
Figure 1.

Part C. Water-Holding Capacity
1. Place each dried soil sample and cloth from Part B in a beaker of water for five to ten minutes or until the soil is saturated.

13

2. Remove the wrapped samples from the beakers and allow excess water to drain from them through the cloths. Find and record the masses of the saturated samples.

3. Calculate the water-holding capacity of each sample as a percentage of the dry mass.

$$\text{percentage water-holding capacity} = \frac{\begin{array}{c}\text{mass of}\\\text{saturated soil}\\\text{and cloth}\end{array} - \begin{array}{c}\text{mass of}\\\text{dried soil}\\\text{and cloth}\end{array}}{\begin{array}{c}\text{mass of dried}\\\text{soil and cloth}\end{array}} \times 100$$

DATA AND OBSERVATIONS

Table 1.

Soil Particle Size Data					
	Amount of each particle type (in mm)				
Soil location	Gravel	Coarse sand	Fine sand	Silt	Clay
1.					
2.					
3.					

Table 2.

Water Content and Water-holding Capacity					
Soil location	Mass of soil and cloth	Mass of dried soil and cloth	Mass of saturated soil and cloth	Percentage water content	Percentage water-holding capacity
1.					
2.					
3.					

ANALYSIS

1. Which type of soil particle made up:

   a. the greatest amount of each soil sample? _____

   b. the least amount? _____

2. Which type of soil particle was:

   a. most closely packed? _____

   b. least closely packed? _____

3. How does the type of soil particles affect water drainage? _____

_____

FURTHER EXPLORATIONS

1. Research the procedure for calculating the organic-matter content of a soil sample.

2. Prepare a chart showing the predominant soil types in various parts of the United States. Show how soil types affect the commercial activities of an area.

The Lesson of the Kaibab

3–2
LAB

 The environment may be altered by forces within the biotic community, as well as by relationships between organisms and the physical environment. The carrying capacity of an ecosystem is the maximum number of organisms that an area can support on a sustained basis. The density of a population may produce such profound changes in the environment that the environment becomes unsuitable for survival of that species. Humans can interfere with these natural interactions and have either a positive or a negative effect.

OBJECTIVES

- Graph data on the Kaibab deer population of Arizona from 1905 to 1939.
- Analyze the methods responsible for the changes in the deer population.
- Propose a management plan for the Kaibab deer population.

MATERIALS

colored pencils (1 green and 1 red)

PROCEDURE

Before 1905, the deer on the Kaibab Plateau in Arizona were estimated to number about 4000 on almost 300 000 hectares of range. The average carrying capacity of the range, was estimated then to be about 30 000 deer. On November 28, 1906, President Theodore Roosevelt created the Grand Canyon National Game Preserve to protect the "finest deer herd in America."

Unfortunately, by this time the Kaibab forest area had already been overgrazed by sheep, cattle, and horses. Most of the tall perennial grasses had been eliminated. The first step to protect the deer was to ban all hunting. In addition, in 1907, the Forest Service tried to exterminate the predators of the deer. Between 1907 and 1939, 816 mountain lions, 20 wolves, 7388 coyotes, and more than 500 bobcats, all predators of the deer, were killed.

1. Using the green pencil, draw and label a straight horizontal line across the graph in Data and Observations to represent the average carrying capacity of the range.
2. Using the red pencil, graph the data in Table 1.
3. Answer Analysis questions 1–4.

Signs that the deer population was out of control began to appear as early as 1920—the range was beginning to deteriorate rapidly. The Forest Service reduced the number of livestock grazing permits. By 1923, the deer were reported to be on the verge of starvation, and the range conditions were described as "deplorable."

Table 1.

Deer Population from 1905 to 1924	
Year	Deer population
1905	4 000
1910	9 000
1915	25 000
1920	65 000
1924	100 000

A Kaibab Deer Investigating Committee recommended that all livestock not owned by local residents be removed immediately from the range and that the number of deer be cut in half as quickly as possible. Hunting was reopened, and during the fall of 1924, 675 deer were killed by hunters. However, these deer represented only one-tenth the number that had been born that spring.

4. Using the red pencil, plot the data in Table 2 on your graph. Label the completed graph.
5. Answer Analysis questions 5–6.

Today, the Arizona Game Commission carefully manages the Kaibab area with regulations geared to specific local needs. Hunting permits are issued to keep the deer in balance with their range. Predators are protected to help keep herds in balance with food supplies. Tragic winter losses can be checked only by keeping the number of deer near the carrying capacity of the range.

6. Answer Analysis questions 7–11.

Table 2.

Deer Population from 1925 to 1939	
Year	Deer population
1925	60 000
1926	40 000
1927	37 000
1928	35 000
1929	30 000
1930	25 000
1931	20 000
1935	18 000
1939	10 000

DATA AND OBSERVATIONS

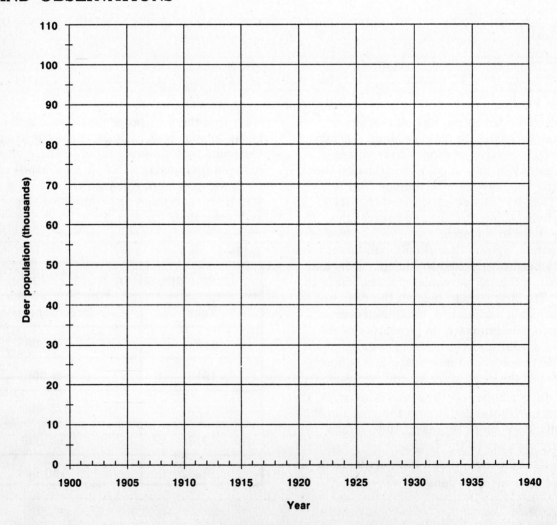

ANALYSIS

1. In 1906 and 1907, what two methods did the Forest Service decide to use to protect the Kaibab deer?

   _____

2. How many total predators were removed from the preserve between 1907 and 1939? _____

3. What was the relationship of the deer herd to the carrying capacity of the range:

   in 1915? _____

   in 1920? _____

   in 1924? _____

4. Did the Forest Service program appear to be successful between 1905 and 1924? _____

   Explain your answer. _____

   _____

5. Why do you suppose the population of the deer declined in 1925 although the predators were being

   removed? _____

   _____

6. Do you think any changes had occurred in the carrying capacity of the range from 1900 to 1940? _____

   Explain your answer. _____

   _____

7. Why do you suppose the population of deer in 1900 was 4000 when the range had an estimated carrying

   capacity of 30 000? _____

   _____

   _____

8. Without the well-meaning interference of humans, what do you think would have happened to the deer

   population after 1900? _____

   _____

   _____

9. What major lessons were learned from the Kaibab deer experience? _____

   _____

   _____

   _____

10. If the lessons learned from the Kaibab deer studies had been known then, what recommendations would

you have made in 1915? _____

_____

in 1923? _____

_____

in 1939? _____

_____

11. What future management plan would you suggest for the Kaibab deer herd? _____

_____

FURTHER EXPLORATIONS

1. Many forests have been endangered by gypsy moth caterpillars. Research how they came to the United States and the methods that have been proposed to control them.
2. Mosquitoes are a great annoyance to many people. Obtain some books from your teacher or librarian that contain information on mosquitoes. Decide whether or not you would try to eliminate them. Justify your answer.

What Organisms Make Up a Microcommunity?

4–1
LAB

INVESTIGATION

▶ Usually, when a community and the organisms associated with it are described, observable organisms such as trees, grass, pigeons, and squirrels are listed. These observable organisms are called macroorganisms. However, a community also has many microorganisms. Microorganisms can make up their own microcommunities. These microcommunities are small and inconspicuous. With a microscope, you can examine communities that would ordinarily go undetected. It is also possible to identify and classify many of the organisms found in microcommunities.

OBJECTIVES

- Observe and identify the organisms found in the microcommunities in bean water and pond water.
- Determine if each organism is motile or sessile.
- Identify each organism as a producer or a consumer.
- Make a hypothesis about the relationship of an organism as a consumer or producer to its motility.

MATERIALS

water
pond water
dropper
spring-type clothespin
bean water (tap water in which beans
          have soaked for several days)

microscope slides (3)
coverslips (2)
Bunsen burner
striker
laboratory apron
crystal violet stain

100-mL graduated cylinder
filter paper
microscope
small beaker or paper cup
funnel
250-mL beaker

PROCEDURE

Part A. Bean-Water Microcommunity

1. Place a drop of bean water on a microscope slide, spreading it into a thin film the size of a nickel.

2. Attach a clothespin to one end of the slide. Holding the slide by the clothespin, quickly pass the slide with the bean-water film through a low Bunsen burner flame several times. **CAUTION:** *Always be careful around open flames. Secure loose hair and clothing to keep them away from the flame.* Warm the slide until the bean water has dried on the slide. This process is called "fixing." Fixing sticks the cells to the slide. See Figure 1.

3. Add several drops of crystal violet stain to the dried film of bean water. Staining the slide will make microorganisms on the slide easier to see. See Figure 2.

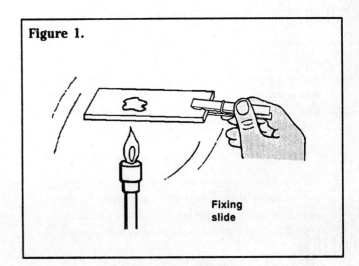

Figure 1.

Fixing slide

Figure 2.

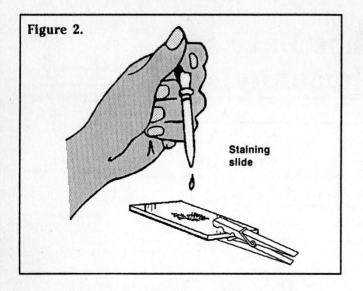

Staining slide

Figure 3.

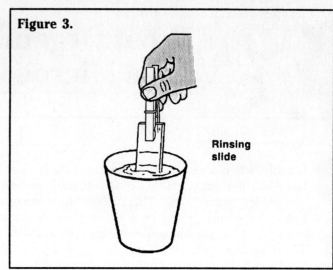

Rinsing slide

Figure 4.

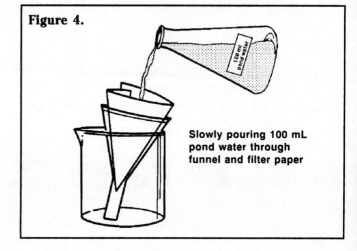

Slowly pouring 100 mL pond water through funnel and filter paper

4. After one minute, rinse the stain off with water. This rinsing is best done by dipping the slide into a beaker or paper cup filled with water. Handling the slide with the clothespin will keep you from staining your fingers. See Figure 3.

5. Allow your slide to air dry. After the slide has dried, examine the bean-water microcommunity under the high power of your microscope. No coverslip is needed.

6. Prepare a wet mount of bean water. This time a coverslip is needed. Examine the wet mount under the high power of your microscope.

7. Using the diagrams in Figure 5 for comparison, identify the organisms in your bean-water microcommunity. Note particularly the shapes of the organisms to aid in your identifications. Record in Table 1 the names of the organisms observed in the bean water.

8. By examining the wet mount, determine whether each organism is motile or is sessile. Record your findings in Table 1.

9. By examining the wet mount, determine whether each organism is a producer or a consumer. Producers usually will contain colored pigments ranging from yellow to green to blue-green. Consumers usually are colorless. Record your findings in Table 1.

10. Make a **hypothesis** to explain the relationship between producers and consumers and their motility. Write your hypothesis in the space provided.

Part B. Pond-Water Microcommunity

1. Line a funnel with filter paper. Filter 100 mL of pond water into a beaker. See Figure 4. This procedure will concentrate any organisms present in the water and make them easier to find.

2. Remove the filter paper from the funnel after the last of the pond water has drained through. Turn the filter paper inside out and touch to a glass slide the moist end that used to be the tip of the paper cone.

3. Add a small drop of water to the slide, if it is dry, and add a coverslip. Observe this wet mount under low and high powers of your microscope. Using the diagrams in Figure 5 for comparison, identify the organisms in your pond-water microcommunity.

4. Record in Table 1 the names of the organisms observed in the pond water. Determine and record in Table 1 whether each organism is motile or sessile. Determine and record in Table 1 whether each organism is a producer or a consumer.

HYPOTHESIS

_____

_____

Figure 5.

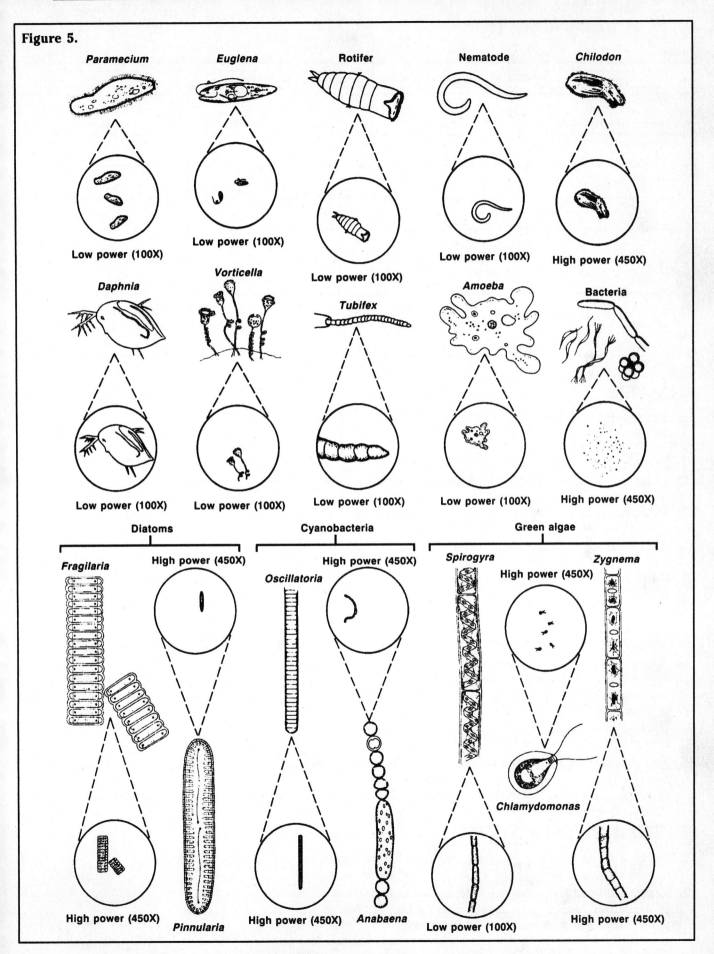

Paramecium

Euglena

Rotifer

Nematode

Chilodon

Low power (100X)

Low power (100X)

Low power (100X)

High power (450X)

Low power (100X)

Daphnia

Vorticella

Tubifex

Amoeba

Bacteria

Low power (100X)

Low power (100X)

Low power (100X)

Low power (100X)

High power (450X)

Diatoms

Cyanobacteria

Green algae

Fragilaria

High power (450X)

Oscillatoria

High power (450X)

Spirogyra

High power (450X)

Zygnema

Chlamydomonas

High power (450X)

Pinnularia

High power (450X)

Anabaena

Low power (100X)

High power (450X)

DATA AND OBSERVATIONS

Table 1.

Organisms in Microcommunities			
Name of organism	Community (bean water or pond water)	Motile or sessile	Producer or consumer

ANALYSIS

1. List several possible sources of the organisms in the bean water. _____

_____

2. **a.** Do you think most of the microorganisms in bean water are producers or consumers? _____

   b. What evidence do you have that may support your answer? _____

_____

3. Explain the color difference between producers and consumers in a pond-water microcommunity. _____

_____

4. Did you find a relationship between motility and whether an organism is a consumer or a producer?

Explain. _____

_____

CHECKING YOUR HYPOTHESIS

Was your **hypothesis** supported by your data? Why or why not?_____

_____

FURTHER INVESTIGATIONS

1. Examine other microcommunities, such as that in standing rainwater. Try to identify the organisms you see.
2. Read Anton van Leeuwenhoek's descriptions of various micro-communities he observed in rainwater and on the surfaces of his teeth. Write a report based on your reading.

INVESTIGATION

How Does the Environment Affect an Eagle Population?

5–1

LAB

▶With unlimited resources and ideal environmental conditions, a population would continue to increase in size. This rarely happens, however, because resources are limited and environmental conditions are not ideal. The carrying capacity of an ecosystem is the maximum number of organisms that an area can support. In nature, many populations remain below the carrying capacity because of a combination of both nonliving (abiotic) and living (biotic) factors. These factors include climate, habitat, available food, water supply, pollution, disease, and interactions between species, including predation, parasitism, and competition. The interactions between a population and the many components of an ecosystem are very complex. In this Investigation, you will make a simplified model of the effects of some abiotic and biotic environmental factors on a bald eagle population.

OBJECTIVES

• Make a model to show how abiotic and biotic factors affect a bald eagle population.

• Hypothesize how biotic factors affect a bald eagle population.

MATERIALS

index card
uncooked rice grains dyed red
   with food coloring (75)

20 cm × 20 cm pieces of paper (2)
uncooked white rice grains (150)
metric ruler

scissors
graph paper
colored pencils (6 colors)

PROCEDURE
Part A.

Eagles mate for life. Only one pair of eagles occupies, defends, and hunts a well-defined territory.

1. Mark off the edges of the two sheets of paper in centimeters. On each sheet, draw a grid of 400 1-cm squares as shown in Figure 1. The grid represents a 4-km² lake (10 cm = 1 km) where the eagles hunt for fish. This will be their only source of food.

2. Cut two 1-cm squares from the index card. Label one of the squares M for male and the other F for female.

3. Lay the two grids near each other on a flat surface. Scatter 150 grains of white rice over one of the grids. Each grain represents a large fish in the lake. Eagles eat only the large fish.

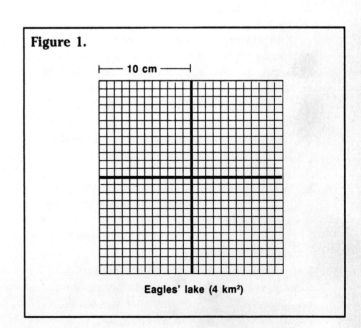

Figure 1.

├── 10 cm ──┤

Eagles' lake (4 km²)

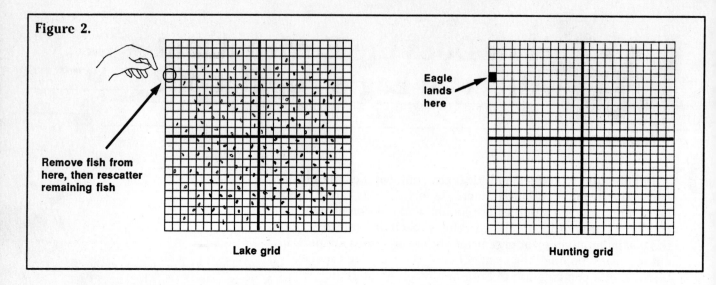

Figure 2.

Remove fish from here, then rescatter remaining fish

Lake grid

Eagle lands here

Hunting grid

4. The other grid is the hunting grid. Hold the M (male eagle) square about 30 cm over the hunting grid and drop the M square onto the grid.

5. Note which grid square is most completely covered by the M square. Remove all of the rice from the corresponding square on the lake grid. Do the same thing with the F square. This process represents the eagles catching fish.

6. Each adult eagle hunts twice a day. Rescatter the remaining rice and repeat steps 4 and 5. Total the number of fish caught by each eagle on Day 1 and record the data for Day 1 in Table 1.

7. Repeat steps 4 to 6 nine times and complete Table 1. This observation is taking place in the fall when the fish population does not increase.

8. An eagle will share the fish it catches with its mate, but it will feed itself first. If an eagle does not eat a total of nine fish in any three-day period, it grows too weak to hunt and dies. Be sure to examine the data for each three-day period as you continue. If one eagle dies, continue hunting with only one eagle for the remaining days.

9. Graph the daily total number of fish from Table 1, using a colored pencil. Record the days on the horizontal axis and the number of fish on the vertical axis. Answer questions 1 and 2 in the Analysis section.

10. Make a **hypothesis** to describe what might happen if two ospreys (other birds of prey) also hunted in the lake, each averaging a take of three fish per day. Write your hypothesis in the space provided.

11. Return all the white rice to the lake grid. Repeat steps 4 to 8, but randomly remove an additional six fish per day that the ospreys catch. Record the data in Table 2.

12. Graph the daily total number of fish from Table 2, using a different colored pencil. Answer question 3 in the Analysis section.

Part B.

1. Rescatter all the rice as before.

2. Other factors in the ecosystem affect the fish population and, thus, the eagle population. Choose Factor D and any three other factors listed below. Follow directions for each factor. Record your data in Tables 3, 4, 5, and 6, labeling them properly. Be sure to rescatter the remaining rice before each hunting trip and scatter all the rice when you begin a new factor.

3. Graph the data from Table 3 through 6. Use a different colored pencil for each table.

Factors That Might Affect an Ecosystem

A. A drought occurs that causes the water level of the lake to fall. This causes one quarter of the fish to die. Remove 38 fish from the lake. Repeat steps 4 to 6 from Part A ten times. Answer Analysis question 4.

B. It is spring. The fish in the lake are spawning. Double the number of fish (rice grains). Repeat steps 4 to 6 from Part A ten times. Answer Analysis question 5.

C. It is winter. The lake is frozen and the eagles cannot hunt fish. Answer Analysis question 6.

D. Insecticides are sprayed on land near the eagles' territory. Some insects that have ingested the insecticide are eaten by small fish in the lake. The small fish are, in turn, eaten by large fish. When eagles eat fish contaminated with insecticide, they lay eggs that do not hatch. Remove half of the fish (75 of the white rice grains) and replace them with fish contaminated with insecticide (75 red rice grains). Repeat steps 4 to 6 from Part A ten times. Answer Analysis questions 7 and 8.

E. Phosphate pollution causes the algae in the lake to grow out of control. The algal growth reduces the amount of dissolved oxygen in the lake water and causes three quarters of the fish to die. Remove 112 fish. Repeat steps 4 to 6 from Part A. Answer Analysis question 9.

F. The eagles have two offspring. The adults have to catch two additional fish a day. Repeat steps 4 to 6 of Part A. Answer Analysis question 10.

HYPOTHESIS

_____

_____

DATA AND OBSERVATIONS

Table 1.

Day	1	2	3	4	5	6	7	8	9	10
No. of fish										

Table 2.

Day	1	2	3	4	5	6	7	8	9	10
No. of fish										

Table 3.

Day	1	2	3	4	5	6	7	8	9	10
No. of fish										

Table 4.

Day	1	2	3	4	5	6	7	8	9	10
No. of fish										

Table 5.

Day	1	2	3	4	5	6	7	8	9	10
No. of fish										

Table 6.

Day	1	2	3	4	5	6	7	8	9	10
No. of fish										

ANALYSIS

1. How might eagle predation affect the fish population over time? _____

_____

2. What effect, if any, might a small scale decrease in the fish population have on the eagle population?

_____

3. Ospreys and eagles compete for food. What effect, if any, might the competition have on the eagle

population? _____

_____

4. Explain how a climate change might or might not indirectly affect the eagle population? _____

_____

5. What effect, if any, does an increase in the fish population have on the eagle population? _____

_____

6. How can a seasonal change affect the eagle population? _____

_____

7. What effect, if any, does the insecticide have on the fish population? _____

on the eagle population? _____

8. Can a pollutant, such as an insecticide, affect one population in an ecosystem and not another? _____

9. Explain how pollution can indirectly affect the eagle population. _____

_____

10. How does an increase in the eagle population affect the fish population? How is this situation similar to

the competition from ospreys? _____

_____

CHECKING YOUR HYPOTHESIS

Was your **hypothesis** supported by your data? Why or why not? _____

_____

FURTHER INVESTIGATIONS

1. Perform an experiment to test the effect of competition among bean plants. Germinate about 25 bean seeds. Select and plant one seedling in a pot three quarters full of soil. This serves as a control. Plant about 18 seedlings in another pot of the same size and filled with the same amount of soil. Measure the growth of the seedlings.

2. Design and carry out an experiment to test the effect of overcrowding in a yeast population. Since yeast reproduces by budding, a population increases rapidly. Observe changes in the number of yeast cells in the population by making a wet-mount slide and counting the number of cells in the field of view. How does light and temperature affect a yeast population?

Recycling Garbage

EXPLORATION

▶ The United States presently generates 160 million tons of garbage per year. This is approximately equivalent to two-thirds of a ton of garbage per person per year. One means of managing this waste problem is to recycle. Items such as plastic, paper, aluminum, and glass can all be recycled. Items like food, rubber, textiles, and wood are not recyclable. Of the 160 million tons of garbage in the United States, 80 percent goes to landfills, 10 percent is recycled, and 10 percent is incinerated. The Environmental Protection Agency (EPA) recommends recycling more of our garbage and decreasing our use of landfills.

OBJECTIVES

- Identify the major components of a bag of selected garbage.
- Determine which types of garbage can be recycled.
- Construct a graph that compares a bag of selected garbage with the garbage generated by the average United States household.
- Calculate how many aluminum cans would have to be recycled to pay a typical electric bill.

MATERIALS

individual bag of selected garbage
metric rulers
graph paper

colored pencils (red, blue, green)
newspapers
household electric bill

PROCEDURE

1. Spread out several layers of newspapers on your work area.

2. Inspect your bag of garbage. Separate the garbage into five separate piles: paper, glass, metal, plastic, and one for food, rubber, wood and other materials.

3. Count the number of items in each pile and record these numbers in Table 1. Note: Garbage is usually weighed to determine percentages, but counting items will be used as an easier method of comparison.

4. Dispose of your garbage in the receptacles designated for recyclable and non-recyclable items. Wash your hands after handling the garbage.

5. Calculate the percentage of your garbage each group of items represents. Use the formula (A/B) × 100, where A equals the number of items in each group and B equals the total

number of your items. Record these percentages in Table 1.

6. On a piece of graph paper, draw X and Y axes. Label the X axis with the names of the five garbage groups and the Y axis "% of total garbage." Allow enough space on the X axis to draw three bars for each garbage group. Construct a bar graph of the groups and their percentages. Use a red pencil.

7. Pool your data with the data from the rest of the class and record the totals in the fourth column of Table 1.

8. Using figures for the entire class, calculate the percentage of total garbage each group of items represents. Use the formula (C/D) × 100, where C equals the total number of items in each group and D equals the total number of all trash items for the class. Record these percentages in the last column of Table 1.

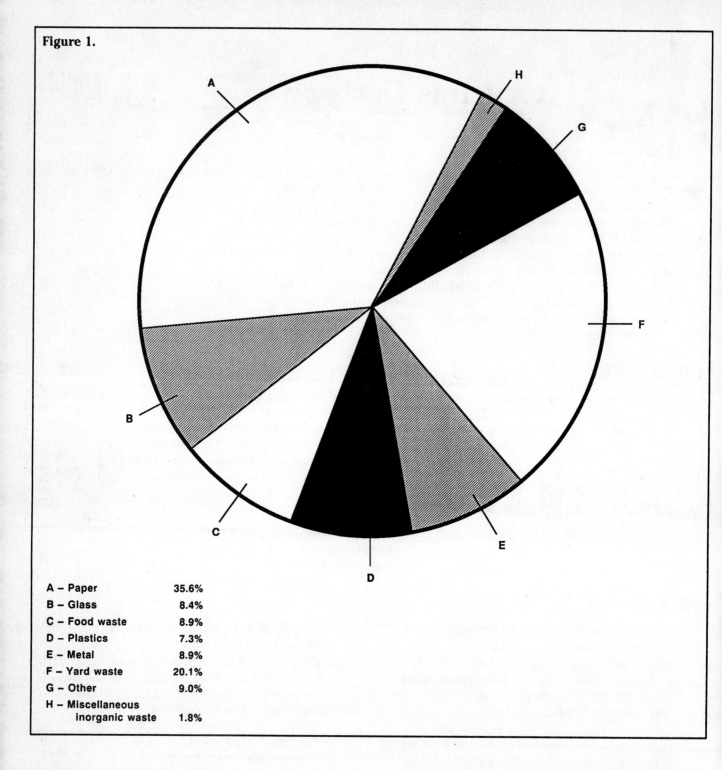

Figure 1.

A – Paper   35.6%
B – Glass   8.4%
C – Food waste   8.9%
D – Plastics   7.3%
E – Metal   8.9%
F – Yard waste   20.1%
G – Other   9.0%
H – Miscellaneous
       inorganic waste   1.8%

9. On the same piece of graph paper, add a second bar for each group, showing the percentages of each group for the class garbage. Use a blue pencil.

10. Examine Figure 1. This pie chart indicates the average percentages of various groups of garbage items for the United States. Plot the percentages of the items in this pie graph on your bar graph also. Use a green pencil for this third set of bars.

11. Recycling one aluminum can saves enough electrical energy to light a 100-watt light bulb for 3.5 hours. Using this information, determine how many cans would have to be recycled to pay the electric bill you have been given. Note that the bill measures kilowatt-hours, the energy used by 1000 watts for one hour. Use your result to answer question 6 in the Analysis.

Name _____ Date _____

DATA AND OBSERVATIONS

Table 1.

Components of Garbage					
Garbage groups	Number of your items = A	% of your total (A/B) × 100		Total number of items for class = C	% of class totals (C/D) × 100
Paper					
Glass					
Metal					
Plastic					
Food					
Rubber					
Wood					
Other					
Totals	B =			D =	

ANALYSIS

1. How does your garbage sample compare with the average U.S. garbage? _____

_____

_____

_____

2. How do the class data compare with the average U.S. garbage? _____

_____

_____

_____

3. What percentage of the class garbage can be recycled? _____

4. What percentage of the class garbage cannot be recycled? _____

5. It takes 30 to 40 years on average for a two-liter plastic soft drink container to decompose. An aluminum can takes between 80 and 100 years to decompose in a landfill. What are the advantages of recycling

these items rather than dumping them in a landfill? _____

_____

_____

6. How many aluminum cans would have to be recycled to pay the electric bill you were given? Remember that recycling one aluminum can saves enough electrical energy to light a 100-watt light bulb for 3.5

hours. _____

FURTHER EXPLORATIONS

1. Develop and conduct a survey to determine how many families in your class or school are currently involved in recycling. Ask them about what they recycle and how they do it. Follow up this survey by devising a way to collect and recycle items from your school.

2. Identify specific products that can be recycled. Locate recycling stations in your community. Publicize all this information at your school using posters and the school paper.

How Does Detergent Affect Seed Germination?

6–2 LAB

INVESTIGATION

▶ Synthetic detergents may contain phosphates and other chemicals that soften the water and prevent minerals and dirt from being redeposited onto the clothing during washing.

Phosphates are nutrients required by plants in small amounts. They exist naturally even in clear lakes in very low concentrations. When many people began using phosphate detergents, problems resulted. The large amounts of phosphates dumped into bodies of water from household wastes greatly increased the concentration of phosphates in lakes and streams. The balance of living things was disrupted and ecological damage occurred.

Today, phosphates have been removed from many detergents to prevent such ecological damage. In fact, phosphate detergents are banned in areas around the Great Lakes and Chesapeake Bay. There are, however, many additives in detergents that accomplish what phosphates did. These detergents without phosphate are referred to as biodegradable. Biodegradable substances can be decomposed by bacteria in water. Although these new detergents do not seem to cause major damage to the ecology, they may affect plants in other ways, such as growth rate and seed germination.

OBJECTIVES

- Hypothesize what effect different concentrations of detergents labeled *phosphate-free* and *biodegradable* will have on seed germination.
- Determine the effect of different concentrations of liquid detergent on seed germination.
- Graph the data collected in the experiment.

MATERIALS

masking tape
scissors
petri dishes (3)
wax marking pencil
toothpicks
paper towels
colored pencils (3)

radish seeds (90)
distilled water
50-mL graduated cylinder
graph paper
1% liquid dishwashing detergent solution
10% liquid dishwashing detergent solution
metric ruler

PROCEDURE

1. Label the lids of three petri dishes CONTROL, 1%, and 10%. Label each dish with your name.

2. Cut six circles of paper toweling to fit the bottom of the dishes. Place two circles in the bottom of each dish as in Figure 1.

3. Distribute 30 radish seeds evenly over the bottom of each dish.

4. Before using the graduated cylinder, make sure it is completely free of detergent by rinsing thoroughly with tap water, then with distilled water. Measure and pour 10 mL distilled water into the dish labeled CONTROL.

5. Measure and pour 10 mL of the 1% detergent solution into the dish labeled 1%.

31

Figure 1.

Paper towel

Petri dish

6. Measure and pour 10 mL of the 10% detergent solution into the dish labeled 10%. Rinse the graduated cylinder with tap water, then with distilled water.

7. Use a toothpick to reposition your seeds if necessary.

8. Replace the lids on the petri dishes and seal with two short pieces of masking tape on opposite sides of the dish.

9. Place all the dishes in a warm, dark place.

10. Make a **hypothesis** that will predict the effects of different concentrations of detergents on seed germination. Write your hypothesis in the space provided.

11. Examine your petri dishes daily for five days. Count the total number of seeds germinated each day in each dish. Germination has occurred if the root is visible. Record your counts in Table 1. Observe and measure the roots of some of the germinated seedlings as in Figure 3. Record your observations and measurements in Table 2.

12. Do not allow the paper towels in the petri dishes to dry out. Add small amounts of water or detergent solution of the proper concentration to the petri dishes if necessary. Be sure you add the same kind of liquid that you originally added to each plate.

13. After collecting data for five days, make a line graph of your data. Place the number of days on the horizontal axis and the number of germinated seeds on the vertical axis. For each treatment, plot the number of seeds germinated over the five-day period. Choose a different colored pencil for each of the three treatments.

Figure 2.

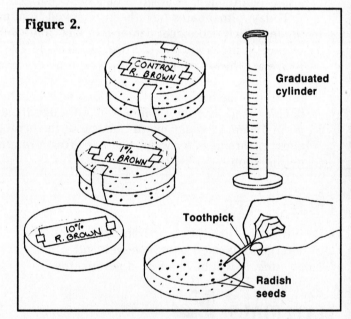

CONTROL R. BROWN

1% R. BROWN

10% R. BROWN

Graduated cylinder

Toothpick

Radish seeds

Figure 3.

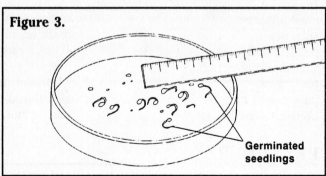

Germinated seedlings

HYPOTHESIS

_____

_____

DATA AND OBSERVATIONS

Table 1.

	Number of Seeds Germinated		
Day	Control	1% detergent solution	10% detergent solution
1			
2			
3			
4			
5			

Table 2.

	Growth of Germinating Seedlings		
Day	Control	1%	10%

ANALYSIS

1. How many of the seeds germinated after five days

   in distilled water? _____ in 1% detergent? _____ in 10 % detergent? _____

2. How was germination of the seeds affected by the detergent? _____

   _____

3. What was the purpose of the control? _____

   _____

4. What were the noticeable differences in growth of the seedlings in the three dishes two days after

   germination? _____

   _____

5. Farmers often irrigate their crops with untreated water from lakes and streams. What would happen to a

   farmer's crop yield if the water used for irrigation contained about 1% detergent? _____

   _____

   _____

6. If a detergent is biodegradable, does that mean it will not harm living things? Explain. _____

   _____

   _____

CHECKING YOUR HYPOTHESIS

Was your **hypothesis** supported by your data? Why or why not? _____

_____

FURTHER INVESTIGATIONS

1. Experiment with a variety of detergent brands (including one with phosphates, if available) and see if they all have the same effects.
2. Conduct the same experiment with more dilute and more concentrated detergents. Try to determine what concentration of detergent, if any, gives the same results as the control. Try to determine what concentration of detergent kills the seeds.

How Does Ionizing Radiation Affect Plant Growth?

6–3

LAB

INVESTIGATION

▶ Organisms are exposed continuously to many forms of radiation from both natural and man-made sources. Most types of radiation are not harmful to organisms and some are even beneficial. Remember that photosynthesis depends upon a form of radiation called visible light. Ionizing radiation has a short wavelength and high energy content, and can severely damage living cells. Ionizing radiation is so energetic that it readily penetrates most forms of matter. Gamma rays and X rays are examples of ionizing radiation. The seeds used in this Investigation have been treated with gamma radiation. Gamma rays are high-energy X rays that are emitted by radioactive isotopes.

OBJECTIVES

- Hypothesize how different amounts of ionizing radiation affect germination and growth of barley seeds.
- Observe the effects of different amounts of ionizing radiation on the germination of barley seeds.

- Observe the effects of different amounts of ionizing radiation on the size and structure of barley plants.
- Compare percent germination, size, and structural changes among control and experimental groups.

MATERIALS

planting flat	metric ruler	plastic bag (transparent, large enough to enclose planting flat)
masking tape	colored pencils (5)	
plastic tie	graph paper (2 pieces)	five kinds of barley seeds (one nonirradiated control group, the rest irradiated with 20 000, 30 000, 40 000 and 50 000 RADs, respectively)
potting soil, moistened	laboratory apron	

PROCEDURE

The barley seeds used in this Investigation have been treated with gamma radiation. The amount of radiation to which the seeds have been exposed is measured in a unit known as a RAD (Radiation Absorbed Dose). The class will be divided into teams of five students. Each student will be responsible for planting and evaluating one kind of seed and for recording data for all five kinds of seeds.

1. Put on a laboratory apron. Obtain a planting flat and fill it with moistened soil.

2. Obtain the five kinds of seeds. Using masking tape, label the flat with your team's name. Divide the flat into five rows and label each row with one radiation dosage.

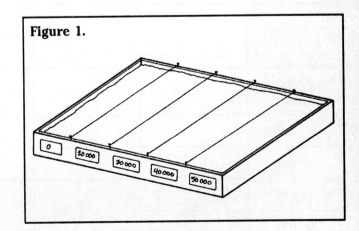

Figure 1.

3. Count the number of each kind of seed and record these numbers in Table 1 as "Number of seeds at start."

4. Plant the barley seeds 5 mm deep and evenly spaced. Cover the seeds with a thin layer of soil. Make sure you plant each kind of seed in its correctly labeled row.

5. Place the flat in a transparent plastic bag and tie the end of the bag with a plastic tie. Set the flat in a well-lighted area at room temperature. Avoid direct sunlight.

6. Make a **hypothesis** that describes how the dosage of radiation will affect barley germination and growth. Write your hypothesis in the space provided.

7. Observe the flat daily. Record the date in Table 1 when each kind of seedling first appears. When seedlings first begin to appear, remove the flat from the plastic bag.

8. Check the seedlings every other day and water as needed once seeds begin to germinate. *DO NOT oversaturate the soil.*

9. Count, once a week, the *total* number of seedlings that have germinated. Calculate the percent germination by dividing the number of germinated seedlings by the total number of seeds planted and multiplying by 100. Record these data in Table 1.

10. Measure, once each week, the height of each of the seedlings in cm and calculate the average height. Record your data in Table 1.

11. Compare the leaves of the seedlings with those of the control seedlings. Record your observations in Table 2.

12. Repeat steps 9, 10, and 11 for three weeks.

13. Average the data from the entire class and make two graphs.

Graph 1. Graph the percent germination (on the vertical axis) against time (weeks, on the horizontal axis). Plot the class data for each kind of seed. Use a different-colored pencil for each kind of seed. You will have five different-colored lines, each representing a different treatment of barley seeds.

Graph 2. Graph the average height (on the vertical axis) against time (weeks, on the horizontal axis). Plot the class data for each kind of seed. Use a different colored pencil for each seed type. Once again you will have five different lines, each representing a different treatment.

HYPOTHESIS

_____

_____

_____

DATA AND OBSERVATIONS

Table 1.

Effects of Ionizing Radiation										
Radiation dosage (RADs)	Number of seeds at start	Germ. date	Number of seeds germinated			Percent germination			Average height (cm)		
			Week			Week			Week		
			1	2	3	1	2	3	1	2	3
No radiation (control)											
20 000											
30 000											
40 000											
50 000											

Table 2.

Observations of Seedlings			
Radiation dosage (RADs)	Week 1	Week 2	Week 3
No radiation (control)			
20 000			
30 000			
40 000			
50 000			

ANALYSIS

1. Why were nonirradiated barley seeds included in this experiment? _____

_____

2. How do increasing amounts of radiation affect germination of barley seeds? _____

_____

3. How does ionizing radiation affect the size and structure of barley seedlings? _____

_____

_____

4. Growth and development are controlled at the molecular level. How might ionizing forms of radiation

affect DNA molecules and so affect growth and development? _____

_____

_____

5. Most of the effects of ionizing radiation are related to the stunting of growth. Ionizing radiation damages the dividing cells in the apical meristem. Why might ionizing radiation be more harmful to seeds than to older, more mature plants? _____

_____

6. The amount of radiation to which organisms are normally exposed is referred to as background radiation. Why does background radiation vary geographically? Remember that the atmosphere protects Earth from radiation. _____

_____

_____

7. How might ionizing radiation be used to preserve food materials? _____

_____

_____

_____

CHECKING YOUR HYPOTHESIS

Was your **hypothesis** supported by your data? Why or why not? _____

_____

FURTHER INVESTIGATIONS

1. Plant irradiated marigold seeds. Grow the plants until they flower. Study the effect of ionizing radiation on flower structure and color.
2. Examine the leaves from each type of seedling under the microscope to determine if ionizing radiation caused damage to the cells.

Tests for Organic Compounds

7–1
LAB

EXPLORATION

▶ Understanding the chemistry of living organisms is an important part of biology. The structures of cells are made up of many different chemical molecules. Cell metabolism involves the production and destruction of many types of molecules. Most of the common molecules found in living things belong to four classes of carbon-containing molecules: carbohydrates, lipids, proteins, and nucleic acids.

OBJECTIVES

- Determine the presence of starch by a chemical test.
- Analyze a glucose solution for the presence of simple reducing sugars.
- Analyze a sample of vegetable oil for the presence of lipids.
- Analyze a sample of gelatin for the presence of protein.

MATERIALS

droppers (9)	wax marking pencil	soluble starch solution
test tubes (3)	vegetable oil	95% ethanol
test-tube rack	biuret reagent	2% gelatin solution
test-tube holder	glucose solution	iodine solution
test-tube stoppers (2)	Benedict's solution	laboratory apron
test-tube brush	hot plate	safety goggles
brown paper	water bath	

PROCEDURE

Part A. Tests for Carbohydrates

Test for Starch

1. Put on goggles and an apron. Label three test tubes 1, 2, and 3. Place them in a test-tube rack.

2. Using a separate dropper for each solution, add 10 drops of soluble starch solution to test tube 1, 10 drops of glucose solution to test tube 2, and 10 drops of water to test tube 3. Record the color of each tube's contents in Table 1.

3. Add 3 drops of iodine solution to each test tube. **CAUTION:** *If iodine is spilled, rinse with water and call your teacher immediately.*

4. Record in Table 1 the color of each tube's contents after addition of the iodine. A blue-black color indicates the presence of starch.

5. Discard the contents of the test tubes according to your teacher's directions. Gently use a test-tube brush and soapy water to clean the three test tubes and rinse with clean water.

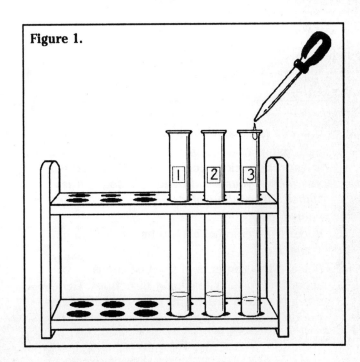
Figure 1.

Test for Simple Reducing Sugars

1. Heat the water bath to boiling on the hot plate.
2. Label three test tubes 1, 2, and 3.
3. Using separate droppers for each solution, add 10 drops of soluble starch solution to test tube 1, 10 drops of glucose solution to test tube 2, and 10 drops of water to test tube 3. Record the color of each tube's contents in Table 2.
4. Add 20 drops of Benedict's solution to each of these three test tubes and place in a boiling water bath for 3 minutes.

Benedict's solution tests for the presence of simple reducing sugars (monosaccharides and some disaccharides, but not polysaccharides). Thus, a color change might or might not occur when Benedict's solution is added to a carbohydrate and heated. A change from blue to green, yellow, orange, or red occurs if a monosaccharide or some disaccharides are present. The original blue color will remain after heating if a polysaccharide or some disaccharides are present.

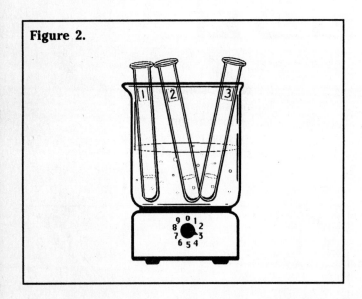

Figure 2.

5. Remove the three test tubes from the water bath using a test-tube holder and place them in a test-tube rack to cool. **CAUTION:** *Be careful not to burn yourself. If Benedict's solution is spilled, rinse with water and call your teacher immediately.*
6. Record the color of each tube's contents in Table 2.
7. Discard the contents of the test tubes according to your teacher's directions. Gently use a test-tube brush and soapy water to clean the three test tubes and rinse with clean water.

Part B. Tests for Lipids
Brown Paper Test for Lipids

1. Place a drop of water on a small piece of brown paper. Place a drop of oil on the same piece of paper. Allow the paper to dry for a few minutes.
2. Hold the piece of paper up to the light. If a semi-transparent (translucent) spot is evident, the sample contains lipids. Record the appearance of the two spots in Table 3.

Solubility Test for Lipids

1. Label two test tubes 1 and 2.
2. Using separate droppers, add 20 drops of 95% ethanol to test tube 1 and 20 drops of water to test tube 2.
3. Add 5 drops of oil to test tubes 1 and 2 and stopper each tube.
4. Shake each tube well, let settle, and record in Table 4 whether the oil is soluble in either solvent.

Lipids are soluble only in nonpolar solvents because lipids, themselves, are nonpolar.

5. Dispose of the contents of the test tubes according to your teacher's directions. Gently use a test-tube brush and soapy water to clean the two test tubes and rinse with clean water.

Part C. Tests for Proteins

1. Label three test tubes 1, 2, and 3.
2. Using separate droppers, add 30 drops of 2% gelatin to test tube 1, 30 drops of glucose solution to test tube 2, and 30 drops of water to test tube 3. Record the color of each tube's contents in Table 5.
3. Add 10 drops of biuret reagent to each test tube. **CAUTION:** *Biuret reagent is extremely caustic to the skin and clothing. If biuret reagent is spilled, rinse with water and call your teacher immediately.*

When biuret reagent is mixed with a protein, it will produce a lavender to violet color.

4. Record in Table 5 the color of each tube's contents after adding biuret reagent.
5. Discard the contents of the test tubes according to your teacher's directions. Gently use a test-tube brush and soapy water to clean the test tubes and rinse with clean water.
6. Fill in the last column of all five tables with the correct interpretation of the test results.

DATA AND OBSERVATIONS

Table 1.

Test for Starch				
Test tube	Substance	Color at start	Color after adding iodine	Starch present (+/−)
1	Starch			
2	Glucose			
3	Water			

Table 2.

Test for Simple Reducing Sugars				
Test tube	Substance	Color at start	Color after adding Benedict's solution	Reducing sugar present (+/−)
1	Starch			
2	Glucose			
3	Water			

Table 3.

Brown Paper Test for Lipids		
Substance	Translucent on brown paper?	Lipids present (+/−)
Water		
Oil		

Table 4.

Solubility Test for Lipids		
Substance	Dissolves?	Lipids present (+/−)
Oil in ethanol		
Oil in water		

Table 5.

Test for Proteins				
Test tube	Substance	Color at start	Color after adding biuret reagent	Protein present (+/−)
1	Gelatin			
2	Glucose			
3	Water			

ANALYSIS

1. What is used to test for the presence of starch? _____

_____

2. How can you tell by using this test that a substance contains starch? _____

_____

_____

3. What is used to test for the presence of simple reducing sugars such as monosaccharides? _____

_____

4. How can you tell by using this test that a substance contains a simple reducing sugar? _____

_____

5. Why was water tested for each chemical? _____

_____

6. What is used to test for the presence of protein? _____

7. How can you tell by using this test that a substance contains protein? _____

_____

_____

8. Biuret reagent will turn the skin brownish-purple. Explain why this occurs. _____

_____

_____

9. a. When greasy food is spilled on clothing, why is it difficult to clean with water alone? _____

_____

   b. What would be better than plain water for removing a greasy food stain? Why? _____

_____

FURTHER EXPLORATIONS

1. Choose a number of common substances that are available and conduct your own tests for the presence or absence of carbohydrates, lipids, and proteins. Cotton, animal hair, fingernails, and various foods such as egg white and cheese are among the many things you might investigate.

2. Investigate the difference between saturated and unsaturated fatty acids. Find recent information concerning these fatty acids and health.

What Is the Action of Diastase?

7–2
LAB

INVESTIGATION

▶ Enzymes are biological molecules that speed up reactions without being changed or used up by the reactions themselves. Every living tissue and its cells depend upon enzymes for all cell reactions. Seeds contain enzymes that digest stored starch and make energy available for the embryo during germination. One seed enzyme, diastase, breaks down starch into the disaccharide maltose early in the germination process. The iodine and Benedict's tests can be used to confirm the activity of the enzyme diastase.

OBJECTIVES

- Prepare an extract of germinating barley seeds.
- Analyze a known solution of diastase for the presence of starch and sugars.

- Hypothesize what changes will occur when barley extract is mixed with starch solution.
- Determine the action of barley extract on starch solution over a period of time.

MATERIALS

droppers (4)
test tubes (7)
test-tube rack
germinating barley seeds,
   5 days old (25 seeds)
hot plate
water bath
Benedict's solution
0.4% starch solution

iodine solution
distilled water
0.1–0.2% diastase solution
mortar and pestle
100-mL beaker
50-mL beaker
clean sand
stirring rods (2)
scoop

plastic spot plate
test-tube holder
10-mL graduated cylinders (3)
white paper
goggles
laboratory apron
clock

PROCEDURE

Part A. Enzyme Preparation

1. Place a small scoop of clean sand in a mortar. Add all the germinating barley seeds and 10 mL distilled water and, using the pestle, grind the seeds until they are fine particles.
2. Pour the contents of the mortar into a 100-mL beaker. Add 40 mL distilled water and allow the mixture to settle for 15 to 20 minutes.
3. Save the barley extract you have made for use in Part B.
4. Set up and begin heating the water bath for Part B.

Figure 1.

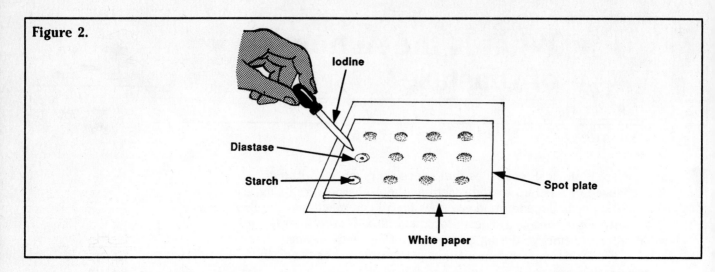

Figure 2.

Labels: Iodine, Diastase, Starch, Spot plate, White paper

Part B. Testing Enzyme Activity

Iodine is used to test for the presence of starch. If starch is present, a blue-black color appears. The Benedict's test is used to test for some sugars such as glucose and maltose. A positive reaction is shown by a color change from blue to orange.

1. Set a clean spot plate on a piece of white paper.

2. Add 3 drops of starch solution to a spot on the spot plate and 3 drops of diastase solution to a second spot. Add 1 drop of the iodine solution to each spot. See Figure 2. **CAUTION:** *If iodine solution is spilled, rinse with water and call your teacher immediately.*

3. Note the color of the mixtures and record the presence or absence of starch in the proper places in Table 1.

4. Place 20 drops of starch solution into a test tube labeled STARCH and 20 drops of diastase solution into a second test tube labeled DIASTASE. Add 2 mL of Benedict's solution to each tube. Mix with two separate stirring rods or by gently tapping each tube against your hand.

5. Use the test-tube holder to place the test tubes into the boiling water bath for 3 minutes. **CAUTION:** *Be careful not to burn yourself. If Benedict's solution is spilled, rinse with water and call your teacher immediately.*

6. Remove the test tubes from the water bath using the test-tube holder and place them in the test-tube rack. Note the color in each tube. Record the presence or absence of sugars in the proper places in Table 1.

7. Make a **hypothesis** to explain the changes that will occur when barley extract is mixed with starch solution. Write your hypothesis in the space provided.

8. Label 5 test tubes Time 0, Time 3, Time 6, Time 9, and Time 12.

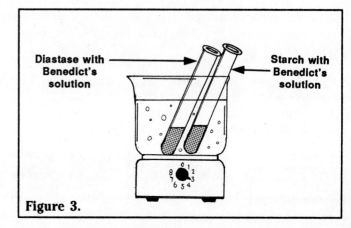

Labels: Diastase with Benedict's solution, Starch with Benedict's solution

Figure 3.

9. Using the graduated cylinder, place 10 mL of the barley extract in a 50-mL beaker and add 10 mL of starch solution. Be careful not to pour off any of the sand.

10. Immediately place 3 drops of this mixture from the beaker on a clean spot on the spot plate and test it for the presence of starch using the iodine solution. Record the results next to Time 0 in Table 2.

11. Quickly place 20 drops of the barley extract-starch mixture from the 50-mL beaker into the clean test tube labeled Time 0. Add 2 mL of Benedict's solution to the test tube and, using the test-tube holder, place the test tube in the water bath for 3 minutes. Record the results of this test next to Time 0 in Table 2.

12. Repeat steps 10 and 11 every 3 minutes for 12 minutes (times 3, 6, 9, 12 minutes). For each repetition, record the results of the two tests in the appropriate places in Table 2.

HYPOTHESIS

_____

_____

Figure 4.

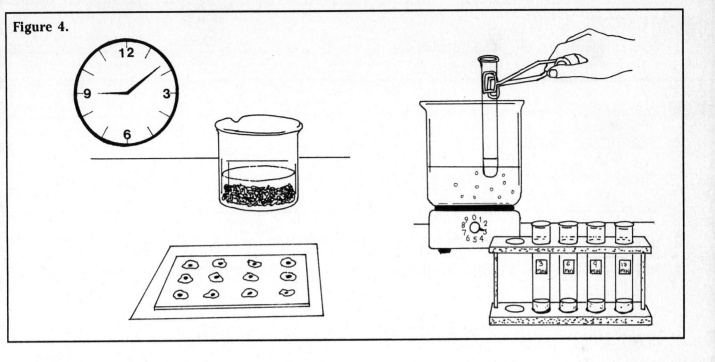

DATA AND OBSERVATIONS

Table 1.

Testing for Substances		
Substance	Iodine test Starch (+/−)	Benedict's test Sugars (+/−)
Starch solution		
Diastase solution		

Table 2.

Testing for Enzyme Action		
Time (Minutes)	Iodine test Starch (+/−)	Benedict's test Sugars (+/−)
0		
3		
6		
9		
12		

ANALYSIS

1. What substance was the iodine solution used to test for? _____

2. What did the iodine test on the starch solution indicate? _____

3. What did the iodine test on the diastase solution indicate? _____

4. What substance was the Benedict's solution used to test for? _____

5. What did the Benedict's test on the starch solution indicate? _____

   _____

6. What did the Benedict's test on the diastase solution indicate? _____

7. What was the purpose of testing the diastase solution with the iodine and Benedict's solution? _____

   _____

8. What did the two chemical tests indicate was in the barley extract-starch mixture at Time 0? _____

   _____

9. a. Over the 12-minute period, what did the iodine test indicate was happening in the barley extract-

   starch mixture? _____

   _____

   b. What did the Benedict's test indicate? _____

   _____

   _____

10. What happened to the starch in the barley extract-starch mixture? _____

   _____

   _____

CHECKING YOUR HYPOTHESIS

Was your **hypothesis** supported by your data? Why or why not? _____

_____

FURTHER INVESTIGATIONS

1. Test for the activity of another common enzyme. Substitute the enzyme amylase, a component of human saliva, for diastase in this experiment. Amylase converts starch to maltose.
2. Design experiments to test the effects of boiling, freezing, dilution, or differences in pH on the activity of diastase.

Use of the Light Microscope

8–1

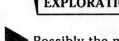
EXPLORATION

▶ Possibly the most important instrument used by biologists is the microscope. A microscope aids scientists by allowing them to investigate worlds that otherwise are too small to be seen. A light microscope magnifies objects up to approximately 400 times their natural size.

Two types of slides are used with the microscope: prepared slides and temporary wet mounts. Prepared slides are permanent and are made to last a long time. Many of the slides you will use in this course will be wet mounts. You will make these slides yourself. As the name temporary wet mount suggests, these slides are not permanent.

OBJECTIVES

- Practice proper handling and use of the light microscope.
- Identify the parts of a light microscope.
- Locate objects under low- and high-power magnification.
- Prepare a wet mount of an insect leg.

MATERIALS

light microscope	coverslip	preserved insect leg
lamp (if needed)	forceps	lens paper
microscope slide	dropper	water

PROCEDURE

Part A. Learning Microscope Parts and Functions

1. Look at Figure 1. Note that the student is carrying the microscope with two hands. Also note that the microscope is carried straight up. Do not tilt the microscope, and carry it with both hands against the body.

2. Position the concave surface (curved surface) of the mirror so that it is turned toward a light source, such as ceiling lights, windows, or a desk lamp. The mirror is attached to most microscopes by means of a swivel joint. If a lamp is built into your microscope, it replaces the mirror and outside light source. **CAUTION:** *Never use direct sunlight as a light source. Direct sunlight will damage your eyes.*

3. Look at Figure 2. Use the diagram that looks more like your microscope to locate microscope parts.

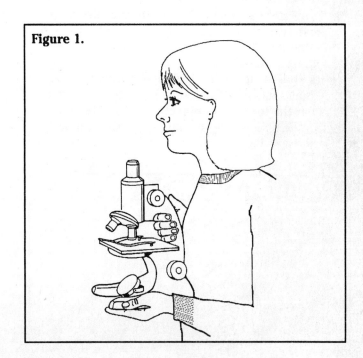

Figure 1.

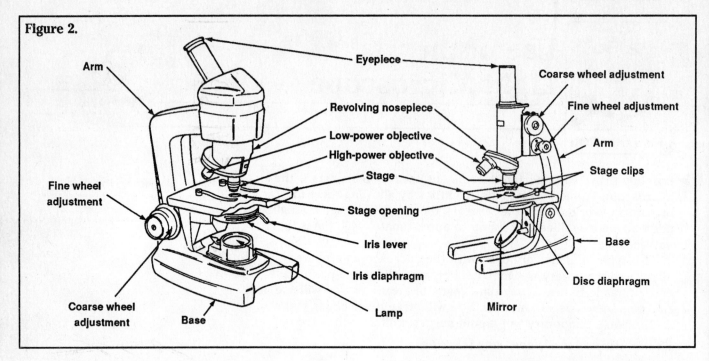

Figure 2.

a. Does your microscope have a lamp, or a

mirror? _____

b. What type of diaphragm does your micro-

scope have? _____

A diaphragm controls the amount of light
entering the microscope. Turning the
diaphragm adjusts the amount of light passing
through the microscope.

4. Use Figure 2 to help you locate the revolving
nosepiece, high-power objective, and low-power
objective on your microscope. The low-power
objective is identified by a 10X marking or by
its short length. The high-power objective
usually has a 43X marking and often is longer
than the low-power objective. The objectives
can be changed by turning the nosepiece as
shown in Figure 3.

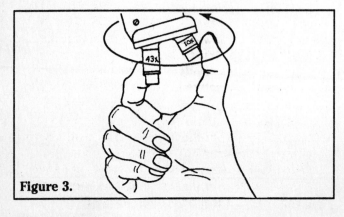

Figure 3.

5. Place a check mark in the square next to each
part of the microscope you have located.
   ☐ diaphragm   ☐ high-power objective
   ☐ lamp or mirror   ☐ low-power objective
   ☐ revolving nosepiece
   *Do not continue with the lab until you know
   where these five parts are located.*

6. Use Figure 2 to help you locate the eyepiece,
coarse wheel adjustment, fine wheel adjustment, stage, and stage opening on your microscope.

7. Place a check mark in the square next to each
part of the microscope you have located.
   ☐ eyepiece   ☐ stage
   ☐ coarse wheel adjustment   ☐ stage opening
   ☐ fine wheel adjustment
   *Do not continue with the lab until you know
   where these five parts are located.*

Part B. Using the Microscope

1. Turn on the lamp or position the mirror toward
the light source.

2. Turn and click the low-power objective so that
it is directly over the stage opening. An objective is in proper viewing position when it is
directly over the stage opening. Most microscopes will "click" when the objective is in
proper viewing position.

3. Look through the eyepiece of the microscope. A
circle of bright light should now be visible.
Keep both eyes open. Keeping both eyes open
will reduce eyestrain.

4. Adjust the mirror and diaphragm to make the
circle of light as bright as possible.

5. Look to the side of the microscope as shown in Figure 4. Slowly turn the coarse wheel adjustment back and forth. DO NOT force the wheel once it stops. When the wheel stops, turn it in the opposite direction. Note the movement of the low-power objective in relation to the stage.

   a. In which direction does the objective move as you turn the coarse wheel adjustment

   toward you? _____

   b. In which direction does the objective move as you turn the coarse wheel adjustment

   away from you? _____

6. The objectives and eyepiece should be cleaned with lens paper at the beginning of each laboratory period. Use one piece of paper and gently wipe each lens. *Always use lens paper to clean lenses. Other types of paper may scratch or smear lenses.*

Part C. Preparation of a Temporary Wet Mount

A temporary wet mount consists of some object such as an insect leg placed in a drop of water on a slide with a coverslip over the object. Use the following steps in preparing your wet mount.

1. Add a small drop of water to a slide as shown in Figure 5A.
2. Place the insect leg to be viewed in the water drop. Use one insect leg only.
3. Use forceps to position a coverslip as shown in Figure 5B. Use of forceps prevents fingerprints getting on the coverslip.
4. Lower the edge of the coverslip down slowly over the water drop and object. This procedure will prevent the trapping of air under the coverslip.

Part D. Locating an Object Under the Microscope

1. Click the low-power objective into viewing position. **NOTE:** *Always locate an object first with low-power magnification even if a higher magnification is required for better viewing.*
2. Place the wet mount of the insect leg on the stage of your microscope. Position the slide on the stage so the insect leg is directly over the center of the stage opening. Secure the slide in place with the clips.
3. Look to the side of your microscope as shown in Figure 4. Slowly lower the low-power objective by turning the coarse wheel adjustment

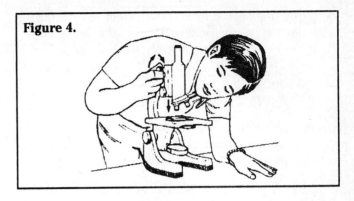

Figure 4.

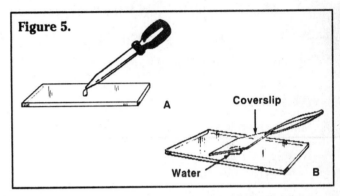

Figure 5.

A

Coverslip

Water   B

until the objective almost touches the glass slide. Some microscopes have an automatic stop that prevents lowering the objective onto the slide. Other microscopes do not. **CAUTION:** *Never lower the objective toward the stage while looking through the eyepiece.*

4. While looking through the eyepiece with both eyes open, slowly turn the coarse wheel adjustment so the objective rises, or moves away, from the stage. The insect leg should soon come into view.
5. Bring the insect leg into sharp focus by turning the fine wheel adjustment.

Part E. Increasing the Magnification of the Microscope

1. Any object to be viewed under high-power magnification *is always located first under low power and focused.* Locate and center the insect leg under low power of your microscope.
2. Move the low-power objective out of viewing position. Look first to the side of the microscope and then revolve the nosepiece. Click the high-power objective into viewing position.
3. Look through the eyepiece. The insect leg should be visible. However, it may need to be focused. Use only the fine wheel adjustment to sharpen the focus. **CAUTION:** *Never use the coarse wheel adjustment for focusing with high power. Damage to the lens and slide may result if the coarse wheel adjustment is used.*

4. If you are unable to find the insect leg, do the following: while looking through the eyepiece, move the glass slide slightly to the left, right, away from, or toward you. These movements may help to reposition the insect leg directly in the center of the high-power objective.

5. Repeat Parts D and E if you are unable to locate the object under high power.

DATA AND OBSERVATIONS

The eyepiece contains a glass lens that magnifies 10 times (10X). The low-power objective also contains a lens that magnifies 10 times (10X). Therefore, the total magnification of an object viewed under low power is 100X. Total magnification is calculated by multiplying the magnification of the objective by that of the eyepiece.

1. What is the total magnification of your microscope under low power? (Use the numbers printed on your low-power objective and eyepiece if present.) _____

2. What is the total magnification of your microscope under high power? (Use the numbers printed on your high-power objective and eyepiece.) _____

ANALYSIS

1. Match the microscope parts with their functions. Write the letter of the function in front of the correct part.

_____ diaphragm
_____ stage opening
_____ mirror or lamp
_____ eyepiece
_____ low-power objective
_____ high-power objective
_____ revolving nosepiece
_____ coarse wheel adjustment
_____ fine wheel adjustment
_____ stage

a. allows light to pass through stage
b. brings objects into rapid but coarse focus
c. regulates amount of light entering microscope
d. is attached to revolving nosepiece and contains a lens capable of 10X magnification
e. contains a lens capable of 43X magnification
f. supports slide
g. directs light into microscope
h. turns to change from one power to another
i. contains a lens capable of 10X magnification
j. brings objects slowly into fine focus

2. Answer the following statements as true or false.

_____ a. Total magnification of a microscope is determined by adding the eyepiece-lens magnification to the objective-lens magnification.

_____ b. The fine wheel adjustment must be used to sharpen focus when using high-power magnification.

_____ c. Always look to the side of a light microscope when lowering the objective.

_____ d. The eyepiece of a microscope is marked 10X. The high-power objective is marked 50X. The total magnification is 500X.

FURTHER EXPLORATIONS

1. Collect standing water from a variety of sources such as tree stumps, abandoned automobile tires, or tin cans. With a microscope, see how many different organisms you can find in the water. Check your library for books that might be helpful in identifying the microscopic organisms.

2. Research the construction of early microscopes, such as the simple instruments made by van Leeuwenhoek. Determine how a simple microscope is different from a compound microscope and draw diagrams showing the optics of each.

How Can a Microscope Be Used in the Laboratory?

8–2 LAB

INVESTIGATION

▶ Several important techniques and ideas can be mastered in the use of a light microscope. A microscope is used for more than magnifying things. Knowing how and why your microscope works will enable you to make better observations. The techniques and hints presented in this Investigation will help you use your microscope correctly.

OBJECTIVES

- Compare the position of an object when viewed through a light microscope with its position on the microscope stage.
- Use stains to aid in viewing objects.
- Hypothesize how the field of view using low power compares with that under high power.
- Compare the depth of field under low and high powers.

MATERIALS

light microscope	absorbent cotton	dropper	iodine solution
microscope slide	single-edged razor blade	lens paper	tissue paper
coverslip	black thread	magazine page	peeled potato
scissors	white thread	forceps	pond water
laboratory apron			

PROCEDURE

Part A. Position of Objects When Viewed with a Microscope

1. Prepare a wet mount (see Exploration 8–1) of a lowercase letter "e" from a magazine page.
2. Place the wet mount of the letter "e" onto your microscope stage. Position the slide on the stage so the "e" faces you as it would on a magazine page.
3. Observe the letter "e" using low power on your microscope. (Review the procedure in Exploration 8–1 if necessary.) Focus the "e" with the fine adjustment. How does the orientation of the "e" viewed through the eyepiece compare with its orientation on the

   stage? _____

4. While looking through the eyepiece, move the slide slowly from left to right. In what direction does the letter move as seen through the

   microscope? _____

5. While looking through the eyepiece, move the slide slowly toward you. In what direction does the letter move when viewed through the

   microscope? _____

Part B. Use of the Diaphragm

1. Prepare a wet mount of a few strands of absorbent cotton. (Follow the procedure outlined in Exploration 8–1.)
2. Observe the cotton fibers with low power. While looking through the microscope, change the amount of light entering the microscope by adjusting the diaphragm. Under what diaphragm setting (maximum, medium, or little light) are the cotton fibers sharpest?

   _____

3. Change to high power and observe the cotton fibers. Again, readjust the amount of light

entering the microscope. Under what diaphragm setting do the cotton fibers appear

sharpest? _____

Part C. Depth of Field

1. Cut very short lengths of some black and white thread.
2. Add a drop of water to a microscope slide. Cross the two strands to form an X in the water drop before adding a coverslip.
3. Locate the strands under low power. Center the slide so you are looking at the point where the strands cross. Adjust the diaphragm for proper lighting. Can both strands be observed clearly

at the same time under low power? _____

4. Change to high power and observe both strands at the point where they cross. Can both strands be observed clearly at the same time

under high power? _____

5. The lens system of your microscope allows you to see clearly only one depth at a time under high power. In order to see objects at different depths, turn the fine wheel adjustment back and forth by a quarter of a turn while looking through the microscope. This movement will give a three-dimensional view of the object. Try this technique while looking at the crossed threads. Using Figure 1 as a guide, note that first one strand is in focus, then the other.

Part D. An Aid to Finding Proper Depth

Locating the proper depth of a wet mount is not always easy. Proper depth is especially hard to find when attempting to view moving organisms or when locating objects so small that they cannot be seen on the slide with the unaided eye. It is easy to focus at the wrong depth of the wet mount and waste valuable time looking at the top surface of the coverslip. The following technique will help you locate the proper depth of a wet mount.

1. Use a dropper to transfer a drop of pond water to a glass slide.
2. Add a hair from your head to the pond water. Add a coverslip.
3. Observe the wet mount under low power by first locating the strand of hair. Focus the hair and adjust the light. The slide can now be moved to find organisms in the pond water. No

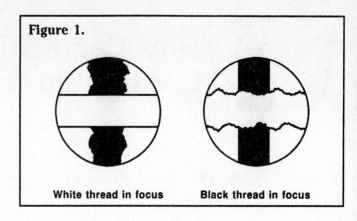

Figure 1.

White thread in focus   Black thread in focus

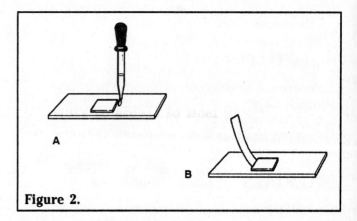

Figure 2.

further adjustment is necessary with the coarse adjustment. You are at the proper depth for finding organisms because the hair and organisms are in the same plane.

4. Locate organisms in the pond water. Attempt to follow a moving organism.

Part E. Stains as an Aid to Microscope Work

Many objects observed with a microscope are colorless. Thus, they appear almost transparent and are difficult to see. Stains often are used in microscope work to color objects for easier and more detailed observation. Stains can be added to a wet mount without disturbing the slide.

1. With a razor blade, gently scrape the edge of a peeled potato. **CAUTION:** *Scrape away from your fingers*.
2. Add a drop of water to a glass slide. Mix the potato scrapings with the water. Add a coverslip.
3. View the wet mount with low power. You are looking at starch grains.
4. Diagram several grains in the circle marked "Unstained" in Data and Observations.

5. Remove the slide from the microscope.

6. Add a drop of iodine solution to your slide along one edge of the coverslip as shown in Figure 2. Do not get any iodine on top of the coverslip. **CAUTION:** *If iodine spillage occurs, wash with water and call your teacher immediately.*

7. Place a piece of tissue paper along the edge of the coverslip opposite the iodine solution. Allow the tissue paper to touch the water of the wet mount as shown in Figure 2. Water will soak into the tissue paper, drawing the iodine stain under the coverslip and into contact with the starch grains.

8. Observe the stained starch grains with low power.

9. Diagram several stained grains in the circle marked "Stained" in Data and Observations.

Part F. A Comparison of Fields of View

Field of view is the area seen through a microscope. Is the field of view with the low power greater than with high power, or are they the same? This exercise will help you answer this question.

1. Make a **hypothesis** to describe how the field of view under low power compares with that under high power. Write your hypothesis in the space provided.

2. Move the slide to a less crowded area of starch grains near the outer edge of the coverslip.

3. Examine the stained starch grains with low power. Count and record the number of grains

   observed under low power. _____

4. Without moving the slide, examine the stained starch grains under high power.
   a. Count and record the number of grains

      observed under high power. _____

   b. How does the number of grains observed under low power compare with the number

      under high power? _____

When using low power, the total area of your field of view is *greater* than when using high power. The width of your low-power field is usually 4 times greater than that of your high-power field. For example, if low power has a magnification of *10X* and high power has *40X*, divide 40 by 10. Your answer, 4, shows the difference in width of these two lenses. Low power has a width that is 4 times larger than that observed under high power.

   c. Calculate the number of times greater the width of low power is than high power for your microscope. Remember that

$$\frac{\text{High-power objective}}{\text{Low-power objective}} = \begin{array}{l}\text{number of times low-power} \\ \text{width is greater than high-} \\ \text{power width}\end{array}$$

Do not confuse width observed with total magnification. As magnification increases, width observed decreases.

   d. Did you observe more or fewer starch grains under low power as compared with high

      power? _____

   e. Did you observe more or less area under low power as compared with high power?

      _____

   f. Your answers to questions 4a and 4b should show about 16 times more starch grains viewed under low power as compared with

      high power. Explain. _____

      _____

      _____

HYPOTHESIS

_____

_____

_____

DATA AND OBSERVATIONS

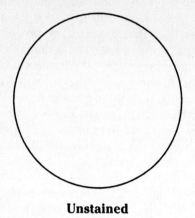

Unstained

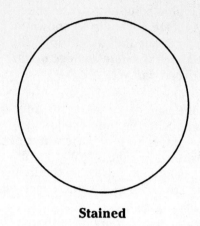

Stained

ANALYSIS

Answer each of the following as true or false.

_____ **1.** Objects viewed under the microscope appear upside down.

_____ **2.** When moving the slide toward the left, objects viewed through the microscope will move toward the left.

_____ **3.** The diaphragm is used to adjust the amount of light entering the microscope.

_____ **4.** All objects in different depths appear in focus at the same time while using high power.

_____ **5.** Stains are used to help make clear objects appear lighter under the microscope.

_____ **6.** Low power shows more area than high power.

_____ **7.** High power shows more detail than low power.

_____ **8.** Observers see about 10 times more width under low power than under high power.

_____ **9.** Your depth of field under high power is less than while observing under low power.

CHECKING YOUR HYPOTHESIS

Was your **hypothesis** supported by your data? Why or why not? _____

_____

FURTHER INVESTIGATIONS

1. Practice using the microscope by inspecting insects that are available locally. Examine the small hairs on the bodies of the insects as well as the eyes, wings, and antennae. Note the effects of changing the diaphragm opening and objective lenses and of moving the slide around.
2. Discuss with your teacher how other stains and methods for staining specific types of plant tissues are used. Practice staining wet mounts from various common garden vegetables and fruits.

Normal and Plasmolyzed Cells

9–1

LAB

EXPLORATION

▶ Diffusion of water molecules across a cell's outer membrane from areas of high water concentration to areas of low water concentration is called osmosis. This movement of water may be harmful to cells. If too much water is lost from the cell, the cell membrane and the cell contents shrink away from the cell wall. This is called plasmolysis. Plasmolysis may lead to death of the cell. Most cells live in an environment where movement of water in and out of the cell is about equal. Therefore, there are no harmful effects to the cell.

OBJECTIVES

• Prepare a wet mount of an *Elodea* leaf.
• Observe plasmolysis in the cell as salt solution is added to the wet mount.

• Observe the reversal of plasmolysis as the salt solution is diluted.
• Compare and diagram normal cells in tap water with plasmolyzed cells in salt solution.

MATERIALS

microscope
microscope slide
paper towel

coverslip
forceps

Elodea (water plant)
droppers (2)

tap water
6% salt solution

PROCEDURE

1. Prepare a wet mount of an *Elodea* leaf as follows. Use Figure 1 as a guide.
2. Use a dropper to place one or two drops of tap water on a microscope slide.
3. Place one leaf, taken from the top whorl of leaves on a sprig of *Elodea*, in the drop of water. Cover the leaf with a coverslip.
4. Observe the leaf under both low and high powers. Note the location of chloroplasts in relation to the cell wall.
5. Diagram a normal cell in the space provided in Data and Observations. Label the cell wall, cell membrane, and chloroplasts.
6. Use a clean dropper to add a drop of 6% salt solution along one edge of the coverslip. Place a piece of paper towel along the opposite edge of the coverslip. The tap water will soak into the paper towel, drawing the salt solution under the coverslip.
7. Observe the leaf under both low and high powers. Again note the location of the

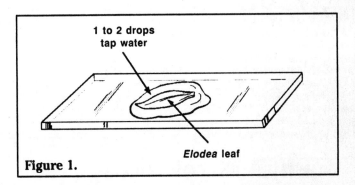

1 to 2 drops
tap water

Elodea leaf

Figure 1.

chloroplasts in relation to the cell wall. If the cell membrane and the cell contents have shrunk away from the cell wall, the cell has plasmolyzed.

8. Diagram a plasmolyzed cell in the space provided in Data and Observations. Label the cell wall, cell membrane, and chloroplasts.
9. Add a drop of tap water to the wet mount, following the procedure in step 6. Observe the appearance of the cells.

DATA AND OBSERVATIONS

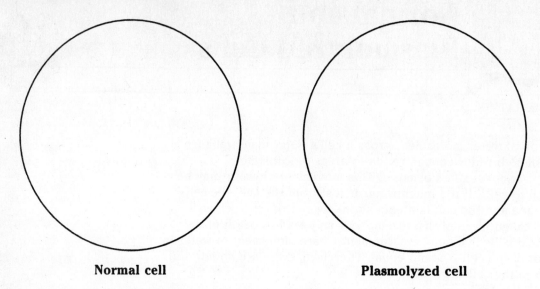

Normal cell   Plasmolyzed cell

ANALYSIS

1. Describe the location of chloroplasts in a normal *Elodea* cell (in tap water). _____

2. Describe the location of chloroplasts in a plasmolyzed cell (in salt solution). _____

3. What did you observe when salt solution was added to the wet mount? _____

_____

4. In which direction did water move when salt solution was added? _____

5. What did you observe when tap water diluted the salt solution? _____

_____

6. In which direction did water move when tap water diluted the salt solution? _____

7. Describe the process of plasmolysis. _____

_____

_____

FURTHER EXPLORATIONS

1. Repeat this Exploration using cells from the epidermis of a purple onion bulb. Compare the results.
2. Repeat this Exploration using 6% glucose solution in place of the salt solution. Compare your results. Explain why the results are different from those in this Exploration.

Chloroplast Pigment Analysis

10–1

LAB

EXPLORATION

▶ When you look at a leaf, the green pigment chlorophyll is usually the only pigment that appears to be present. Actually, chlorophyll is only one of many types of pigments present in the leaf and one of several that are involved in the process of photosynthesis. Once removed from the leaf, the photosynthetic pigments can be separated from one another and identified using a process called chromatography.

Chromatography is a physical process in which several compounds are separated from a solution and from each other. In thin layer chromatography, the solvent is absorbed by a thin layer of silica gel. As the solvent moves upward in the gel, it carries with it the compounds that have been placed on the gel. These compounds each move upward at a specific rate in relation to the moving solvent and can be identified by the distances they move.

OBJECTIVES

- Extract a mixture of plant photosynthetic pigments.
- Separate pigments of spinach leaves by thin layer chromatography.
- Prepare and analyze a silica gel chromatogram.
- Calculate the R_f values for various photosynthetic pigments.

MATERIALS

baby food jar with lid	funnel	capillary tube	mortar and pestle
spinach leaves, dried	cheesecloth	metric ruler	laboratory apron
chromatography solvent	dark-colored bottle or vial	pencil	
thin layer chromatography slide	goggles	ethyl alcohol	

PROCEDURE

Part A. Preparing For Chromatography

1. Obtain a small amount (approximately 5 mL) of the chromatography solvent from your teacher. Pour enough of this solvent into the baby-food jar so that it just covers the bottom of the jar, but is less than 1 mm deep. Screw the cap onto the jar and set aside for later use.

2. Place a pea-sized amount of dried spinach in a mortar. Using the pestle, grind up the spinach for 2 minutes. Add 2 mL of ethyl alcohol to the ground spinach and continue to grind for another 2 minutes as shown in Figure 1. The product should be a deep green fluid. Filter this fluid through a double layer of cheesecloth into a dark-colored bottle or vial. Stopper the bottle tightly until needed.

Figure 1.

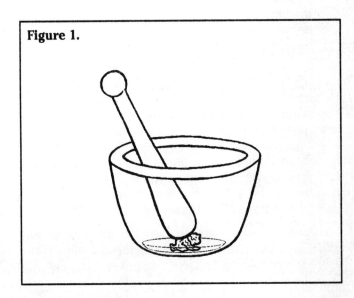

Part B. Making and Analyzing the Chromatogram

1. Select a chromatography slide. Handle it only by its edges. Make a small pencil dot 5 mm from the bottom of the slide. DO NOT use a pen to make the dot.

2. Dip a capillary tube into the pigment-containing fluid in the dark bottle.

3. Lightly touch the filled end of the capillary tube to the dot on the coated slide as shown in Figure 2. Allow a small amount of the fluid to be deposited on the slide, forming a spot 1 mm in diameter. Do not disturb the silica film above the spot. Allow the spot to dry (about 30 seconds).

4. Repeat step 3, applying leaf pigments to the same spot four or five times, being sure to allow the spot to dry each time. This will produce a concentrated spot of pigments.

5. Hold the slide along the outside of the jar to verify that the spot will not be below the level of the solvent. If the solvent is too deep pour a little of it out of the jar into a specially labeled container. Place the slide into the baby food jar on a level surface as shown in Figure 3. Do not allow the spot to contact the solvent at any time. Quickly screw the cap on the jar. Do not move the jar once the slide is placed in the solvent.

6. Watch the slide closely and note the movement of solvent up the film of silica gel. Remove the slide from the baby food jar when the solvent front nearly reaches the top of the slide.

7. Mark the top of the solvent front with a pencil as shown in Figure 4.

8. Make a drawing of your slide in the space provided in Data and Observations. Be sure to indicate the position of the original spot of pigments as well as the locations of pigments anywhere else on the slide. Indicate the relative amounts of pigments in each spot by drawing the spots the same size and darkness as those on your slide.

Carotenes, which are yellow or orange pigments, usually appear near the top of the slide. Lutein is a gray pigment just below the carotenes. Chlorophyll *a* will appear next as a blue-green pigment. Xanthophylls are yellow pigments, and chlorophyll *b* is a yellow-green pigment. They are found together just below chlorophyll *a*. Your chromatogram may or may not have all of these pigments.

9. Measure with a metric ruler the distance in mm from the original spot to the solvent front you marked in step 7. Record this measurement in Table 1.

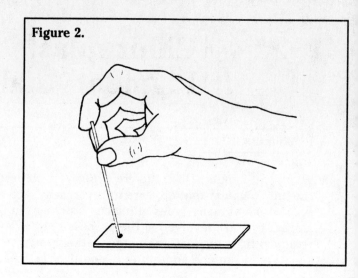

Figure 2.

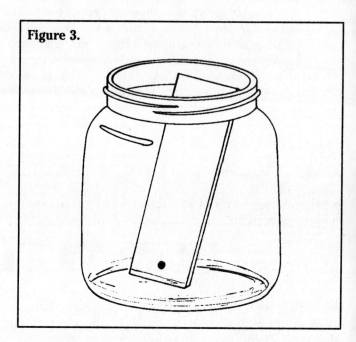

Figure 3.

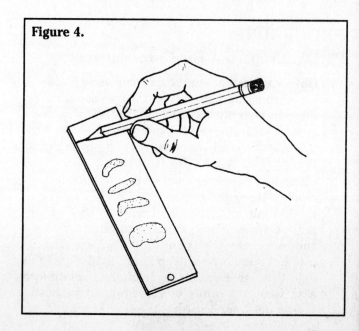

Figure 4.

10. Measure the distance each pigment traveled from the original spot to its final location. Record these data in Table 1.
11. Calculate the R_f value for each pigment spot. The R_f value is the ratio of the distance traveled by the pigment to the distance traveled by the solvent. By comparing R_f values of unknown compounds with the R_f values of known compounds, an unknown substance can be identified.

$$R_f = \frac{\text{distance pigment traveled}}{\text{distance solvent traveled}}$$

12. Record your R_f values in Table 1.
13. Clean your equipment and dispose of your solvents in the designated container.

DATA AND OBSERVATIONS

Table 1.

Chromatography Data		
Substance	Distance from original spot (mm)	R_f value
Solvent front		
Carotenes		
Lutein		
Chlorophyll *a*		
Xanthophylls		
Chlorophyll *b*		

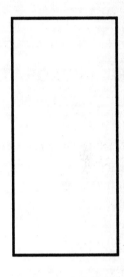

Your thin layer slide

ANALYSIS

1. Which pigments were you able to identify? _____

2. Judging from the darkness of the pigment spots on your chromatogram, which pigment would you say is

   most abundant in spinach leaves? _____

3. Which pigment appeared to travel the fastest? _____ slowest? _____

4. Which pigment had the highest R_f value? _____

5. How do R_f values compare with the rate of travel of the pigment? _____

   _____

6. Why do the pigments travel in the solvent at different speeds? Remember that each pigment is a different

   molecule with its own characteristic size and mass. _____

   _____

   _____

   _____

7. Do you think you would get similar results if you used a different kind of leaf? Explain. _____
_____

8. Why do leaves appear green even though there are other pigments present? _____
_____

9. Many leaves change color in the autumn. How is it possible for this color change to occur? Base your answer on your new knowledge of pigments present in leaves. (HINT: Chlorophyll *a* and chlorophyll *b* are broken down in autumn when day length begins to shorten and temperatures decrease.)
_____

_____

FURTHER EXPLORATIONS

1. Conduct this Exploration using several different plants with differently colored leaves to see how their leaf pigments compare with those of spinach.
2. Separate pigments in spinach leaves by paper chromatography and compare the results with those obtained by thin layer chromatography.

How Does Concentration of Sugar Affect Fermentation? 10-2

LAB

INVESTIGATION

▶ Yeast are unicellular fungi that obtain their energy by fermenting organic materials. Alcoholic fermentation is a process by which yeast break down carbohydrates to produce ethyl alcohol and carbon dioxide. The progress of fermentation can be monitored by examining either ethyl alcohol or carbon dioxide production. Ethyl alcohol can be detected by its distinct odor. Carbon dioxide can be detected by a color change in bromothymol blue.

OBJECTIVES

- Prepare four different growth environments for yeast and measure the rate of respiration in each.
- Make a hypothesis that describes what effect an increase in carbohydrates will have on yeast respiration.

- Draw a graph that compares the rates of respiration for yeast in the four environments.

MATERIALS

colored pencils (4)
molasses solutions (10%, 20%, 50%)
graduated fermentation tubes (4)
bromothymol blue solution

laboratory apron
graph paper
wax marking pencil
distilled water

yeast suspension
stopwatch

PROCEDURE

1. Label the fermentation tubes 1, 2, 3, and 4.
2. Add 1 mL of yeast suspension and 1 mL of bromothymol blue solution to each tube.
3. Add enough distilled water to fermentation tube 1 to bring the fluid level in the closed part of the tube to within 2 cm of the top as shown in Figure 1. Mark with your wax marking pencil the level of fluid in the closed end of the fermentation tube.

Molasses is a syrup that contains a variety of carbohydrates. Your teacher will provide you with molasses solutions of three different concentrations. Use these solutions to carry out steps 4, 5, and 6.

4. Add enough 10% molasses solution to fermentation tube 2 to bring the fluid level in the closed part of the tube to within 2 cm of the top. Mark the fluid level with the wax marking pencil.

The fluid level in the closed end of the fermentation tube will be used to measure the amount of gas produced. As gas accumulates, the fluid level in this part of the tube will drop. You will be monitoring how much the fluid level drops over time.

5. Add enough 20% molasses solution to fermentation tube 3 to bring the fluid level in the closed part of the tube to within 2 cm of the top. Mark the level of the fluid in the closed end of the tube as in step 4.
6. Repeat step 5 using the 50% molasses solution in fermentation tube 4.
7. Make observations and complete the row marked 0 minutes in Table 1 for each of the four fermentation tubes. Begin timing the experiment with a stopwatch.
8. Make a **hypothesis** to predict how increasing the concentration of carbohydrates will affect the rate of respiration in yeast. Write your hypothesis in the space provided.
9. After 10 minutes have elapsed, make observations and complete the row in the table marked 10 minutes for each of the four tubes.

Figure 1.

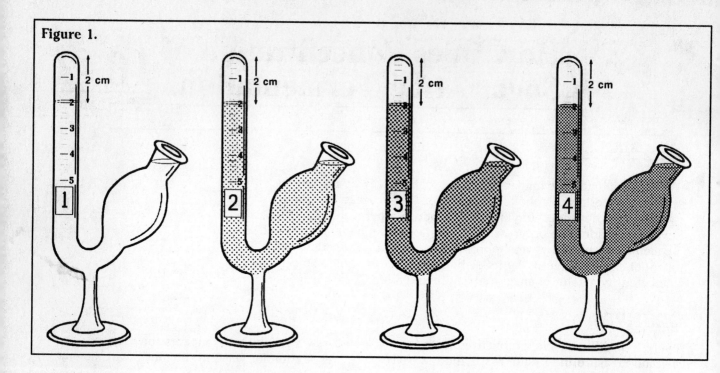

10. Repeat step 9 three more times at 20 minutes, 30 minutes, and 40 minutes after the start of the experiment. Complete the appropriate rows in the table for each of the four fermentation tubes.

11. Place your fermentation tubes in a protected place.

12. At the next class meeting, observe the tubes and complete Table 1.

13. Clean and return experiment materials to their appropriate places.

14. Construct a graph from the data in Table 1. Plot the amount of gas produced over time. The vertical axis represents the amount of gas produced in mL, and the horizontal axis represents the time elapsed in minutes. Use a different-color pencil to draw a line connecting the data points for each fermentation tube.

HYPOTHESIS

_____

_____

DATA AND OBSERVATIONS

Table 1.

Testing For Respiration Rate				
	Time elapsed	Color of solution	Odor of ethyl alcohol (+/−, weak, strong)	Gas production (Volume in mL)
Tube 1	0 min.			
	10 min.			
	20 min.			
	30 min.			
	40 min.			
	24 hrs.			

Name _____ Date _____

Table 1. (continued)

	Time elapsed	Color of solution	Odor of ethyl alcohol (+/−, weak, strong)	Gas production (Volume in mL)
Testing For Respiration Rate				
Tube 2	0 min.			
	10 min.			
	20 min.			
	30 min.			
	40 min.			
	24 hrs.			
Tube 3	0 min.			
	10 min.			
	20 min.			
	30 min.			
	40 min.			
	24 hrs.			
Tube 4	0 min.			
	10 min.			
	20 min.			
	30 min.			
	40 min.			
	24 hrs.			

ANALYSIS

1. In which fermentation tubes did the fluid color change from blue to yellow? _____

   When did the changes take place? _____

   _____

   _____

2. In which tubes did you detect the odor of ethyl alcohol? _____ When did the odor first

   become noticeable? _____

   _____

3. What gas was released in tubes 2, 3, and 4? _____ How might the presence of this

gas be demonstrated? _____

_____

4. What changes took place in tube 1? _____

Explain. _____

_____

5. In which tube was the most gas produced? _____ Why? _____

_____

CHECKING YOUR HYPOTHESIS

Was your **hypothesis** supported by your data? Why or why not? _____

_____

FURTHER INVESTIGATIONS

1. Carry out the same experiment under different conditions. You might keep the molasses concentration constant in the three tubes that contain carbohydrate and vary the amounts of yeast suspension.
2. Devise and carry out an experiment to determine the optimum temperature for yeast fermentation.

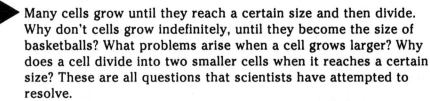

Why Don't Cells Grow Indefinitely?

11–1

LAB

INVESTIGATION

▶ Many cells grow until they reach a certain size and then divide. Why don't cells grow indefinitely, until they become the size of basketballs? What problems arise when a cell grows larger? Why does a cell divide into two smaller cells when it reaches a certain size? These are all questions that scientists have attempted to resolve.

Cell division is a necessary part of the life of any multicellular organism and allows for growth, repair, and formation of cells for reproduction. Growth of an organism occurs mainly by an increase in the number of cells rather than by the enlargement of individual cells. This process seems to be controlled by simple physical laws.

OBJECTIVES

- Make a hypothesis that describes the relationship among surface area, volume, and mass of a cell.
- Determine the relationship between surface area and volume of a model cell.
- Determine the relationship between surface area and mass of a model cell.
- Apply these mathematical relationships to living cells.

MATERIALS

photocopy of 3 cell models
white glue

scissors
balance

coarse sand
small scoop

PROCEDURE

Work with a partner. Obtain a photocopy of the three cell models in Figure 2 for you to cut out.

1. Cut out the three cell models. Fold and glue together all sides of each model. You will have three structures that resemble open boxes, as in Figure 1. Imagine that each cell model has a sixth side and is a closed box. These models represent a cell at three different stages of growth. The youngest stage in growth is represented by the model that is 1 unit to a side. The latest stage in growth is represented by the model that is 4 units to a side.

2. Make a **hypothesis** that predicts what will happen to the surface area-to-volume ratio and the surface area-to-mass ratio as the cell grows larger. The ratio can increase, decrease, or stay the same. Write your hypothesis in the space provided.

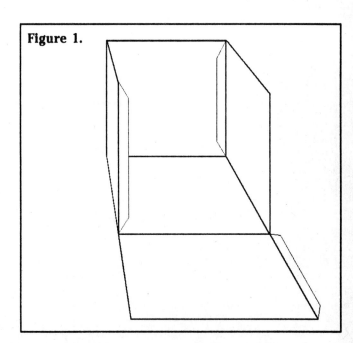

Figure 1.

3. Examine Table 1. Next to the row marked Formulas, write in the mathematical formulas that can be used to calculate the value described in the heading of each column. Use **s** as the length of one side of your model. Once you have written in the formulas, let your teacher check them before you continue the Investigation.

4. Use your formulas to calculate values (except for the last column) for each of the three cell models. Record your values in Table 1.

5. Fill each cell with sand, using the scoop.

6. Determine the mass of each sand-filled model cell by using a balance. Record the masses in the last column of Table 1.

7. Calculate the ratio of total surface area to volume for each model cell. To do this, divide the cell's total surface area by its volume. Place your answers in Table 2.

8. Calculate the ratio of total surface area to mass for each model cell. To do this, divide the cell's total surface area by its mass. Place your answers in Table 2.

HYPOTHESIS

_____

_____

Figure 2.

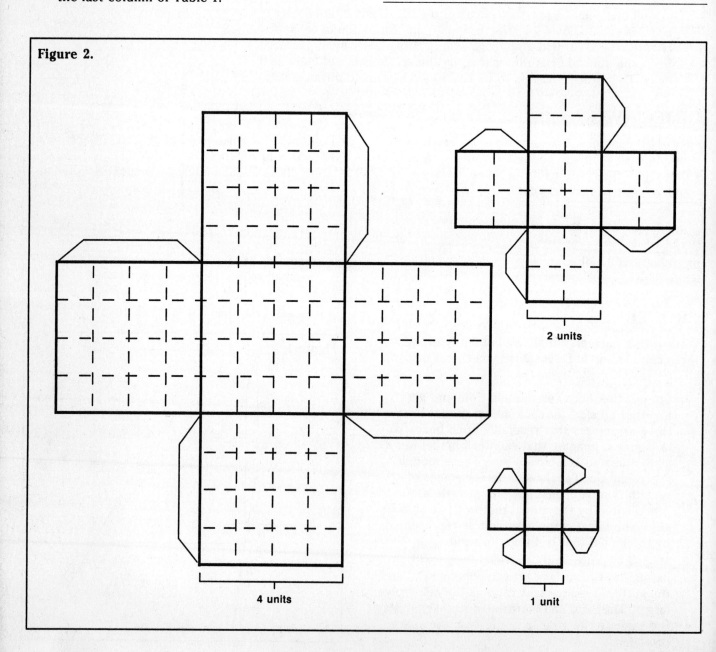

4 units

2 units

1 unit

DATA AND OBSERVATIONS

Table 1.

Measurements of Cell Models					
Formulas					
Cell size (Length of one side)	**Area of one face (Square units)**	**Total surface area of cell (Square units)**	**Volume of cell (Cubic units)**	**Distance from center to edge (Units)**	**Mass of cell (Grams)**
1					
2					
4					

Table 2.

Ratios of Cell-Model Measurements		
Cell size (Length of one side)	**Total surface area to volume**	**Total surface area to mass**
1		
2		
4		

ANALYSIS

1. Anything that the cell takes in, such as oxygen or food, or lets out, such as carbon dioxide, must pass through the cell membrane. Which of the measurements of your model cells best represents the surface

   area of the cell membrane? _____

2. The cell contents, including the nucleus and the cytoplasm, use food and oxygen and produce wastes.

   Which two measurements best represent the contents of one of your model cells? _____

3. As a cell grows larger and accumulates more contents, will it need more or less cell membrane to

   survive? Explain your answer. _____

   _____

4. As a cell grows larger, does the surface area-to-volume ratio get larger, get smaller, or remain the same?

   _____

5. As a cell grows larger, what happens to the surface area-to-mass ratio? _____

6. Which cell model has the greatest surface area-to-volume and surface area-to-mass ratios? _____
_____

7. a. Why can cells not survive when the surface area-to-volume ratio and surface area-to-mass ratio

   become too small? _____
   _____

   b. Which cell model then has the greatest chance of survival? _____

8. How many cells with s=1 would fit into a cell with s=3? _____

9. Which has more total surface area, one cell with s=3 or 27 cells, each with s=1? _____

10. How can the surface area-to-volume and surface area-to-mass ratios be increased in a large cell? _____
_____

11. As the length of a side doubles in a cell, what happens to the distance that nutrients must travel to reach

    the center of the cell? _____
    _____

CHECKING YOUR HYPOTHESIS

Was your **hypothesis** supported by your data? Why or why not? _____
_____

FURTHER INVESTIGATIONS

1. Try calculating surface area-to-volume ratios by using different-shaped cells, such as spheres or rectangles.
2. Investigate actual cell sizes by using a microscope. Use a micrometer to measure the cell diameter, or estimate cell size from the size of the microscope's field of view.

How Does the Environment Affect Mitosis?

11–2

LAB

INVESTIGATION

▶ Mitosis is the division of the nucleus of eukaryotic cells followed by the division of the cytoplasm. If the division proceeds correctly, it produces two cells that are genetically identical to the original cell. Mitosis is responsible for the growth of an organism from a fertilized egg to its final size and is necessary for the repair and replacement of tissue. Anything that influences mitosis potentially has an impact on the genetic continuity of cells and the health of organisms.

   How do environmental factors affect the rate and quality of mitotic division? Scientists are perhaps most keenly interested in this question from the perspective of disease, specifically, the uncontrolled division of cells known as cancer. This investigation will allow you to make a simplified study of the relationship between the environment and mitosis. First, you will observe onion bulbs grow roots by mitotic division. Then you will test the rate of growth of the same onions when exposed to an environmental chemical, caffeine in the form of coffee.

OBJECTIVES

• Prepare squashes of onion root tips to observe mitosis.
• Make a hypothesis to describe the effect of caffeine on mitosis.

• Compare growth of onion roots in water and caffeine.

MATERIALS

onion bulbs (4)
toothpicks (16)
150-mL glass jars (4)
concentrations of caffeine
   (coffee): 0.1%, 0.3%, 0.5%
metric ruler
wax marking pencil
scalpel
paper towels
distilled water

slides (4)
coverslips (4)
Feulgen stain
methanol-acetic acid fixative
3% hydrochloric acid
45% acetic acid in a dropper
   bottle
forceps
microscope

25-mL graduated cylinders (2)
test-tubes (8)
test-tube holder
test-tube rack
thermometer
hot plate
water bath
goggles
laboratory apron

PROCEDURE

Part A. Comparing Rates of Growth

1. Put on a laboratory apron and goggles. Label the small glass jars A, B, C, and D.

2. Insert a toothpick into opposite sides of each onion bulb so that each bulb can be balanced over the mouth of a jar. Then pour water into each jar until just the root area of the bulb is immersed.

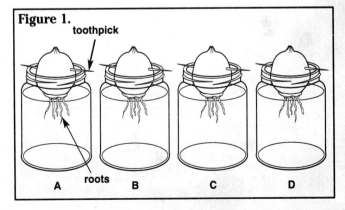

Figure 1.
toothpick

roots

A   B   C   D

3. Examine the bulbs each day. In Table 1 record the number of roots that emerge from each bulb and the average of their lengths.

4. When the roots have grown to 1 cm in length, pour the water out of jars B, C, and D. Your teacher will provide you with caffeine solutions of three different concentrations. Fill jar B with the 0.1% solution, jar C with the 0.3% solution, and jar D with the 0.5% solution. Once again, balance the bulbs over the mouth of jars B, C, and D so that the roots are immersed.

5. Measure the roots for 3 more days, each time recording the average length of the roots for each of the treatments (i.e., water and the three concentrations of caffeine) in Table 2.

Part B. Comparing Phases of Mitosis

Note: READ ALL STEPS BEFORE YOU START.

1. Label 4 test tubes A, B, C, and D to correspond to the treatments to which the onion bulbs are being subjected. Then pour 5 mL of methanol-acetic acid fixative into each of the tubes.

2. Set up and begin heating the water bath to 60°C.

3. Use the scalpel to remove all of the roots from each of the onion bulbs. **CAUTION:** *Use the scalpel with care. Cut away from your fingers.* Then use the scalpel to cut a 3 mm piece from the *tips* of each root. Immediately place these tips from onion bulbs treated in A, B, C, and D jars into the corresponding test tubes containing the methanol-acetic acid fixative.

4. Use the test-tube holder to place test tubes A—D into the water bath at 60° C for 15 minutes.

5. Carefully pour the fixative from each tube into a labeled container to be disposed of by your teacher. Transfer the root tips from each tube to four new test tubes labeled A—D.

6. Pour 5 mL of 3% hydrochloric acid into each of the new test tubes in order to prepare the DNA for staining. **CAUTION:** *Hydrochloric acid is a strong acid and causes burns. Avoid contact with skin or eyes. Flush with water immediately if contact occurs.* Place the test tubes into the water bath at 60°C for 10 minutes.

7. Carefully pour the acid into a labeled empty beaker that your teacher has set aside for the acid. Add enough drops of Feulgen stain into each test tube to cover the roots. **CAUTION:** *The stain can discolor your clothes and skin. Use it with care.* Let the tissues sit in the stain for 15 minutes.

8. From tube A, remove one root tip with a pair of forceps. Place the root tip in the center of a labeled slide. Add one or two drops of acetic acid. Then place a coverslip over the specimen.

9. Place the slide on a paper towel cushion and cover the slide and coverslip with a piece of paper towel. Push down onto the coverslip with the eraser of a pencil. This is called a squash. *Do not press too hard or you will break the coverslip.*

10. Repeat steps 8 and 9 for treatments B, C, and D.

11. Make a hypothesis to describe the effect of caffeine on the stages of mitotic division.

12. Look at your slides under the microscope at low and high power for cells undergoing mitosis. The cells will not be as neatly arranged as they would be on prepared slides. Examine the size, shape, and position of chromosomes in each treatment in order to help you identify phases of mitosis. In comparing treatments, do you notice differences in the number of cells in each phase? In the stronger caffeine solutions, do the chromosomes in any particular phase seem especially distinct? Count and record in Table 3 the number of cells in each phase of the cell cycle.

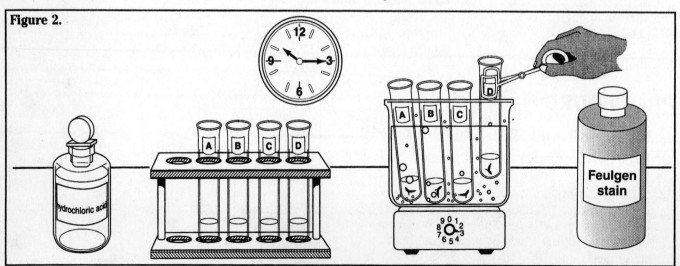

Figure 2.

13. On a sheet of paper, sketch the stages of mitosis observed from roots in each treatment.

HYPOTHESIS

_____

_____

_____

DATA AND OBSERVATIONS

Table 1.

	BULB A		BULB B		BULB C		BULB D	
Day	Number	Av. Length	Number	Av. Length	Number	Av. Length	Number	Av. Length
1								
2								
3								

Number of Roots and Average Length in Water

Table 2.

	BULB A (water)		BULB B (0.1%)		BULB C (0.3%)		BULB D (0.5%)	
Day	Number	Av. Length	Number	Av. Length	Number	Av. Length	Number	Av. Length
1								
2								
3								

Number of Roots and Average Lengths

Table 3.

Treatment	Interphase	Prophase	Metaphase	Anaphase	Telophase	Division of cytoplasm
Bulb A						
Bulb B						
Bulb C						
Bulb D						

Number of Mitotic Phases in Each Treatment

ANALYSIS

1. Identify the control and variable for the experiment.

_____

_____

2. Study Tables 1 and 2. Compare the rate of growth of the roots immersed in water with the rate of growth in the various concentrations of caffeine.

_____

_____

3. Describe any differences in the number of cells in each mitotic phase among the four squashes.

_____

_____

_____

4. How do your observations about mitotic phases in Part B relate to your observations about rate of root growth in Part A?

_____

_____

5. What are some conditions or factors in the environment that might have an effect upon the rate or quality of mitotic division?

_____

_____

CHECKING YOUR HYPOTHESIS

Was your **hypothesis** supported by your data? Why or why not? _____

_____

FURTHER INVESTIGATIONS

1. Design an experiment to test the effects of environmental chemicals on the growth and development of _Ascaris_ eggs.

2. Design an experiment to test the effect of ultraviolet radiation upon the mitotic division of onion, bean, and _Ascaris_ cells.

Observation of Meiosis

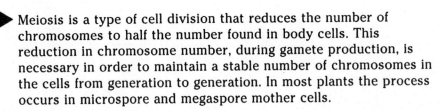

12–1
LAB

▶ Meiosis is a type of cell division that reduces the number of chromosomes to half the number found in body cells. This reduction in chromosome number, during gamete production, is necessary in order to maintain a stable number of chromosomes in the cells from generation to generation. In most plants the process occurs in microspore and megaspore mother cells.

OBJECTIVES

• Observe the stages of meiosis in lily anthers.

• Draw and label the stages of meiosis from lily anthers.

MATERIALS

microscope
prepared slide of lily anthers

drawing paper (optional)
pencil (colored pencils if desired)

PROCEDURE

1. Place a prepared slide of lily anthers on the microscope under low power.
2. Locate cells in the anthers that are undergoing cell division.
3. Observe a cell in meiosis and identify what stage of division the cell is in by comparing it with the stages in Figure 1.

4. In the space provided in Data and Observations, draw the cell and label it with the name of the appropriate stage of meiosis.
5. Continue to observe, identify, and draw cells for as many different stages of meiosis as can be found.

Figure 1.

Prophase I	Prophase I	Metaphase I	Anaphase I
Telophase I	Metaphase II	Anaphase II	Telophase II

DATA AND OBSERVATIONS

ANALYSIS

1. Why do you think lily anthers were chosen for this observation? _____

    _____

2. Which stages of meiosis appeared most frequently? _____

    _____

3. Describe the chromosomes as they appear in the anther cells. _____

    _____

4. What is the overall function of meiosis in lily anthers? _____

    _____

FURTHER EXPLORATIONS

1. Obtain a textbook from your teacher or the library that identifies
   and discusses the meiotic stages in humans. Compare and contrast
   the stages of meiosis in humans with those in flowering plants.
2. Make slides of onion root-tip cells. Compare the cells undergoing
   mitosis with those of lily anthers undergoing meiosis.

Chromosome Extraction and Analysis

13–1
LAB

EXPLORATION

▶ Fruit flies have been used since 1909 as a subject of genetic studies. The cells in the salivary glands of fruit fly larvae have giant polytene chromosomes, which are more than 200 times larger than those in the adult fruit fly. They are formed by multiple chromosome replications without any separation of the replicated strands. As a result, the salivary glands of the larva carry many copies of each gene found in the adult fruit fly. When stained, polytene chromosomes have a banded appearance. Scientists have been able to correlate the size, width, and location of the bands with specific genes. As in the adult fruit fly, the cells from the larva salivary glands have 4 pairs of homologous chromosomes. However, they cannot all be distinguished because the chromosomes in the glands are attached to each other.

    A female fruit fly can lay up to 500 eggs. Each egg hatches into a larva, which grows and molts twice before forming a hard pupal case. Inside its case, a larva changes into an adult fruit fly.

OBJECTIVES

- Extract chromosomes from a fruit fly larva.
- Stain and examine the chromosomes under a microscope.

MATERIALS

fruit fly larvae in a culture
0.7% saline solution
18% hydrochloric acid solution
45% acetic acid solution
aceto-orcein stain
microscope slides and coverslips
paper towels
dissecting probe
forceps, fine-nosed
dissecting microscope or hand lens
compound microscope
metric ruler
laboratory apron
goggles

PROCEDURE

1. Place a drop of saline solution on a microscope slide. Using a forceps, carefully remove a larva from a culture and place it in the drop of solution.

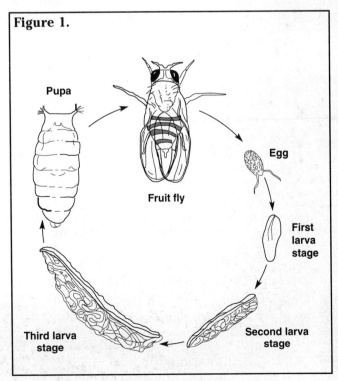

Figure 1.

Pupa

Fruit fly

Egg

First larva stage

Second larva stage

Third larva stage

Figure 2.

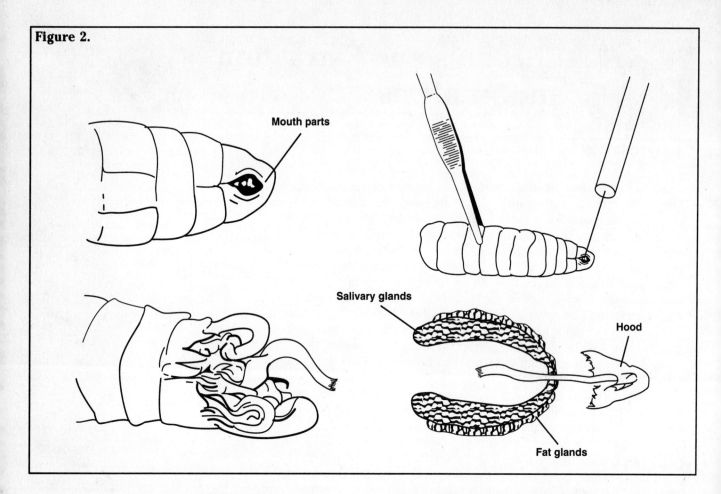

Mouth parts

Salivary glands

Hood

Fat glands

2. Locate the anterior and posterior ends of the larva. The head is the darker area. Also, larvae move in the direction of the anterior end.

3. Use a dissecting microscope or hand lens to examine the anterior end. Use Figure 2 to help you locate the mouth parts.

4. Use the dissecting microscope to perform this step. With a forceps, carefully squeeze the larva about 2/3 of the way back from the mouth parts (see Figure 2). This will force fluid toward the anterior end and make the head region protrude.

5. With a dissecting probe, pierce the head and gently pull straight forward. If done properly, the salivary glands and other internal structures will emerge in strings. The salivary glands are directly behind the head region. Study Figure 2 so that you do not confuse the salivary glands with intestines or fat glands. Essentially, they will appear transparent and cellular. If you were unsuccessful with the procedure, repeat it with another larva.

6. Remove all parts of the larva from the slide except the salivary glands. Drain the solution from the slide, and carefully dab the slide dry with a paper towel. Do not touch the salivary glands with the toweling.

7. Flood the glands with hydrochloric acid solution for 3 minutes and then drain the slide. CAUTION: *If hydrochloric acid is spilled, rinse with water and call your teacher immediately.*

8. Flood the glands with aceto-orcein stain for two minutes. Now tilt the slide and allow the stain to drain to one side. Wipe any excess stain from the slide with a piece of paper toweling, being careful to avoid touching the glands.

9. Look for a dark red stained area. Then place the slide on a paper towel and gently rinse it with acetic acid, being very careful not to wash away the salivary glands. CAUTION: *If acetic acid is spilled, rinse with water and call your teacher immediately.* Drain the excess acid onto a paper towel. Repeat this step until most of the stain has been removed.

10. Carefully lower a coverslip over the salivary glands. Using your thumb, *gently* press the coverslip. Your purpose is to spread the cells into a single layer, at the same time releasing the chromosomes from their nuclei.

11. Observe the slide under low power and high power. Sketch the chromosomes in the space provided.

DATA AND OBSERVATIONS

ANALYSIS

1. Describe the structure of the pairs of chromosomes. Why do they have this appearance?

_____

_____

_____

_____

2. The larval stage is a period of enormous growth in the fruit fly's life cycle. What do you think is the function of having polytene chromosomes with many sets of genes in the larval stage?

_____

_____

_____

_____

3. Suppose that you had grown one group of larvae on an enriched culture and another on a minimal nutrient culture. Would you expect any observable difference in the polytene chromosomes from each group? Why or why not?

_____

_____

_____

_____

4. Compare the chromosomes in this lab with those you observed in mitotic division of onion cells in the BioLab for Chapter 11 of your text. How are they different?

_____

_____

_____

_____

FURTHER EXPLORATIONS

1. Compare the chromosomes in the salivary glands of a variety of insects and describe differences in the number and arrangement of the chromosomes.

2. Athletes are often given a chromosome test prior to competing in the Olympics. Cells from inside the mouth are used as a source of chromosomes. Use reference books to determine the procedure used. Include in your research the purpose, the reliability, and the ethics of the test.

What Phenotypic Ratio Is Seen in a Dihybrid Cross?

14–1

LAB

The fruit fly, *Drosophila melanogaster*, is one of the most important organisms used by geneticists in studying the mechanisms of inheritance. Fruit flies are used for a number of reasons. A very short life cycle allows scientists to study several generations in a short time. Several hundred offspring can be produced by each female under ideal conditions. Small vials and simple media enable the easy culture of the insects. There are many mutations available for study in this small animal, which has only eight chromosomes in its diploid cells.

OBJECTIVES

- Learn to care for and raise two generations of fruit flies.
- Develop hypotheses to describe the results of two dihybrid crosses.

- Determine the results of a cross between two flies showing recessive traits.
- Construct Punnett squares for two dihybrid crosses.

MATERIALS

culture vials with medium and foam plugs (2)
culture of vestigial-winged, normal-bodied
 fruit flies
culture of long-winged, ebony-bodied fruit flies

vial of alcohol
anesthetic
anesthetic wand
white index card

camel-hair brush
wax marking pencil
stereomicroscope or
 hand lens

PROCEDURE

You will need to learn to identify the sexes of fruit flies. Look at the drawing in Figure 1 for help. The dark, blunt abdomen with dark-colored claspers on the underside identifies the male. Males also have a pair of sex combs on the front pair of legs.

In this Investigation, you will study the inheritance of two traits. Both of these traits are easily observable in Figure 2. The allele for long wings (**W**) is dominant to the allele for vestigial wings (**w**). Long wings that permit flight contrast clearly with the vestigial wings that are short and useless for flying. Normal body color (**B**) is dominant to ebony color (**b**). The normal body color of fruit flies is tan, while ebony is a darker brown or black.

The matings in this experiment must begin with unmated (virgin) female flies. This will ensure that only the desired males contribute to the offspring of the cross. A female *Drosophila* can store enough sperm from a single mating to fertilize all the eggs she produces in her lifetime. By using virgin females, the parentage of each generation can be controlled.

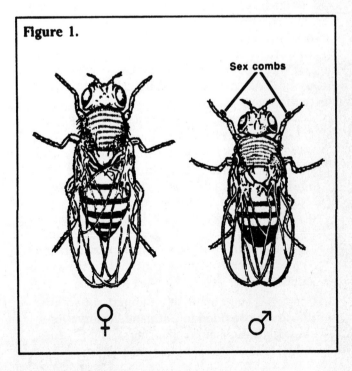

Figure 1.

Sex combs

♀   ♂

Figure 2.

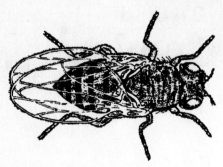

Ebony body color
(bb)

Vestigial wings
(ww)

1. Remove all adult fruit flies from both parental cultures when larvae begin to pupate. Remove the adult flies and place them into a vial of alcohol to kill them. This step will ensure that only virgin females remain in the culture.

2. On the first morning that new adults emerge in the parental cultures, obtain a new sterile culture vial. Label this vial "**F₁**" – ♀ **wwBB** × ♂ **WWbb** and add your name. This vial will be used to raise the first filial generation.

3. Anesthetize and collect five virgin females from the vial that contains vestigial-winged, normal-bodied flies, and collect five males from the vial that contains long-winged, ebony-bodied flies.

To collect flies, tap the vial against a table so that the flies fall to the bottom of the vial. Quickly remove the plug and place a wand containing a few drops of anesthetic into the vial. When the flies are anesthetized, remove the wand from the vial. Use a stereomicroscope or hand lens to aid in identification of males and females. Manipulate the flies with the paintbrush. It is easier to examine flies using a piece of white paper or a white index card as a background. Act carefully, but quickly, before the anesthetic wears off.

4. Place the 10 flies into the vial marked "**F₁**" – ♀ **wwBB** × ♂ **WWbb** and plug the vial with a foam plug. **HINT:** *Always leave the vial on its side until the flies recover from the effects of the anesthetic. This will ensure that they are not injured or killed.* These ten parental flies will mate and produce the first filial generation (**F₁**). The female flies will lay eggs, and larvae will appear in 8 to 10 days. When the larvae begin to pupate, remove the parent flies to the vial of alcohol.

5. Construct and complete a Punnett square in the space provided in Data and Observations, showing the possible offspring of the F₁ cross.

6. Make a **hypothesis** that predicts the numbers and types of F₁ offspring from this cross. Write your hypothesis in the space provided.

7. Record the numbers and type of parental flies in Table 1.

8. Store your vial of living flies according to the instructions of your teacher.

9. Label a new sterile culture vial with your name and "**F₂ cross.**"

10. As new F₁ adults emerge over a period of about two weeks, anesthetize and count the numbers and types of flies that appear in this F₁ generation. Record these data in Table 1.

11. Place several of these F₁ adults, as they are anesthetized and counted, into the vial marked "**F₂ cross.**" These F₁ adults will be used to produce the **F₂** (second filial) generation. Be sure to include both males and females in the **F₂** vial.

12. Construct and complete a Punnett square in the space provided in Data and Observations for this dihybrid cross.

13. Make a **hypothesis** that predicts the numbers and types of F₂ offspring from this dihybrid cross. Write your hypothesis in the space provided.

14. When larvae begin to appear in the F₂ vial, remove the adult F₁ flies to the vial of alcohol. Do not add any more F₁ flies to the F₂ vial.

15. As F₂ adults emerge from their pupae, anesthetize and count the different types of flies. Remove flies after they are counted.

16. Record your data for the F₂ generation in Table 1.

17. Record the totals for F₁ and F₂ below Table 1.

18. Dispose of your cultures as directed by your teacher.

Name _____ Date _____

HYPOTHESIS 1

_____

_____

HYPOTHESIS 2

_____

_____

DATA AND OBSERVATIONS

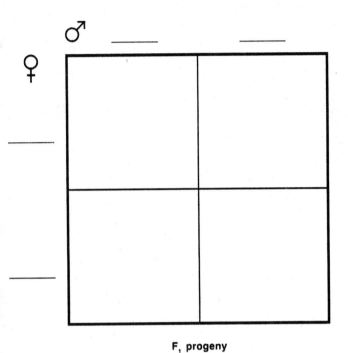

F₁ progeny

F₂ progeny

Table 1.

Results of Dihybrid Crosses				
Generation	Long-winged normal-bodied	Vestigial-winged normal-bodied	Long-winged ebony-bodied	Vestigial-winged ebony bodied
Parental males				
Parental females				
F₁ males				
F₁ females				
F₂ males				
F₂ females				

Total number of F₁ individuals _____   Total number of F₂ individuals _____

ANALYSIS

1. What phenotypes were evident in the F_1 generation? _____

2. What phenotypes were evident in the F_2 generation? _____

    _____

    _____

3. According to your Punnett square, what was the expected F_2 phenotypic ratio? _____

    What was the actual F_2 phenotypic ratio you observed? _____

    _____

4. What conditions would have to be present to produce the expected F_2 phenotypic ratio? _____

    _____

5. Which of Mendel's Laws account for the production of new phenotypes in offspring that were not seen in

    the parents? _____

6. How is Metaphase I of meiosis important in the production of the observed ratio of traits? _____

    _____

    _____

7. Why was it necessary to have virgin females, but not virgin males, for this Investigation? _____

    _____

    _____

CHECKING YOUR HYPOTHESIS

Were your **hypotheses** supported by your data? Why or why not? _____

_____

FURTHER INVESTIGATIONS

1. Perform a similar experiment for two generations with fruit flies
    that have a sex-linked trait, such as white eyes, to determine how
    the ratio of phenotypes changes.
2. The genes for black body, purple eyes, and long wings in *Drosophila*
    are all located on the same chromosome. Devise an experiment that
    will allow you to make a chromosome map for this chromosome.
    Plan crosses that will allow you to measure the frequency of
    crossing-over.

EXPLORATION

Determination of Genotypes from Phenotypes in Humans 15–1

LAB

An organism can be thought of as a large collection of phenotypes. A phenotype is the appearance of a trait and is determined by pairs of genes. The pairs of genes represent the genotype for the trait. If you were told a large enough number of phenotypic traits that belonged to another person, you would be able to recognize that person.

In this Exploration, you will determine some of your own phenotypic traits. From these, you will be able to determine what your genotypes are for some of the traits. If a trait is dominant and you possess that trait, you will not be able to determine your exact genotype because you could be either homozygous or heterozygous for the gene. However, if a trait is determined by incomplete dominance, you can tell if you are homozygous or heterozygous. Genotypes of recessive traits can be identified. By comparing your genotypes and phenotypes with other people in your class, you will see why you are a unique individual. Given the almost limitless number of gene combinations, it is almost impossible that anyone would have all the same traits as you.

OBJECTIVES

- Determine your phenotype for nine different characteristics.
- Determine your possible genotypes for the nine different characteristics.

- Compare your phenotypes and genotypes with those of other students in the class.
- Evaluate your uniqueness as an individual.

MATERIALS

PTC taste paper
untreated taste paper
mirror

PROCEDURE

1. Obtain one piece each of PTC paper and untreated taste paper from your teacher. First, place the untreated paper on your wet tongue to see how it tastes. Then dispose of it in the wastebasket, and place the PTC paper on your wet tongue to see if you can taste phenylthiocarbamide—PTC.

2. PTC is quite bitter and you will notice readily whether or not you have the ability to taste this chemical. If you can taste PTC, enter "taster" in the proper place in the "Your

Phenotype" column in the table. If you cannot taste the chemical, enter "nontaster" in the table. Discard the taste paper in the wastebasket.

3. Now that you have determined your phenotype, enter in the column marked "Your Possible Genotypes" what your genotype could be. Tasters are either **TT** or **Tt**. Nontasters are **tt**.

4. For each of the following traits, observe and record your phenotype in the table. Then record your possible genotypes.

a. hairline—The widow's peak hairline comes to a point in the center of the forehead (**WW** or **Ww**). Individuals that lack the trait are **ww**.

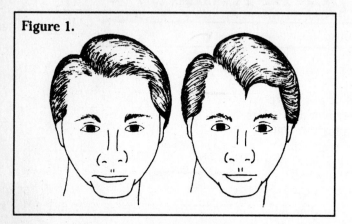

Figure 1.

b. eye shape—Almond-shaped eyes (**AA** or **Aa**) are dominant to round eyes (**aa**).

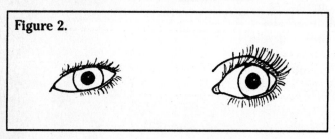

Figure 2.

c. eyelash length—Long eyelashes (**EE** or **Ee**) are dominant to short eyelashes (**ee**).

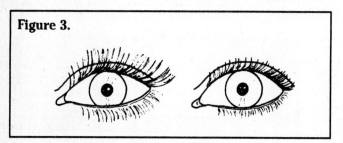

Figure 3.

d. tongue rolling—The ability to roll the tongue (**CC** or **Cc**) is dominant to the lack of this ability (**cc**).

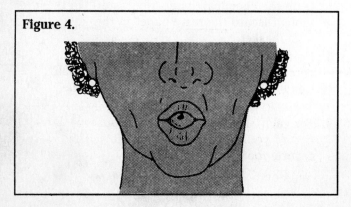

Figure 4.

e. thumb—One whose thumb tip bends backward more than 30 degrees (hitch-hiker's thumb) is dominant (**BB** or **Bb**) to a straight thumb (**bb**).

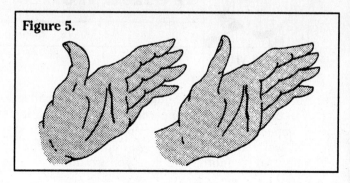

Figure 5.

f. lip thickness—Thick lips (**LL** or **Ll**) are dominant to thin lips (**ll**).

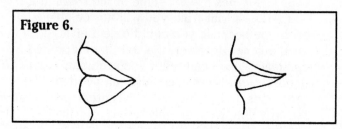

Figure 6.

g. hair texture—Curly hair (**HH**) is incompletely dominant to straight hair (**SS**). Those that have wavy hair are **HS**.

Figure 7.

h. inter-eye distance—The distance between the eyes is an example of incomplete dominance. Close-set eyes are **DD**, eyes set far apart are **FF**, and medium-set eyes are **DF**.

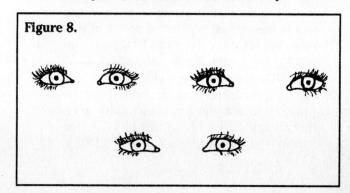

Figure 8.

Name _____ Date _____

i. **lip protrusion**—Protruding lips (**PP**) are incompletely dominant to nonprotruding lips (**NN**). Slightly protruding lips are **PN**.

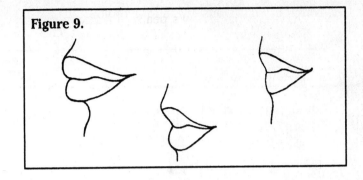

Figure 9.

DATA AND OBSERVATIONS

Table 1.

Human Phenotypes and Genotypes				
	Traits		Your phenotype	Your possible genotypes
	Dominant	Recessive		
PTC taste	Taster		Nontaster		
Hairline	Widow's peak		Straight line		
Eye shape	Almond		Round		
Eyelash length	Long		Short		
Tongue dexterity	Can roll		Unable to roll		
Thumb	Hitchhiker's thumb		Straight thumb		
Lip thickness	Thick		Thin		
Hair texture	Curly	Wavy	Straight		
Inter-eye distance	Close together	Medium distance	Far apart		
Lip protrusion	Protruding	Slightly protruding	Not protruding		

ANALYSIS

1. Which traits do you have that are dominant? _____

2. Which traits do you have that are recessive? _____

3. Which of your traits are governed by incomplete dominance? _____

4. Which of your traits do you share with one or more of your classmates? _____
_____

85

5. Which of your traits are unique to you? _____

   _____

6. If you and a particular classmate shared all of the same traits examined in this Exploration, what traits

   could you describe to prove your uniqueness? _____

   _____

7. What determines your traits? _____

8. With knowledge of the phenotype of a human, how can a person's genotype be determined? _____

   _____

   _____

   _____

   _____

9. Why was untreated paper used in the PTC taste test? _____

   _____

FURTHER EXPLORATIONS

1. Books from the library or your teacher on human genetics will discuss many other human characteristics. Identify some other characteristics that you or your classmates have and try to determine the genotypes that cause them.

2. Calculate the percentage of the class that has each phenotype and compare these figures with national averages. Suggest reasons why your class might differ from the national percentages of some phenotypes.

How Can Karyotype Analysis Explain Genetic Disorders? 15–2

LAB

INVESTIGATION

▶ A karyotype is a picture in which the chromosomes of a cell have been stained so that the banding pattern of the chromosomes appears. Cells in metaphase of cell division are stained to show distinct parts of the chromosomes. The cells are then photographed through the microscope, and the photograph is enlarged. The chromosomes are cut from the photograph and arranged in pairs according to size, arm length, centromere position, and banding patterns. Karyotypes have become of increasing importance to genetic counselors as disorders and diseases have been traced to specific visible abnormalities of the chromosomes.

OBJECTIVES

- Construct a karyotype from the metaphase chromosomes of a fictitious insect.
- Analyze prepared karyotypes for chromosome abnormalities.

- Identify the genetic disorders of six fictitious insects by using the insects' karyotypes.
- Hypothesize how karyotype analysis can be used to explain the presence of a genetic disorder.

MATERIALS

photocopies of metaphase chromosomes
  from six insects (2 pages)

scissors
rubber cement

PROCEDURE

For this Investigation, assume that a new species of insect has been discovered. This insect has three pairs of very large chromosomes. Researchers have been able to trace four genetic disorders to specific

chromosomal abnormalities in this insect. Study the karyotypes and phenotypes of normal male and female insects as illustrated in Figure 1.

Figure 1.

Normal karyotype – male

Normal phenotypic male

Normal karyotype – female

Normal phenotypic female

Note that the normal male insect has a pair of sex chromosomes similar to those of the human male, one large and one small. In the same way, the female has a pair of sex chromosomes similar to those of the human female, both large. These sex chromosomes make up the third pair of chromosomes.

The disorder known as size reduction disorder appears when there is a monosomy of the sex-chromosome pair. A single large chromosome produces a small female insect. A single small chromosome produces a small male insect. This disorder is shown in Figure 2.

Clear wing disorder, as shown in Figure 3, appears to result from trisomy of the chromosomes of the second pair. The extra chromosome of the second pair produces sterile insects that lack coloring in their wings. Since sterility always results, the clear wing disorder is not passed on to progeny.

A duplication of a portion of a chromosome from pair 1 produces an insect with a double head. This duplication also produces banding on the wings and additional body segments. See Figure 4.

The deletion of a short segment of the large sex chromosome results in a loss of body segmentation and a reduction of body size. This disorder is shown in Figure 5.

1. Obtain copies of the metaphase chromosomes of six insects from your teacher.
2. Write a **hypothesis** to describe how karyotype analysis can be used to explain the presence of a genetic disorder. Write your hypothesis in the space provided.
3. Cut out the chromosomes for insect 1 from the photocopy and place them along the line for insect 1 in Data and Observations. Arrange similar chromosomes together as shown in the normal karyotypes in Figure 1. Match up similar chromosomes by comparing chromosome size, length of the arms of each chromosome, centromere position, and banding patterns. Be sure to line up chromosomes that resemble the first pair of the normal karyotype above the number 1, those that resemble the second pair above the number 2, and those that resemble the third pair (sex chromosomes) above the number 3.
4. Once the chromosomes are positioned, paste their centromeres to the straight line using rubber cement. This represents the karyotype for one insect.
5. Repeat steps 3 and 4 for each of the fictitious insects.

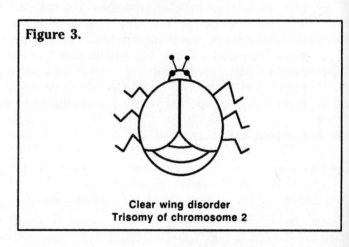

Figure 2.

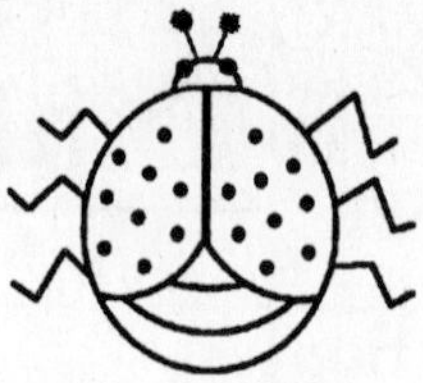

Size reduction disorder
Monosomy of chromosome 3

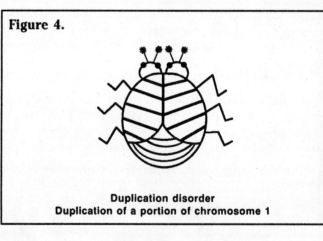

Figure 3.

Clear wing disorder
Trisomy of chromosome 2

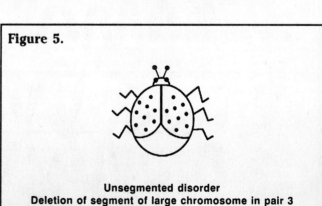

Figure 4.

Duplication disorder
Duplication of a portion of chromosome 1

Figure 5.

Unsegmented disorder
Deletion of segment of large chromosome in pair 3

6. Compare your karyotypes with the karyotypes of the normal insects and with the descriptions of the genetic disorders.
7. Complete the Analysis for this Investigation.

HYPOTHESIS

_____

_____

DATA AND OBSERVATIONS

Insect 1

Insect 2

_____

_____

1   2   3   1   2   3

Insect 3

Insect 4

_____

_____

1   2   3   1   2   3

Insect 5

Insect 6

_____

_____

1   2   3   1   2   3

ANALYSIS

1. Identify the sex, genetic disorder, and chromosome error for each of the fictitious insects.

	Sex	Genetic disorder	Chromosome error
Insect 1	_____	_____	_____
Insect 2	_____	_____	_____
Insect 3	_____	_____	_____
Insect 4	_____	_____	_____
Insect 5	_____	_____	_____
Insect 6	_____	_____	_____

2. Which type of chromosome abnormality is the most difficult to detect by means of a karyotype?

   _____ the easiest? _____ Why? _____

   _____

3. How can duplication of a chromosome of the first pair produce a double head and, at the same time,

   affect the wing pigmentation and body segmentation? _____

   _____

4. What kind of information would be required if karyotype analysis were to be used to detect the genetic

   disorders of real organisms? _____

   _____

CHECKING YOUR HYPOTHESIS

Was your **hypothesis** supported by your observations? Why or why not? _____

_____

FURTHER INVESTIGATIONS

1. Any book on human genetics will have discussions of genetic disorders and karyotype analysis. Borrow a book from your teacher or the library and investigate some of the common abnormal karyotypes in humans.
2. Obtain additional copies of metaphase chromosomes from the six fictitious insects. Construct new karyotypes from the various chromosomes. Imagine how they might look. Draw pictures of the new fictitious insects.

DNA Sequencing

EXPLORATION

▶ DNA sequencing is the process of identifying the order of nitrogen bases within a DNA macromolecule. It begins with cutting the chromosome using restriction enzymes, and making many copies. The strands are separated and the end of one specific strand is labeled with a radioactive probe. The multiple copies of this strand are divided into equal quantities and placed in four test tubes, each of which contains a different chemical that selectively destroys one kind of nitrogen base. The concentration of each chemical is such that it does not destroy the given nitrogen base each and every place it occurs on the DNA strand. As a result, each chemical treatment produces DNA segments of different lengths. The segments are then sorted by size and charge in a gel electrophoresis. The segments with radioactive labels will show up on the X ray and can be "read" to determine the sequence of nucleotides in the gene under study. From this sequence, a scientist can determine the type of protein that would be assembled by information from the gene.

OBJECTIVES

- Learn to read a gel electrophoresis.
- Model the process of DNA sequencing through gel electrophoresis.
- Convert the DNA sequence into a protein by using an mRNA codon chart.

MATERIALS

photocopy of strands of DNA (24)
colored pencils
scissors
meter stick
large poster board (approx. 24" × 36")
tape
plastic tray or rectangular lid, small (4)
wax marking pencil
marker

PROCEDURE

Part A. Reading a Gel Electrophoresis

1. Figure 1 represents an X ray of the gel from the electrophoresis of segments of a DNA strand. Each letter at the top represents one of the four bases in a nucleotide of a DNA molecule. The marks under each of the bases represent segments of DNA that migrated through the gel. (The radioactive probes attached to the segments

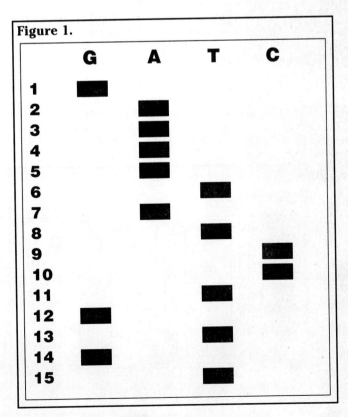

Figure 1.

Figure 2.

radioactive probe

ACGGGACTACCATGGGCCTTA

"burn" these marks into the X ray film when it is exposed to the gel.) The numbers represent the relative distances traveled by the segments, with "1" being the shortest distance. The smaller the segment, the longer the distance it migrates. The DNA sequence is read *from the bottom of the X ray to the top,* that is, from the smallest segment of the DNA strand to the largest.

2. "Read" the DNA sequence shown in Figure 1, and record it in Data and Observations.

Part B. Modeling DNA Sequencing

1. Cut out the 24 photocopied DNA strands as shown in Figure 2. Color the radioactive probe at the left end of each copy of the strand.

2. Place 6 of the strands into each of the plastic trays. The trays represent test tubes containing different chemical treatments. As shown in Figure 3, label the trays left to right with the letters G, A, T, and C to show the specific nitrogen base that will be destroyed.

3. On your posterboard, use your marker to draw a diagram of an electrophoresis gel with wells at the top where you will place segments of the DNA strands. Label the wells G, A, T, C, from left to right, as shown in Figure 4. Number the left side of the gel 1-21 (to represent the 21 nitrogen bases you are sequencing), making sure to space the numbers equidistantly. Construct a grid on your diagram, as shown in

Figure 4, to help you position the DNA segments that "migrate" through the gel.

4. In "test tube" G, cut diagonally through a G on each strand to "destroy" it. *Make sure to cut the G at a different location on each strand so that you will have segments of DNA of different lengths, each with a radioactive probe attached.* Place the segments in the appropriate well of your gel template. Repeat this procedure with the DNA strands in test tubes T and C, cutting T and C, respectively.

5. In test tube A, cut one nitrogen base A on each of the strands. (In the laboratory, if the first base after the radioactive label is destroyed, the DNA segment will not move. Scientists "glue on" a starter DNA sequence to avoid this problem.) Place the segments in their well.

6. From well G, move the segments through the gel according to their size. The smallest piece will move the farthest; the longest piece will move the shortest distance. (For example, a

Figure 4.

	G	A	T	C
1				
2				
3				
4				
5				
6				
7				
8				
9				
10				
11				
12				
13				
14				
15				
16				
17				
18				
19				
20				
21				

Figure 3.

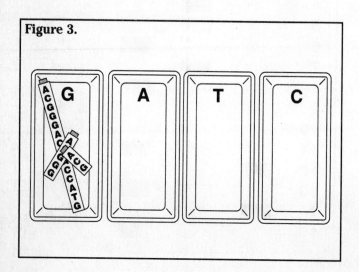

segment with one base would move to position 21, a segment with three bases would move to position 19, and so forth.) Do not use the segments without the radioactive label. These segments would move through the gel but would not register on the X ray film. Uncut segments will remain in the wells. Tape each segment in place.

7. Repeat step 6 for the segments from wells A, T, and C.
8. "Read" the DNA sequence of your gel electrophoresis in the same way you read the diagram of the X ray in Figure 1. Record your results.

DATA AND OBSERVATIONS

Part A
DNA sequence of Figure 1:

_____

Part B
DNA sequence of your gel electrophoresis:

_____

ANALYSIS

Second Base in Code

First Base in Code	A	G	U	C	Third Base in Code
A	Lysine Lysine Asparagine Asparagine	Arginine Arginine Serine Serine	Isoleucine Methionine Isoleucine Isoleucine	Threonine Threonine Threonine Threonine	A G U C
G	Glutamic acid Glutamic acid Aspartic acid Aspartic acid	Glycine Glycine Glycine Glycine	Valine Valine Valine Valine	Alanine Alanine Alanine Alanine	A G U C
U	STOP STOP Tyrosine Tyrosine	STOP Trytophan Cysteine Cysteine	Leucine Leucine Phenylalanine Phenylalanine	Serine Serine Serine Serine	A G U C
C	Glutamine Glutamine Histidine Histidine	Arginine Arginine Arginine Arginine	Leucine Leucine Leucine Leucine	Proline Proline Proline Proline	A G U C

1. What is the mRNA sequence that can be transcribed from the DNA sequence in Part A?

_____

2. Refer to the mRNA codon chart above. Give the amino acid sequence that is coded by the mRNA sequence from Question 1.

_____

3. How does the polypeptide coded for in Figure 1 differ from most proteins?

_____

_____

4. What is the mRNA sequence that can be transcribed from the DNA sequence in Part B?

_____

5. Refer to the mRNA codon chart above. Give the amino acid sequence that is coded by the mRNA sequence from Question 4.

_____

6. How are the strands of "DNA" you studied in this lab different from an actual gene?

_____

_____

Further Explorations

1. Research and write an essay on the use of DNA sequencing to identify human genetic disorders such as cystic fibrosis, Down syndrome, and hemophilia.

2. Research the topic of genetic markers and design an experiment for identifying markers using gel electrophoresis.

Analyzing Fossil Molds

17–1
LAB

EXPLORATION

▶ Brachiopods are solitary, bivalved animals. All modern brachiopods live in the sea and, based on the fossil record, it is likely that extinct species were also marine organisms. In this lab you will use two sheets of plastic molds of brachiopods to model the analysis of fossil molds found in sedimentary rock. The molds will be labeled A and B to distinguish two populations. From measurements of the molds, you will speculate about whether the fossils belong to the same or different populations and whether they would have been found in the same layers of sedimentary rock.

OBJECTIVES

- Measure brachiopod molds and record data.
- Prepare a graph of the data from the two sets of molds.

- Analyze the graphs.
- Interpret results relative to geologic time.

MATERIALS

plastic sheets of brachiopod molds, A and B
metric ruler

colored pencils
graph paper

PROCEDURE

1. The length of the brachiopods will be used as an indicator of brachiopod size. Measure the length of each mold. Record each different size in Table 1 as well as the number of molds corresponding to that size. If you need more room for your measurements, continue the table on a separate sheet of paper.

2. Use data from Table 1 to make a line graph or bar graph. The x-axis will represent the length in millimeters; the y-axis will represent the number of molds. Use a different colored pencil for each set of molds. For assistance in graphing, refer to *Organizing Information* in the *Skill Handbook*, pages 1145–1148.

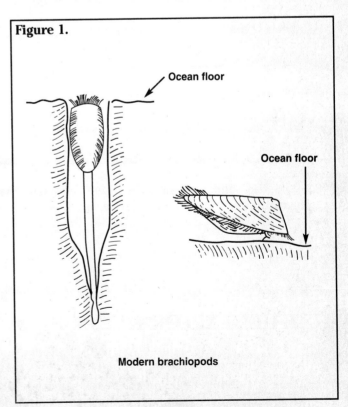

Figure 1.

Ocean floor

Ocean floor

Modern brachiopods

DATA AND OBSERVATIONS

Table 1.

SHEET "A" BRACHIOPODS		SHEET "B" BRACHIOPODS	
Length in mm	Number of molds	Length in mm	Number of molds

ANALYSIS

1. What is the mean size for the molds on sheet A? on sheet B? _____

_____

2. Describe the shape of the graphs for the A molds and the B molds. _____

_____

3. Do you think the brachiopods on sheet A represent the same population as the brachiopods on

sheet B? Explain your answer. _____

_____

4. Do you think the molds on sheet A came from the same time period as the molds on sheet B? Explain your answer.

_____

_____

5. Using the concept of relative dating and assuming for the purpose of analysis that the average size of brachiopods increased over time, which sheet of brachiopods would have been found in a more recent layer of rocks?

_____

_____

FURTHER EXPLORATIONS

1. Conduct research to find out why fossils of marine animals, such as brachiopods, are more useful in relative methods of dating sedimentary rocks than land animals.

2. There are about 260 living species of brachiopods but more than 30 000 known fossil species. Explore possible reasons for the extinction of so many species.

How Is Camouflage an Adaptive Advantage?

18–1

LAB

INVESTIGATION

▶ Natural selection can be described as the process by which those organisms best adapted to the environment are more likely to survive and reproduce than are those organisms that are poorly adapted. Organisms have developed many different kinds of adaptations that have helped them survive in their environments. These include adaptations that help an organism find food, such as keen night vision in nocturnal animals, as well as adaptations that help an organism avoid being eaten. Some organisms use camouflage as a way to escape predation from other organisms. Camouflage allows them to blend in with the background.

OBJECTIVES

- Use an artificial environment to demonstrate the concept of natural selection.
- Construct bar graphs to show the results of the Investigation.

- Hypothesize what will happen if natural selection acts on organisms exhibiting camouflage over a period of four generations.
- Compare this example of artificial selection to natural selection.

MATERIALS

hand hole punch
colored paper (purple, brown, blue, green, tan, black, orange, red, yellow, and white) (1 sheet of each color)

plastic film cannisters or petri dishes (10)
piece of brightly-colored, floral fabric (80 cm × 80 cm)
graph paper (2 sheets)

PROCEDURE

1. Work in a group of four students.
2. Punch 20 dots of each color from the sheets of colored paper and place each color in a different plastic container.
3. Spread out the floral cloth on a flat surface.
4. Spread 10 dots of each color randomly over the cloth. See Figure 1.
5. Select a student to choose dots. That student must look away from the cloth, turn back to it, and then immediately pick up the first dot he or she sees.
6. Repeat step 5 until 10 dots have been picked up. Be sure the student looks away before a selection is made each time.
7. Record the results in Table 1. Return the 10 collected dots to the cloth in a random manner.

Figure 1.

Spread 10 dots of each color

Assume that the dots represent individual organisms that, if allowed, will reproduce more of their own type (color). Also assume that selection of dots represents predation.

8. Write a **hypothesis** to predict what will happen over time if selected dots are not returned to the cloth and the remaining dots "reproduce." Write your hypothesis in the space provided.

9. Each student in the group must, in turn, pick up 20 dots following the method in steps 5 and 6. Place the dots in their original containers. Remember to look away each time a selection is made.

10. After each student has removed 20 dots, shake the remaining 20 dots off the cloth onto the table. See Figure 2.

11. Count and record in Table 2 the number of dots of each color that remains.

12. Give each of the "surviving" dots four "offspring" of the same color by adding dots from the containers. You may need to punch out more of certain colors. Return all of the dots to the cloth in a random manner. This will bring the total number of dots on the cloth back to 100. See Figure 3.

13. Repeat steps 9–12 three more times. Each repetition represents the survival and reproduction of a single generation. Continue to record the results of each repetition in Table 2.

14. Make a bar graph to show the number of dots of each color that were on the cloth at the beginning of the investigation. Label the horizontal axis with the names of the 10 colors and the vertical axis with the number of dots.

15. Make a second bar graph to show the number of dots of each color that were on the cloth at the end of the fourth generation. Label the axes as on the first graph.

HYPOTHESIS

_____

_____

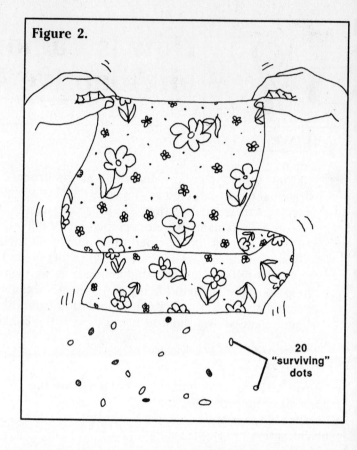

Figure 2.

20 "surviving" dots

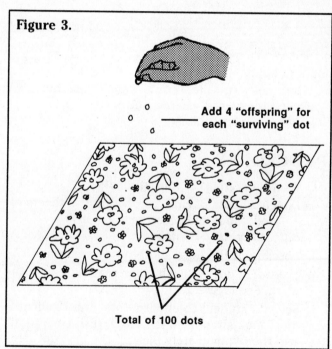

Figure 3.

Add 4 "offspring" for each "surviving" dot

Total of 100 dots

DATA AND OBSERVATIONS

Table 1.

Selection of Dots	
Color	**Number of dots selected**
Purple	
Brown	
Blue	
Green	
Tan	
Black	
Orange	
Red	
Yellow	
White	

Table 2.

Color	Number remaining after generation			
	1	**2**	**3**	**4**
Purple				
Brown				
Blue				
Green				
Tan				
Black				
Orange				
Red				
Yellow				
White				

Table title: **Number of Dots Remaining After Each Generation**

ANALYSIS

1. Which colors were picked up from the floral background? _____

_____

2. Which colors, if any, were not picked up? Why not? _____

_____

_____

3. If the dots represent food to a predator, what is the advantage of being a color that blends in with the

background? _____

4. Give two examples in nature where camouflage helps an organism avoid predation. _____

_____

5. As dots on the cloth pass through several generations, what trends in predation and survival of colors do

you observe? _____

_____

6. What variations could occur in this artificial environment to simulate changes in a natural environment? For example, what would happen if the predator were color blind? _____

_____

7. How would the outcome of this Investigation have been affected if dots that were subject to predation (those picked up) tasted bad or were able to harm the predator in some way, such as by stinging it?

_____

_____

_____

8. Describe a possible example of natural selection that would be similar to the example of artificial selection in this Investigation. _____

_____

CHECKING YOUR HYPOTHESIS

Was your **hypothesis** supported by your data? Why or why not? _____

_____

FURTHER INVESTIGATIONS

1. Certain species of fish are dark colored on the dorsal side of their bodies and light-colored on the ventral side. Research this adaptation and show how it is an example of camouflage.
2. Design an experiment to show that a chameleon changes its color in response to the color of its background.

Plant Survival

EXPLORATION

▶ Predator and prey relationships are not confined to the animal kingdom. Enormous numbers of insects and other animals prey on plants, which they damage or destroy in the process of feeding on them. Since plants cannot run away, they have had to evolve a variety of other defense mechanisms in order to avoid decimation by hungry herbivores. Chemical defenses are common. Nearly all species of plants produce chemicals that are distasteful or dangerous to their potential enemies. Many of these chemicals were probably, at first, metabolic waste products that accumulated in plants. In this Exploration you will observe the effect of plant chemicals on the behavior of the land snail, which normally consumes a wide variety of plant tissues.

OBJECTIVES

- Prepare a feeding tray containing plant tissues soaked in various chemicals.
- Observe the feeding behavior of a herbivore.

- Analyze the effect of different chemical defenses on a land snail.

MATERIALS

land snail
plant extracts prepared by your
   teacher
large leaf lettuce (firm) or
   Chinese cabbage

plastic feeding container
   (approx. 20 cm in width) with
   perforated lid
graph paper

scissors
stereomicroscope or hand lens
wax marking pencil

PROCEDURE

1. Cut a piece of graph paper 3.5 cm square. Use it as a template for cutting 4 pieces of lettuce or cabbage. Avoid cutting the leaf along its veins.

2. Label the bottom of the feeding container with the numbers 1–4 evenly spaced along the width of the container, as shown in Figure 1.

3. Soak one square of the lettuce or cabbage in water for 5 minutes. Remove the square and allow the excess water to drip onto a paper towel. Then place the leaf square on the bottom of the feeding container below "1."

4. Soak each of the remaining leaf squares in a different extract. Place each square below a number in the feeding container. Record the location of each extract in Table 1.

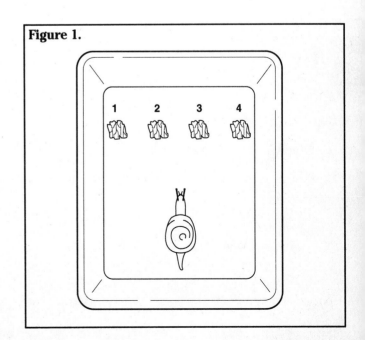

Figure 1.

5. Introduce a snail to each container. Cover the container with its lid and place it in an area of the room designated by your teacher. After 24 hours, remove the remains of each leaf and place it on graph paper. Count and record the number of small squares, or grids, consumed. Divide this figure by the number of grids in 3.5 cm^2 of graph paper to find the percent of each leaf that was consumed.

DATA AND OBSERVATIONS

Table 1.

Type of Extract	Number of squares consumed	Percent of squares consumed
1. water		
2.		
3.		
4.		

ANALYSIS

1. Which leaves were not touched by the snail? Which leaves were eaten the most? _____

_____

_____

2. What is the evolutionary significance of plant chemical defenses? _____

_____

_____

_____

3. What are some nonchemical plant defenses? _____

_____

_____

FURTHER EXPLORATIONS

1. Monarch butterfly caterpillars feed on milkweeds, consuming large quantities of alkaloids which remain in the body of the adult during metamorphosis. The alkaloids make the butterfly distasteful to birds. Design an experiment to test whether monarch butterflies raised on other plants are equally distasteful to birds.

2. Research the coevolution of cabbage butterflies and plants of the mustard family. Discover how these butterflies have overcome the defensive chemicals that are toxic to most insects. Explain the evolutionary advantage of the cabbage butterfly's specialized diet.

Gene Frequencies and Sickle Cell Anemia

19-1
LAB

▶ Sickle cell anemia, a potentially fatal disease, results from a mutant gene for hemoglobin, the oxygen-carrying chemical in red blood cells. There are two alleles for the production of hemoglobin. Individuals with two Hemoglobin A alleles (AA) have normal red blood cells. Those with two mutant Hemoglobin S alleles (SS) develop the disease. And persons who are heterozygous for the sickle cell allele (AS) have some normal red cells and some red cells with the potential to sickle. Heterozygous individuals carry the mutant allele but do not suffer from its debilitating effects.

In the United States about 1 in 500 African-Americans develops sickle cell anemia. But in Africa about 1 in 100 individuals develops the disease. Why is the frequency of a potentially fatal allele so much higher in Africa?

The answer is related to another potentially fatal disease, malaria. Individuals with an AA genotype have a significantly greater risk of contracting malaria and may die from the disease. This results in the removal of A alleles from the gene pool. The SS genotype, which results in sickle cell anemia, is usually fatal before the age of twenty. This results in the removal of S alleles from the population. A person with an AS genotype does not develop sickle cell anemia and has less chance of contracting malaria. Such a person is able to survive and reproduce in a malaria-infected region. Therefore both the individual's A alleles and S alleles remain in the population. The frequency of the hemoglobin S allele in malaria-infected regions of Africa is 16%. But in the United States, where malaria has been eradicated, the allelic frequency is 4%.

OBJECTIVES

- Determine the change in allelic frequencies of the Hemoglobin A and S alleles.

- Explain the process of natural selection as a force affecting allelic frequencies.

MATERIALS

75 red beans and 25 black beans
wax marking pencil

plastic food containers, tall (5)
blindfold

PROCEDURE

1. Using the wax marking pencil, label one of the containers AA, a second AS, and a third SS, to represent the three possible genotypes for the sickle cell trait. Label a fourth container "Non-Surviving Genes." Leave the fifth container unlabeled. It will be used to represent a small population in Africa living in a malaria-infected region.

2. Place 75 red beans in the unlabeled container. These beans represent gametes carrying the Hemoglobin A allele.

3. Add 25 black beans to the unlabeled container. These beans represent gametes carrying the Hemoglobin S allele.

4. Place a cloth or plastic wrap over the top of the container and secure it with a rubber band or tightly knotted string. Shake the container of beans until they are well mixed.

5. You will now simulate the fusion of gametes and record the resulting genotype of each offspring using all of the beans in the unlabeled container. You will also simulate the effect of being homozygous or heterozygous for the hemoglobin gene in a malaria-infected region.

   Wearing a blindfold to ensure a random choice, one team member will select two beans from the unlabeled container 50 times. After each selection, a second team member will identify and record the genotype of the "offspring" in Table 1 by making a slash under the appropriate column head. The same team member (or another) then places each pair of beans in container AA, AS, or SS, depending on the genotype.

   During the periods when the blindfolded team member is making a selection, a third team member will randomly call out the word "malaria" a total of 25 times. (This represents a 50% malaria infection rate.) If the genotype of the selected pair of gametes is AA, that offspring will contract malaria and die. Therefore place that pair of alleles in the container labeled "Non-Surviving Genes" and put a circle around the slash (i.e., the recorded genotype) in Table 1. If the genotype is AS, the individual will survive. Put a circle around the slash in Table 1, and place the pair of beans in the container labeled AS. If the genotype is SS, the individual will die. As with the individual homozygous for the normal Hemoglobin gene, place the beans in the container labeled "Non-

Surviving Genes" and draw a circle around the recorded genotype.

6. After you have completed the cycle for one generation, you are ready to begin the next. All of the beans in the containers labeled AA and AS should be emptied into the unlabeled container. However, place all the beans from the container labeled SS into the container for "Non-Surviving Genes." (Since individuals who are homozygous SS will not usually live long enough to have children, you will not use the SS gametes for tallying the next generation.)

7. Count the number of red beans and black beans in the unlabeled container and record the individual and combined totals in the appropriate spaces under Table 1. The combined total represents the total number of alleles for hemoglobin in the population. Calculate the allelic frequencies as shown and record your results.

8. To determine the 50% malaria infection rate for this generation of the population, take the total number of A and S alleles remaining in the population and divide it by 4. This is the number of times you will call "malaria" in the next round of fusion of gametes. (For example, if you had 47 A and 17 S genes, you would have a total of 64 alleles in the population. Divide this by 4 and you would get the number 16. This would be 50% of the next generation of people produced by the gamete fusions.)

9. Repeat the procedure you followed in step 5 and then step 6. Use Table 2 to record your data.

10. Repeat the procedure you followed in step 7. Record your data and calculations in the spaces provided under Table 2.

Figure 1.

DATA AND OBSERVATIONS

Table 1.

Second Generation		
AA genotypes	AS genotypes	SS genotypes

a. How many A alleles are remaining in the population? _____

b. How many S alleles are remaining in the population? _____

c. What is the total number of alleles in this population? $A + S =$ _____

d. What is the frequency (percent) of the A allele? $\frac{A}{A + S} \times 100 =$ _____

e. What is the frequency (percent) of the S allele? $\frac{S}{A + S} \times 100 =$ _____

Table 2.

Third Generation		
AA genotypes	AS genotypes	SS genotypes

a. How many A alleles are remaining in the population? _____

b. How many S alleles are remaining in the population? _____

c. What is the total number of alleles in this population? $A + S =$ _____

d. What is the frequency (percent) of the A allele? $\frac{A}{A + S} \times 100 =$ _____

e. What is the frequency (percent) of the S allele? $\frac{S}{A + S} \times 100 =$ _____

ANALYSIS

1. **a.** What was the frequency of the A allele in the original population? _____

    b. What was the frequency of the S allele in the original population? _____

    c. What was the frequency of the A allele in the second generation? _____

    d. What was the frequency of the S allele in the second generation? _____

    e. What was the frequency of the A allele in the third generation? _____

    f. What was the frequency of the S allele in the third generation? _____

    g. Explain your findings. _____

    _____

2. Since few people with sickle-cell anemia are likely to survive to have children of their own, why hasn't

    the mutant allele been eliminated by natural selection? _____

    _____

    _____

3. Why is the frequency of the sickle-cell allele so much lower in the United States than in Africa?

    _____

    _____

4. Scientists are working on a vaccine against malaria. What impact would the vaccine have on the

    frequency of the sickle-cell allele in Africa? _____

    _____

    _____

5. In 5 out of 100 million individuals in a population, the allele for Hemoglobin A will spontaneously undergo mutation into the allele for Hemoglobin S. Will such mutations cause major changes in the allelic

    frequencies of a population? Explain your answer. _____

    _____

    _____

FURTHER EXPLORATION

1. Genetic engineering holds the potential for altering genes in human gametes. Write an essay speculating on the impact of genetic engineering on human evolution.
2. Use the library to research a human population that has been geographically or socially isolated over the course of many generations and determine how the allelic frequencies have been affected over time.

Can a Key Be Used to Identify Organisms?

20–1
LAB

INVESTIGATION

▶ Classification is a way of separating a large group of closely related organisms into smaller subgroups. Identification of an organism is easy with a classification system. The scientific names of organisms are based on the classification systems of living organisms. To identify an organism, scientists often use a key. A key is a listing of characteristics, such as structure and behavior, organized in such a way that an organism can be identified.

OBJECTIVES

- Use a key to identify fourteen shark families.
- Examine the method used in making statements for a key.
- Construct your own key that will identify another group of organisms.
- Hypothesize how organisms can be identified with a key.

MATERIALS

PROCEDURE

1. Make a **hypothesis** to describe how sharks can be identified using a key. Write your hypothesis in the space provided.
2. Use Figure 1 as a guide to the shark parts used in the key on page 109.
3. Read statements 1A and 1B of the key. They describe a shark characteristic that can be used to separate the sharks into two major groups. Then study Shark 1 in Figure 2 for the characteristic referred to in 1A and 1B. Follow the directions in these statements and continue until a family name for Shark 1 is determined. For example, to key a shark that has a body that is not kite shaped, and has a pelvic fin, and six gill slits, follow the directions of 1B and go directly to statement 2. Follow statement 2B to statement 3. At statement 3A, identify the shark as belonging to Family Hexanchidae.
4. Continue keying each shark until all have been identified. Write the family name on the line below each animal in Figure 2.
5. Have your teacher check your answers.

HYPOTHESIS

_____

_____

Figure 1.

First dorsal fin

Dorsal (top) side

Second dorsal fin

Caudal fin

Anal fin

Mouth back along underside

Gill slits

Pelvic fin

Pectoral fin

Ventral (bottom) side

DATA AND OBSERVATIONS

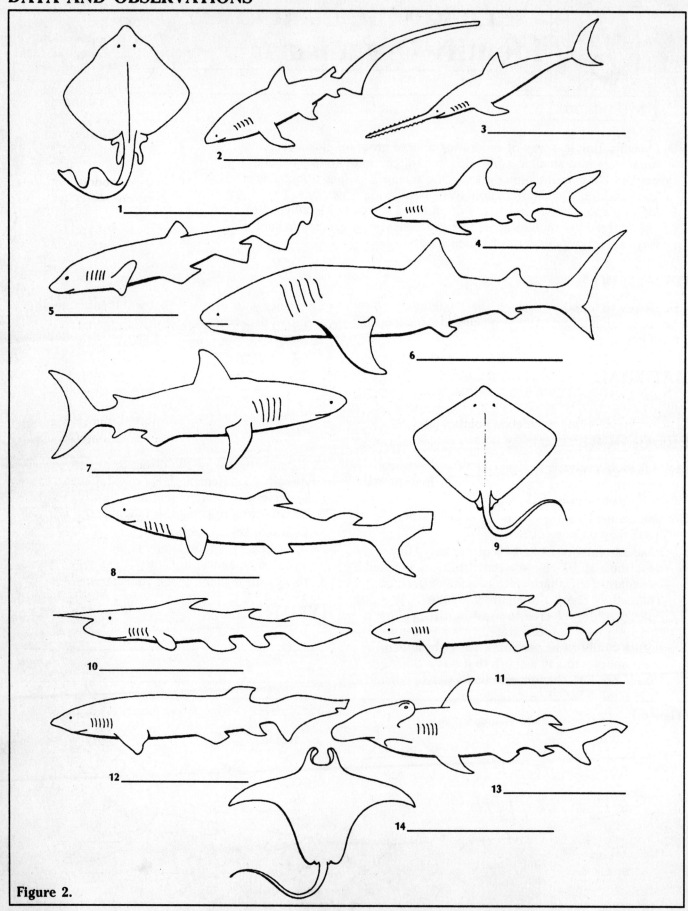

1 _____

2 _____

3 _____

4 _____

5 _____

6 _____

7 _____

8 _____

9 _____

10 _____

11 _____

12 _____

13 _____

14 _____

Figure 2.

108

1. A. Body kitelike in shape (if viewed from above) .... Go to statement 12
    B. Body not kitelike in shape (if viewed from above) .... Go to statement 2

2. A. Pelvic fin absent and nose sawlike .... Family Pristophoridae
    B. Pelvic fin present .... Go to statement 3

3. A. Six gill slits present .... Family Hexanchidae
    B. Five gill slits present .... Go to statement 4

4. A. Only one dorsal fin present .... Family Scyliorhinidae
    B. Two dorsal fins present .... Go to statement 5

5. A. Mouth at front of head rather than back along underside of head .... Family Rhinocodontidae
    B. Mouth back along underside of head .... Go to statement 6

6. A. Head expanded on side with eyes at end of expansion .... Family Sphyrnidae
    B. Head not expanded .... Go to statement 7

7. A. Top half of caudal fin exactly same size and shape as bottom half .... Family Isuridae
    B. Top half of caudal fin different in size and shape from bottom half .... Go to statement 8

8. A. First dorsal fin very long, almost half total length of body .... Family Pseudotriakidae
    B. First dorsal fin length much less than half the total length of body .... Go to statement 9

9. A. Caudal fin very long, almost as long as entire body .... Family Alopiidae
    B. Caudal fin length much less than length of entire body .... Go to statement 10

10. A. Nose with long needlelike point on end .... Family Scapanorhynchidae
    B. Nose without needlelike point .... Go to statement 11

11. A. Anal fin absent .... Family Squalidae
    B. Anal fin present .... Family Carcharhinidae

12. A. Small dorsal fin present near tip of tail .... Family Rajidae
    B. Small dorsal fin absent near tip of tail .... Go to statement 13

13. A. Hornlike appendages at front of shark .... Family Mobulidae
    B. Hornlike appendages not present at front of shark .... Family Dasyatidae

ANALYSIS

1. What is a biological key and how is it used? _____

    _____

2. List four different characteristics that were used in the shark key. _____

    _____

3. a. Which main characteristic could be used to separate Shark 4 from Shark 8? _____

    _____

    b. Which main characteristic could be used to separate Shark 4 from Shark 7? _____

    _____

4. Prepare your own key for the five fish in Figure 3. Use the same format as on page 109. The family names to be used are the numbers I, II, III, IV, and V. Your key should correctly use traits that will lead to each fish family. To help you get started, the first statements are given. Statement 1 divides the five fish into two main groups, based on body shape. Next, choose another characteristic that will divide the fish not having a tubelike body into two groups. Continue to choose characteristics that will separate a group into smaller groups. Write your key in the space below.

      1. A. Fish with long tubelike body
           B. Fish with body shape not tubelike

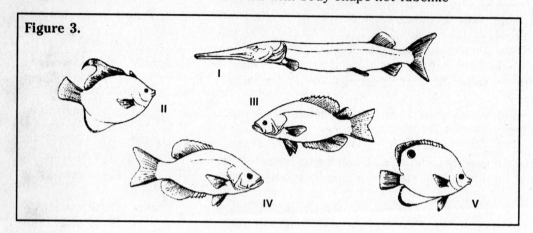

Figure 3.

Key

1. A. _____

    B. _____

2. A. _____

    B. _____

3. A. _____

    B. _____

4. A. _____

    B. _____

CHECKING YOUR HYPOTHESIS

Were you able to identify sharks using the key? Did your **hypothesis** describe the method correctly? _____

_____

FURTHER INVESTIGATIONS

1. Exchange keys with one of your classmates. Work through it to identify the fish. Is the key correct?

2. The library will have many books that include simple keys to different plants and animals, as well as to rocks, fossils, and stars. Select a book that includes keys to local plants or animals. Take a walk and practice keying some of the organisms that live around you.

Virus Replication

EXPLORATION

▶ Viruses are very successful at invading the cells of organisms. Even though the protein coat of the virus has antigens that trigger an organism's immune system to fight off the invader, many viruses can change their coats (called capsids), or hide inside the coats of other viruses. After a virus particle attaches itself to the outside of a cell, a viral enzyme damages the cell membrane and allows the virus to enter. Some viruses inject their nucleic acid into the cell and leave their coat outside the cell; other viruses still have their coats when they enter the host. Once viruses are inside a cell, they insert virus genes into the host DNA and take over the cell's genetic machinery. The viral genes direct replication and cause the cell to make many new copies of viral genes. At the same time, the cell's protein synthesis machinery is directed to make many new viral coats and enzymes. The virus particles are assembled, and the new viruses escape from the cell either by exocytosis or by bursting out of the cell.

OBJECTIVES

- Trace the steps of viral reproduction in cells.
- Describe each step of viral reproduction.
- Construct an analogy for viral replication in cells.

MATERIALS

glue
scissors
photocopy of viruses and virus parts

PROCEDURE

1. Study Figures 2 and 3 so that you become familiar with virus structure and with the structure of the model viruses used in this Exploration.
2. Cut out the drawings representing viruses and virus parts that your teacher will give you.

Each of the drawings corresponds to one of the stages of virus replication listed at the end of this Procedure.

3. Using Figure 3 as a key, identify the drawing on your photocopy that represents each stage of virus replication. Begin with stage 1 and identify all ten drawings.

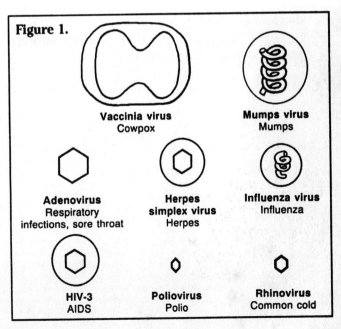

Figure 1.

Vaccinia virus
Cowpox

Mumps virus
Mumps

Adenovirus
Respiratory
infections, sore throat

Herpes
simplex virus
Herpes

Influenza virus
Influenza

HIV-3
AIDS

Poliovirus
Polio

Rhinovirus
Common cold

Figure 2.

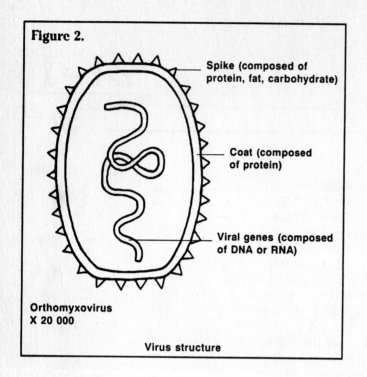

Spike (composed of protein, fat, carbohydrate)

Coat (composed of protein)

Viral genes (composed of DNA or RNA)

Orthomyxovirus
X 20 000

Virus structure

Figure 3.

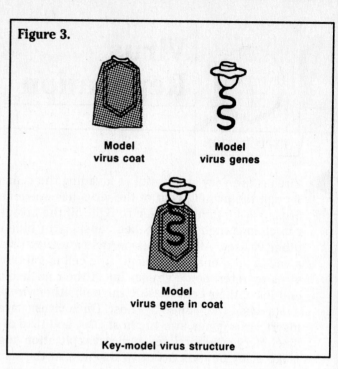

Model virus coat

Model virus genes

Model virus gene in coat

Key-model virus structure

4. Sequence the stages in the virus life cycle. Label the small tab of each drawing with the number of the stage of replication it represents, using the numbers in the "Stages of Virus Replication" list.

5. Examine Figure 4 in Data and Observations. This is a diagram demonstrating the ten stages of virus replication within a cell. The location of each stage is labeled. Copy your numbers from the drawings onto this diagram in the spaces provided.

6. Place the drawings, each representing a stage of replication, under its proper label on the diagram in Figure 4. When complete, have your diagram checked by your teacher for accuracy.

7. Once the drawings are in their proper position on the diagram, glue them in place.

Stages of Virus Replication

1. The virus is attached to the cell membrane.

2. A viral enzyme damages the host cell membrane; the virus invades the cell and is carried inside.

3. The protein coat of the virus is removed.

4. Viral genes are activated as they become a part of the cell's machinery.

5. a. The activated virus directs replication of viral genes.

   b. The activated virus directs translation, the making of new viral coats.

6. a. New genes are completed.

   b. New coats are completed.

7. The new viruses are assembled. Genes are inserted into the protein coats.

8. The new viruses escape from the cell.

DATA AND OBSERVATIONS

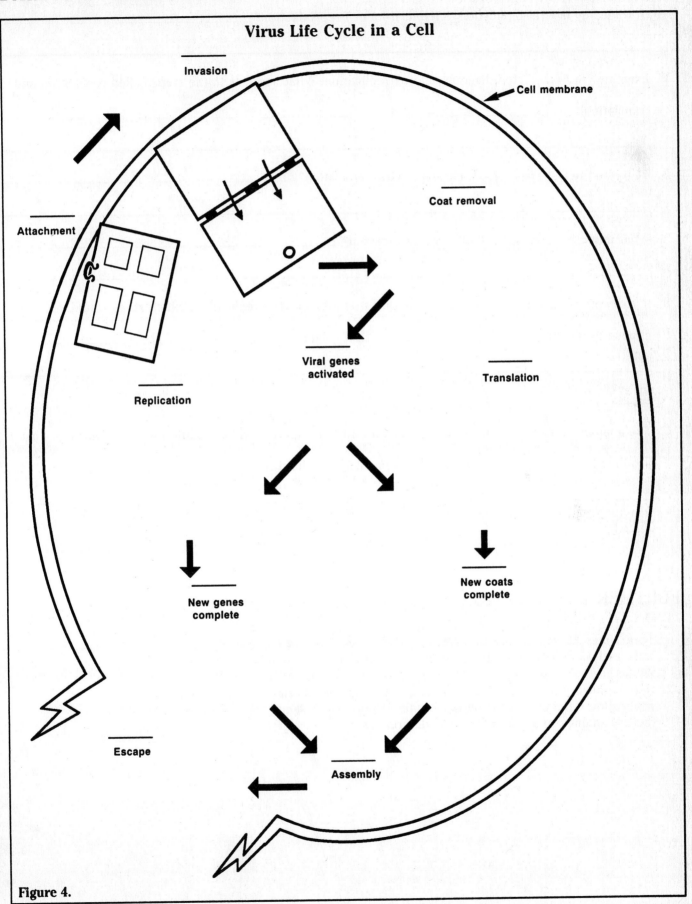

Figure 4.

ANALYSIS

1. In what way is viral replication different from cell reproduction? _____

_____

2. Examine your completed diagram of virus replication. What happens in the steps called replication and

translation? _____

_____

3. Describe the last stage before escape in viral replication? _____

_____

4. What might happen to a cell after viruses leave it? _____

_____

5. If you were a scientist developing a drug that would block viral replication, which steps would you

choose to block? _____ Explain. _____

_____

_____

6. The analogy you are working with in this Exploration compares viral replication with the making of a

product in a factory. In what ways does the analogy not hold true? _____

_____

_____

_____

FURTHER EXPLORATIONS

1. Read the article "Viruses Revisited," by Robin Marantz Henig, *New York Times Magazine*, Nov 13, 1988, pp. 70–72. Write an essay or have a class debate about the pros and cons of altering human genes.
2. Use the library to research retroviruses. Determine how their replication differs from that shown in this Exploration. Change the factory analogy to account for these differences.

How Are Bacteria Affected By Heat?

21–2

LAB

INVESTIGATION

▶ Bacteria comprise much of the dry weight of feces and can contaminate lakes, streams, and ground water through untreated sewage or drainage from cattle farms. Because of this potential for contamination, water is usually treated with chlorine. But milk, another potential source of bacteria, cannot be treated with chlorine. Raw milk—which is milk taken directly from the cow, then filtered and consumed—was once a cause of illness. But a technique for destroying various bacteria in milk by heating was developed in the late 1800s by the chemist Louis Pasteur. Modern-day pasteurization consists of heating milk to 145 degrees Fahrenheit (about 63° C) for 30 minutes. This kills many of the bacteria present in the milk. The milk is then cooled and kept at a temperature below 50 degrees Fahrenheit to prevent remaining bacteria from spoiling the milk. There is a "shelf life" for milk, after which it can spoil due to a variety of bacteria. In this lab you will investigate how heat affects *Escherichia coli* (or *E. coli*), one of the types of bacteria found in milk.

OBJECTIVES

- Practice sterile techniques for handling bacterial cultures and inoculating agar plates.
- Hypothesize how heat affects the growth of bacteria.

- Compare the effects of different temperatures on the growth of bacteria.

MATERIALS

sterile petri dishes containing hardened bacto-methylene blue agar (2)
sterile cotton swabs (10)
broth culture of *Escherichia coli*
test tube containing 10 mL of *E. coli* culture
test tube with 10 mL of milk inoculated with *E. coli* and autoclaved

hot plate
water bath
thermometer
test tubes (3)
test-tube holder
test-tube rack
sterile pipette
wax marking pencil
incubator

refrigerated milk
paper towels (2)
tape, masking
250-mL beaker with 100 mL of alcohol
25-mL graduated cylinder
laboratory apron
goggles

PROCEDURE

CAUTION: *Do not touch eyes, mouth, or any other part of your face while doing this lab. Wear your laboratory apron and goggles. Wash your work surface with disinfectant solution, using a paper towel, both before and after doing the lab.*

Part A. Preparing the Treatments

1. Prepare a hot water bath by heating the water to 63°C. One person should monitor the hot plate and thermometer while the rest of the team does the following activities.

2. Place the two petri dishes containing bacto-methylene blue agar on the counter in front of you. Turn the dishes upside down, being careful not to allow them to open. With a wax marking pencil, divide each dish into thirds as shown in Figure 1. On dish A, number the sections 1, 2, and 3. On dish B, number the sections 4, 5, and 6. These numbers refer to the following treatments:

   1. Control
   2. Refrigerated milk
   3. Refrigerated milk inoculated with *E. coli*
   4. Refrigerated milk inoculated with *E. coli* and autoclaved
   5. *E. coli* culture
   6. Refrigerated milk inoculated with *E. coli* and heated on the hot plate

3. Label 3 test tubes with numbers 2, 3, and 6 to correspond to the treatments listed in step 2, and place them in a test-tube rack. To each test tube, add 10 mL of refrigerated milk from the same milk source.

4. Label with the number 4 the test tube of milk that has been inoculated and autoclaved. Label with the number 5 the test tube containing 10 mL of *E. coli* culture.

5. With a sterile pipette, add 1 mL of *E. coli* culture to test tubes 3 and 6. Place the contaminated pipette in the container of alcohol on the counter. Carefully roll the test tubes between your palms to mix the *E. coli* into the milk. Do not tip the test tubes. Put the test tubes back into the rack after each mixing.

6. Using a test-tube holder, place test tube 6 in the hot water bath. Insert the thermometer into the test tube, and heat the tube for 30 minutes or until the milk is 63°C. Using the test tube holder, remove test tube 6 from the heat and place it into the test-tube rack. Place the contaminated thermometer in the container of alcohol.

Part B. Inoculating the Agar Plates

1. Dip a sterile cotton swab into test tube 2. Press the swab against the inside of the tube to remove excess milk. Lift the lid off petri dish A, and swab the surface of the agar in section 2 using a light back-and-forth motion. Do not gouge the agar. Replace the lid. Discard the swab in the beaker of alcohol.

2. Dip a sterile cotton swab into test tube 3. Press the swab against the inside of the test tube to remove excess milk. Lift the lid off petri dish A and touch the agar surface in section 3 with the cotton swab. Discard the swab in the beaker of alcohol. **CAUTION:** *Do not let the cotton end of the swab touch the outside of the dish or any other surface. If any of the liquid spills from test tube 3, call your teacher immediately.* Using another sterile swab, lightly spread the culture in section 3 using a tight s-shaped pattern. Replace the lid on the petri dish and discard the cotton swab. Secure the lid of the petri dish with tape, label the tape with the name of your team, and turn the dish upside down.

3. You will now inoculate petri dish B. Dip a cotton swab into test tube 4. Press the swab against the inside of the tube to remove excess milk. Lift the lid, and touch the agar surface in section 4 with the cotton swab. Discard the swab in the alcohol. Using another sterile swab, spread the treatment throughout section 4. Replace the lid on the petri dish and discard the swab.

4. Repeat step 3 with test tubes 5 and 6, using sterile cotton swabs to apply and spread cultures in sections 5 and 6 of petri dish B. Exercise special caution when inoculating from test tube 5. When you have completed the inoculations, secure the lid of the petri dish with tape. Label the tape with the name of your team, and turn the dish upside down.

Figure 1.

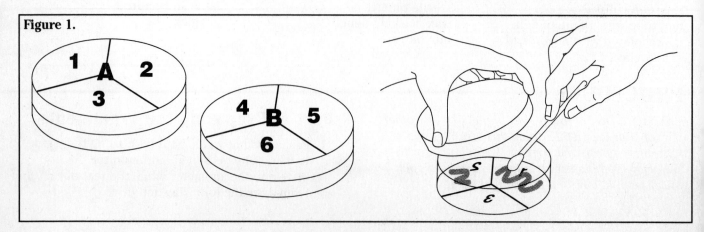

5. Place both petri dishes in the incubator overnight at 35–37°C. Incubate the dishes upside down to prevent moisture from accumulating on the surface of the agar. Wash your hands thoroughly.

6. Make a hypothesis to describe how heat will affect the bacteria.

7. On the following day or at the next lab period, examine the two petri dishes and record your observations in Table 1. Any *E. coli* colonies will appear dark and button-like, often with concentric rings and a greenish metallic sheen. Other bacteria may also be present but will not have the same appearance.

HYPOTHESIS

_____

_____

_____

DATA AND OBSERVATIONS

Table 1.

Treatment	Observations
#1 Control	
#2 Refrigerated milk	
#3 Refrigerated milk + *E. coli*	
#4 Refrigerated milk + *E. coli*, autoclaved	
#5 *E. coli*	
#6 Refrigerated milk + *E. coli*, heated	

ANALYSIS

1. What was the purpose of the control section on the agar? _____

_____

_____

2. What was the function of section 2 on petri dish A? _____

_____

_____

3. What was the function of the pure *E. coli* culture? _____

_____

_____

4. Explain your results using the milk inoculated with *E. coli* test tube. _____

_____

_____

5. Explain your results with test tube 6. _____

_____

_____

6. Explain your results with the autoclaved milk. _____

_____

_____

CHECKING YOUR HYPOTHESIS

Was your **hypothesis** supported by your data? Why or why not? _____

_____

FURTHER INVESTIGATIONS

1. Obtain some ground hamburger from the grocery store and design an experiment to test for the presence of *E. coli* in the meat.
2. Go to the library and research the use of *E. coli* in Recombinant DNA experiments. When *E. coli* was first used for this type of research, many people feared its use. Explain why people thought that and why they should not have been worried.

Observing Algae

EXPLORATION

22–1
LAB

Algae are simple plant-like organisms belonging to the Kingdom Protista. Algae contain chlorophyll and are therefore capable of carrying out photosynthesis. The chlorophyll is contained in green organelles called chloroplasts. Specialized structures within the chloroplasts of some algae are important for starch formation and storage. These structures are known as pyrenoids. The cellular organization of algae varies. Some are unicellular, while others are multicellular. The multicellular algae may be in the form of seaweeds, threads called filaments, or groups called colonies. Colonies vary in size from three or four cells to thousands of cells, depending on the species.

OBJECTIVES

- Prepare wet mounts of algae for microscopic examination.
- Locate and identify chloroplasts, holdfasts, and pyrenoids of various species of algae.
- Distinguish among unicellular, filamentous, and colonial forms of organization.

MATERIALS

microscope
microscope slides (6)
coverslips (6)
droppers (6)
iodine solution
paper towel
laboratory apron

Living specimens of
Closterium
Oedogonium
Scenedesmus
Synedra
Ulothrix
Volvox

PROCEDURE

Part A. Unicellular Algae

1. Remove a small sample of *Synedra* from the specimen jar with a dropper. Place a drop of the culture on a clean microscope slide and cover with a coverslip.

2. Locate the alga under low power of the microscope. Look for an elongated cell as shown in Figure 1.

3. Switch to high power and note the cell wall. Label the cell wall of the cell in Figure 1.

4. Record in Table 1 the cellular organization of this alga.

5. Prepare a wet mount, as in step 1, of *Closterium* using a clean slide and coverslip.

Figure 1.

Synedra

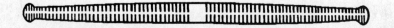

6. Observe first under low power and look for a yellow-green crescent-shaped cell. The cell has a constriction in its middle called an isthmus.

7. Add a drop of iodine solution to the edge of the coverslip. Using a piece of paper towel as in Figure 2, carefully draw the iodine across the slide. The iodine will stain pyrenoids blue-black.

8. Switch to high power and examine the alga again. Label the cell wall, pyrenoid, and isthmus in Figure 3.

9. Record in Table 1 the cellular organization of this alga.

Part B. Filamentous Algae

Oedogonium is a pale yellow-green alga. It may be floating or attached to surfaces in the specimen jar by means of a holdfast cell located at its base.

1. Locate the algae in the jar and remove a small sample with a clean dropper.

2. Place the specimen on a clean slide and cover with a coverslip.

3. Observe the alga under low power. Look for an unbranched filament as shown in Figure 4.

4. Stain the alga with iodine using the technique in step 7 of Part A.

5. Observe the stained alga again under low power. Note the net-like chloroplast and the many pyrenoids.

6. Switch to high power and observe the cell wall, chloroplast, pyrenoid, and holdfast cell. Label these structures in Figure 4.

7. Estimate the number of individual cells that compose the filament and record your observation in Table 1. Record the cellular organization of this alga.

Ulothrix is an alga usually found attached to the sides of a culture jar or to substances in the jar by its holdfast cell.

8. Locate the alga and remove a small sample from the jar with a clean dropper . Make a wet mount of this alga with a clean slide and coverslip.

9. Look for an unbranched filament as shown in Figure 5. Observe it first under low power. Note the ribbon-shaped chloroplast.

10. Add a drop of iodine and follow the procedure as in step 7 of Part A.

11. Switch to high power. Find the cell wall, chloroplast, pyrenoid, and holdfast. Label these structures in Figure 5.

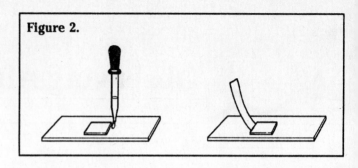

Figure 2.

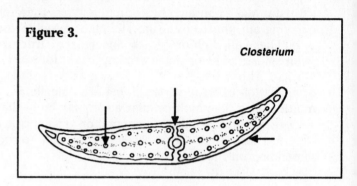

Figure 3.

Closterium

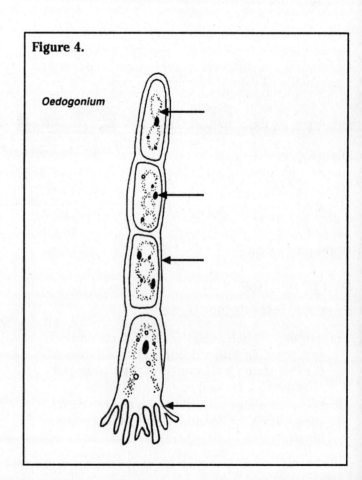

Figure 4.

Oedogonium

12. Estimate the number of individual cells that compose the filament and record your observation in Table 1. Record this alga's cellular organization.

120

Figure 5.

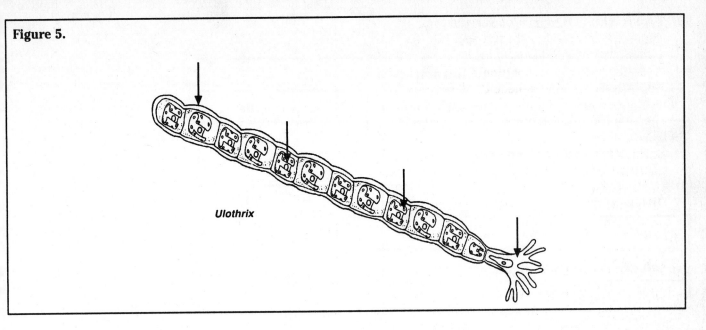

Ulothrix

Part C. Colonial Algae

Volvox can be seen as tiny green spheres spinning through the water. A *Volvox* colony consists of a single layer of 500 to 60 000 cells arranged in a hollow sphere.

1. Locate and remove a colony of *Volvox* using a clean dropper. Make a wet mount of this alga using a clean slide and coverslip.

2. Look for a large, green spherical colony as shown in Figure 6. Observe first under low power.

3. Switch to high power. Locate a daughter colony inside the sphere. Daughter colonies are groups of cells within the colony that eventually break away to form new colonies. Label the flagella and daughter colonies in Figure 6.

4. Estimate the number of individual cells that make up the *Volvox* colony and record the number in Table 1. Record this alga's cellular organization.

Scenedesmus can be found floating throughout the water in the specimen jar.

5. Prepare a wet mount of *Scenedesmus*. Be sure to use a clean dropper, slide, and coverslip.

6. Look for a colony of oval or crescent-shaped cells with short spines as shown in Figure 7. Observe first under low power.

7. Switch to high power and find the cell wall, chloroplasts, and spines. Label these structures in Figure 7.

8. Estimate and record the number of cells that make up this colony. Record this information in Table 1 and record this alga's cellular organization.

Figure 6.

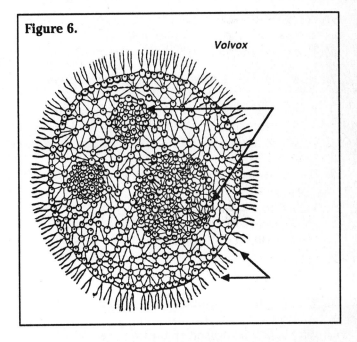

Volvox

Figure 7.

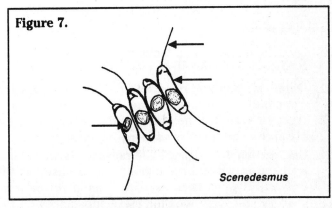

Scenedesmus

9. When all observations have been recorded, clean up and wash all your equipment.

DATA AND OBSERVATIONS

Table 1.

Algal Characteristics		
Algal specimen	Cellular organization	Number of cells
Synedra		
Closterium		
Oedogonium		
Ulothrix		
Volvox		
Scenedesmus		

ANALYSIS

1. Which algae were unicellular? _____

2. How many cells were present in *Scenedesmus*? _____ Did all of the specimens have the same

   number of cells? _____ Explain. _____

   _____

3. Why was it difficult to count the number of cells in *Volvox*? _____

   _____

   What structure(s) allow this colony to move? _____

4. Which algae possessed pyrenoids? _____

5. What is the function of pyrenoids? _____

6. Why was iodine used to locate pyrenoids? _____

   _____

FURTHER EXPLORATIONS

1. Obtain a sample of pond water. Prepare a slide for microscopic examination and locate algae. Look carefully since many different types of algae will be present. Make a drawing of each type of algae observed and identify as many as possible.

2. Carageenan and alginate are polysaccharides derived from the cell walls of brown and red algae. They are used as thickeners in ice cream and some cheeses, as well as in cosmetics. Visit a local grocery store and look for these substances on the labels of various products. Make a list of the products and the algae-derived substances they contain.

How Can Digestion Be Observed in Protozoans?

22-2

LAB

INVESTIGATION

▶ In the process of physical digestion, ingested food is broken down into small particles. Chemical digestion is a complex series of chemical reactions, each of which is controlled by a different enzyme. Each enzyme works best under a different set of conditions. As the pH of the food changes during the digestion process, enzymes with different optimal pH ranges become active. As food is broken down in the food vacuoles of *Paramecium* or other ciliates, the pH of the vacuole contents changes. If the vacuole contents are stained with a pH indicator, the progress of digestion can be followed.

OBJECTIVE

- Prepare an indicator-stained food source for *Paramecium*.
- Observe the feeding behavior of *Paramecium*.

- Hypothesize the changes in the pH of the food vacuole as food is digested by *Paramecium*.

MATERIALS

small beaker
cream or whole milk
wood splint
Congo red indicator
microscope slide

coverslip
toothpick
petroleum jelly
droppers (3)
methyl cellulose

culture of *Paramecium*
microscope
clock or stopwatch

PROCEDURE

Part A. Preparing Materials

1. Obtain a small amount of milk in a beaker.

2. Use a wood splint to transfer a few grains of Congo red to the milk. Stir the mixture with the wood splint. Congo red is used to stain structures that contain fats or oils. Congo red is also an indicator of pH. The indicator is red at a pH of 5.0 or greater. It is blue at a pH of 3.0 or less.

3. Make a **hypothesis** to explain how the color of the indicator will change as food is digested. Write your hypothesis in the space provided.

4. Use a toothpick to outline a square of petroleum jelly on the center of a microscope slide. The square should be the size of a coverslip.

5. Use a dropper to place one drop of methyl cellulose in the center of the square. This substance will slow the movement of the paramecium.

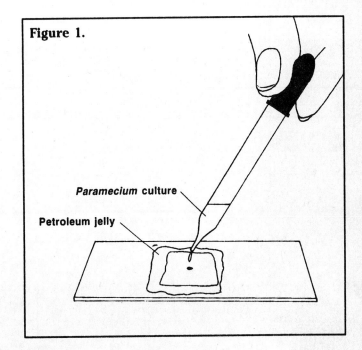

Figure 1.

Paramecium culture

Petroleum jelly

6. Use a clean dropper to add one drop of *Paramecium* culture to the slide.
7. Use a clean dropper to add one drop of stained milk to the slide.
8. Add a coverslip to the slide.

Part B. Observing Digestion

1. Observe your slide under low power. Locate a paramecium that is near some of the stained fat droplets.
2. Change to high power, and look for food vacuoles in the cell. Record the color of the contents of the food vacuoles. This is an indication of pH. Write your observations of the paramecium's behavior and movement of food vacuoles in Table 1.

3. In the space provided, make a drawing of the paramecium. Label the oral groove and any food vacuoles that you see. Use your textbook for more information on the structures of *Paramecium*.
4. Continue to observe the paramecium. Every ten minutes for one-half hour, record the color of the food vacuole contents, your observations, and make a drawing of the paramecium. Be careful to draw the positions of the food vacuoles accurately.

HYPOTHESIS

_____

_____

DATA AND OBSERVATIONS

Table 1.

Time (min)	Color of food vacuole contents	Other observations
0		
10		
20		
30		

Drawings of *Paramecium*

0 min.	10 min.	20 min.	30 min.

ANALYSIS

1. What color was the mixture of Congo red and milk? _____ What does this tell you about the

   pH of the milk? _____

   _____

   _____

2. When you first observed the paramecium, did it contain any stained food vacuoles? _____

   If not, when did you first see stained food vacuoles? _____

   _____

   _____

3. What color was the contents of the food vacuoles when the paramecium first began to ingest the

   milk? _____ What does this indicate about the pH of the food at the beginning of

   digestion? _____

   _____

   _____

   _____

4. At what time did the stained food vacuole change color? _____

What color did it become? _____ What does this indicate about the pH of the food vacuole

contents as digestion progresses? _____

_____

_____

_____

5. Did the stained food vacuole change color a second time? _____ What does the change or lack of

change indicate about the pH of the food vacuole contents late in digestion? _____

_____

_____

_____

6. Thymol blue is an indicator that is red at a pH of 1.2 or less. It is yellow between 2.8 and 8.2. It is blue at 10.0 or higher. How might repeating the experiment with this indicator give additional information about

the pH of food as it is digested in the food vacuole? _____

_____

_____

_____

CHECKING YOUR HYPOTHESIS

Was your **hypothesis** supported by your data? Why or why not? _____

_____

FURTHER INVESTIGATIONS

1. Repeat the experiment with another ciliate, such as *Blepharisma*, or *Bursaria*.
2. Repeat the experiment with the addition of a drop of *Didinium*. This ciliate is a predator that feeds on *Paramecium*. See if the *Didinium* shows color changes as it digests the stained *Paramecium*.

Identification of Common Molds

23-1

LAB

EXPLORATION

Fungi and molds are classified mainly by means of their reproductive structures. Zygomycota produce spores in a sporangium. The sporangium of the common black bread mold, *Rhizopus*, resembles a black lollipop. Ascomycota produce spores in a sac known as an ascus. They may also produce spores in conidia. The conidia of some fungi resemble either bony fingers or dandelions. Basidiomycota produce spores in a club-shaped structure called a basidium. Basidia are arranged within gills in the cap of the fungus. Deuteromycota have only an asexual reproductive phase, but they may be reclassified in one of the other divisions of fungi if a sexual reproductive phase is discovered.

OBJECTIVES

- Observe the color and appearance of fungus colonies.
- Examine the reproductive structures of various types of fungi.

- Draw, label, and identify the four different types of fungi.

MATERIALS

cultures of fungi (4)
cellophane tape

microscope
microscope slides (4)

PROCEDURE

1. Examine a fungus culture and record in Table 1 the color and appearance of the colony.

2. Gently touch the reproductive surfaces of one of your fungus samples with the adhesive side of a piece of cellophane tape. Do not crush the reproductive structures. Your teacher will demonstrate where on the fungus spores are located.

3. Carefully place the tape, adhesive side up, on a microscope slide.

4. Observe the reproductive structures under low power on the microscope.

5. Draw and label the structures that you see in the space provided in Table 1.

6. Identify the division of fungi by using your textbook or any other available reference materials.

7. Repeat this procedure for each of the fungi provided.

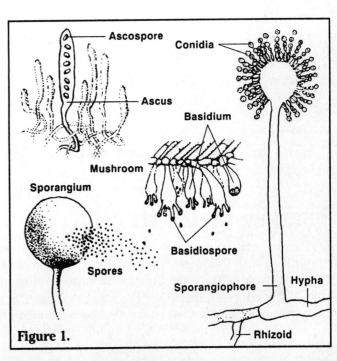

Figure 1.

DATA AND OBSERVATIONS

Table 1.

Identification of Common Molds			
Fungus	**Appearance of colony**	**Drawings of structures**	**Division**

ANALYSIS

1. Write a general description of the spores seen in this Exploration. Include relative numbers, shape, number of cells, and size. _____

_____

_____

2. Why are different fungi identified by using entire reproductive structures rather than by using only spores? _____

3. List the four divisions of fungi and give an example of each division. _____

_____

FURTHER EXPLORATIONS

1. Bring some fungus specimens from home, and use the same procedure to identify their division. Good places to look for fungi are cheese, fruit, and the spaces between tiles in the shower.

2. Using spores from one of the fungus cultures in this Exploration, inoculate a petri dish containing nutrient medium. Study the growth of the fungus and determine its life cycle.

Which Foods Can Bread Mold Use for Nutrition?

23-2
LAB

INVESTIGATION

▶ Molds are organisms that belong to Kingdom Fungi. Molds therefore have characteristics different from organisms in the protist and moneran kingdoms. First, fungi are usually multicellular and often can be seen without a microscope. Fungus cells have nuclei. Fungi contain no chlorophyll and so are not able to make their own food. Organisms in this kingdom are either saprophytes or parasites. Saprophytic fungi obtain their food by feeding off once-living material, whereas parasitic fungi obtain their food by feeding off living material.

OBJECTIVES

- Examine bread mold by using a hand lens or stereomicroscope.
- Hypothesize the conditions a bread mold needs for growth.

- Determine if bread mold can use a variety of food sources.
- Determine if bread mold needs moisture.

MATERIALS

hand lens or stereomicroscope
petri dish containing living bread mold
small jars with covers (6)

water
labels or wax marking pencil
cardboard from a box

dehydrated potato flakes
raisins
cotton swabs (6)

PROCEDURE

Part A. Observing Bread Mold

1. Use a hand lens or stereomicroscope to observe bread mold growing in a petri dish.
2. Note the following parts.
   (a) Tiny black round structures called sporangia are located on top of long stalks. Reproductive spores form in these structures.
   (b) Clear stalklike parts called sporangiophores hold up the sporangia.
   (c) A mass of threadlike structures spread along the surface of the mold's food supply. These structures are called hyphae, and they enable the mold to secure its food.
   (d) Hyphae that penetrate the food supply are rhizoids.
3. Label the following parts on Figure 1: *sporangium, sporangiophore, rhizoid, hypha*.

Part B. Testing Different Possible Sources of Food

1. Label small jars with your name and the numbers one to six.
2. Prepare the jars as follows:
   Jar 1: Cover the bottom with dehydrated (dry) potato flakes.
   Jar 2: Cover the bottom with potato flakes and add enough water to make a paste.
   Jar 3: Cover the bottom with dry raisins.
   Jar 4: Cover the bottom with dry raisins and add enough water to soak them thoroughly.
   Jar 5: Stand a small piece of cardboard upright in the jar.
   Jar 6: Stand a small piece of cardboard upright in the jar and add a small amount of water to the bottom of the jar.
3. Rub a damp cotton swab over the surface of the bread mold studied in Part A. Rub the swab over the surface of the contents of jar 1. Repeat this procedure for the remaining five jars, using a different swab for each jar.

4. Cover the jars and let them sit for several days at room temperature.

5. Make a **hypothesis** to describe the conditions a mold needs to grow. Write your hypothesis in the space provided.

6. After several days, examine each jar for the presence or absence of bread mold growth.

7. Record your observations in Table 1.

8. Dispose of your jars according to your teacher's instructions.

HYPOTHESIS

_____

_____

DATA AND OBSERVATIONS

Figure 1.

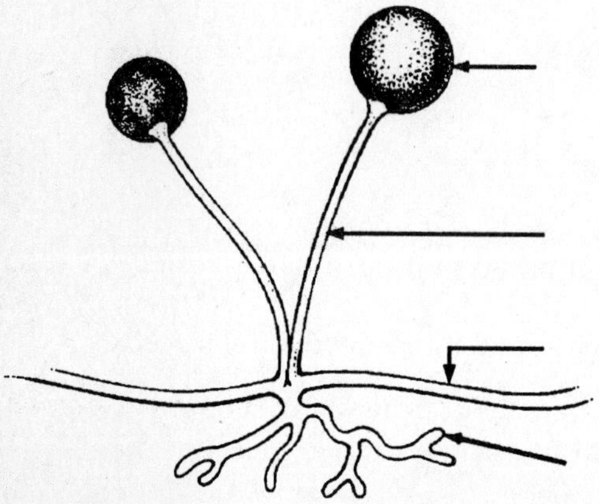

Table 1.

	Results of Mold Growth	
Jar	Contents	Mold growth?
1	Dry potato flakes	
2	Wet potato flakes	
3	Dry raisins	
4	Wet raisins	
5	Dry cardboard	
6	Wet cardboard	

ANALYSIS

1. In which jars did bread mold grow? _____

2. What growth conditions were supplied to the fungus in the jars in which you observed bread mold growth? _____

3. How does moisture affect the growth of bread mold? _____

CHECKING YOUR HYPOTHESIS

Was your **hypothesis** supported by your data? Why or why not? _____

_____

_____

FURTHER INVESTIGATIONS

1. Set up two additional jars with the same contents as the jar that had the best growth of fungus. Label the jars 7 and 8. Place one in a dark closet and leave one in the light. Examine both after several days.

2. Design experiments to show the effects of temperature and chemicals such as table salt on the growth of bread mold. Carry out the experiments under your teacher's supervision.

130

Name _____ Date _____

Lichens

EXPLORATION

23–3

LAB

▶All organisms classified in the same kingdom have some
similarities. Thus, lichens, which contain a fungus, have some
characteristics in common with other fungi. But lichens are unique
in that they are a combination of two different organisms: a green,
unicellular organism and a fungus. These two organisms exist
together in a symbiotic relationship. Symbiosis is the living
together in close association of two different organisms. In lichens,
the green organism is either a cyanobacterium or a green alga.
Through photosynthesis, this organism provides food for the
lichen. The green organism is surrounded by the mycelium of the
fungus, which provides moisture, protection, and possibly some
minerals to the lichen. Because both organisms benefit from their
association, a lichen is an example of mutualism.

OBJECTIVES

- Determine the specific lichen type of three lichen
  samples.
- Diagram the macroscopic appearance of three
  lichen samples.

- Observe and diagram the microscopic appearance
  of a typical lichen.

MATERIALS

microscope
microscope slide

coverslip
dropper

water
Cladonia (reindeer moss)

three lichen samples
labeled A, B, and C

PROCEDURE

Part A. Lichen Types and Appearance

1. Examine the lichen samples provided. Each
   sample is labeled with the letters A, B, or C for
   identification.

2. Using your observations and the information
   given in Table 1, complete Table 2.

Table 1.

Lichen Type		
Lichen type	**Description**	**Appearance**
Crustose	Flat or crusty, forms a mat on rocks or bark	(Natural size)
Foliose	Spreading leaflike lobes, has a papery appearance	(Natural size)
Fruticose	Stalked vertical growth or branching, may appear hairlike or as branching threads	(2X natural size)

Part B. Microscopic Appearance of Lichens

1. Make a wet mount of a small piece of *Cladonia* (the size of a dime). Before adding the coverslip, mash the lichen with the eraser end of a pencil. Add one or two more drops of water if necessary.

2. Observe the lichen under low and high powers of your microscope. Look for small round green cells and long thin colorless hyphae. Diagram and label the parts of the lichen in the space provided in Data and Observations.

DATA AND OBSERVATIONS

Table 2.

Lichen Characteristics			
Sample	General description	Diagram	Type
A			
B			
C			

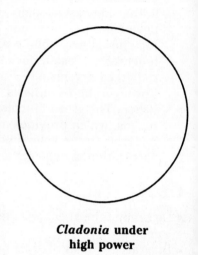

Cladonia under high power

ANALYSIS

1. Define

   a. lichen. _____

   b. symbiosis. _____

2. How are lichens symbiotic? _____

_____

3. Microscopically, how do the green organism and fungus of a lichen differ? _____

_____

FURTHER EXPLORATIONS

1. Examine a wet mount of one of the other lichen samples and compare it with what you saw in the *Cladonia* sample.

2. Lichens are known as pioneer organisms. Use the library to find out how lichens help prepare the way for other organisms in barren areas.

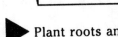

Roots and Stems

24–1
LAB

Plant roots and stems are living tissues that show a close relationship between their function and structure.

Plant roots absorb water and minerals from the soil and transport them upward through the stem. Roots also store food and anchor plants into the soil.

One of the major functions of stems is support. Stems also are pathways for transporting food and water. Woody stems can be used to determine the age of trees.

OBJECTIVES

- Identify and label root tissues.
- Describe the functions of root tissues.

- Identify and label the stems of one-, two-, and three-year-old trees.
- Explain the functions of stem tissues.

MATERIALS

cross section of parsnip root
single-edged razor blade
microscope slide

iodine stain
dropper
coverslip

microscope
laboratory apron

PROCEDURE

Part A. Root Anatomy

Your teacher has prepared cross sections of parsnip roots as shown in Figure 1.

1. Examine your cross section. You can see two different regions. These regions are marked A and B in Figure 1.
2. Prepare your root tissue for microscopic viewing as shown in Figure 2 below.
   A. Cut root section in half. **CAUTION:** *The blade is sharp. Cut away from your fingers.*
   B. Cut the root section in half again in the opposite direction to form a longitudinal section as shown.

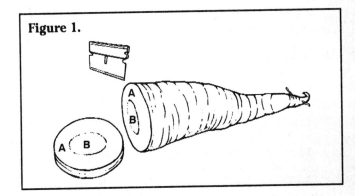

Figure 1.

C. Slice off a thin section from one edge of the remaining root section.

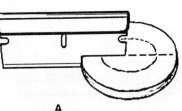 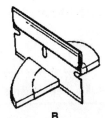

Figure 2.

A   B   C

3. Place your thin section on a microscope slide. With a dropper, add iodine stain. Then cover with a coverslip. **CAUTION:** *Iodine is a poison. If spillage occurs, immediately wash with water and call your teacher.*

4. Examine the slide under the low-power magnification of your microscope. Slowly, move the slide across the field of view so that you observe all tissues of the root section. You will observe two different kinds of cells. One kind of cell looks like a railroad track. This cell is a transporting cell. The other kind of cell is rounded or squared and is packed closely together in groups. This cell is a storage cell. Storage cells are much smaller than transporting cells. Starch grains in the storage cells will appear blue from the iodine stain.

5. Draw in Table 3 a few of the two different kinds of cells in regions A and B as seen under low power.

6. Complete Table 4 by giving the functions for the two regions. Note that the table gives the names for these regions.

7. Look at Figure 3, which shows a thin section of parsnip root seen under low power. The thin section was prepared as shown in Figure 4. This thin section is different from the first one you observed. This one is a cross section.

8. Identify and label the root structures in Figure 3 by using the descriptions in Table 1. Areas A and B (cortex and central cylinder) are labeled.

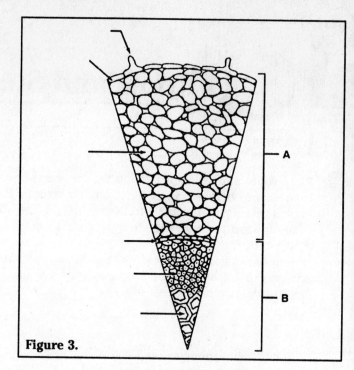

Figure 3.

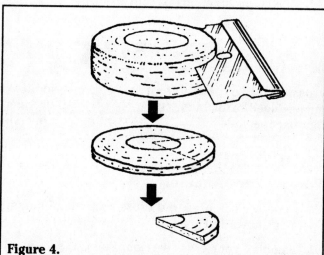

Figure 4.

Table 1.

Root Parts and Functions	
Tissue	**Description**
Xylem	Thick-walled cells in Area B at tip end of slice; part of central cylinder, transports water
Phloem	Thin-walled cells in Area B found in groups next to xylem; part of central cylinder, transports food
Epidermis	Outermost layer of root; protective covering; one cell thick
Root hairs	Fingerlike projections on some epidermal cells; increases surface area for water absorption
Endodermis	Single layer of cells, ringlike, separating Area A from Area B; protective covering
Cortex	Widest area of root; stores food; makes up most of Area A

Part B. One-Year-Old Tree Stem

Figure 5 shows how a cross section of a tree stem is made.

1. Examine Figure 6, which shows the cross section of a one-year-old tree stem viewed through a microscope.

2. Study the descriptions of stem tissues in Table 2. Then label each kind of tissue in Figure 6.

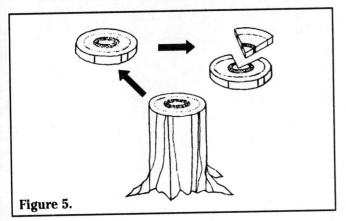

Figure 5.

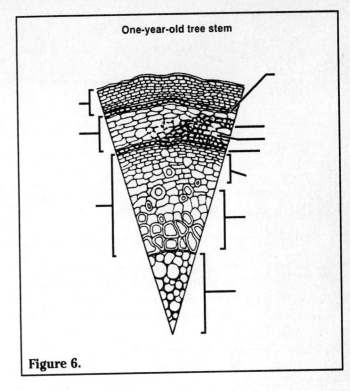

One-year-old tree stem

Figure 6.

Table 2.

	Stem Tissues
Tissue	**Description**
Cork	Outermost layer, about eight cell layers thick; protects against water loss
Cork cambium	Single layer of cells just inside the cork layer; produces new cork cells
Cortex	First layer inside cork cambium, about ten cells thick; cells larger and with thinner cell walls than cork cells; stores food
Pith	Tissue at center of stem (pointed end of wedge diagram); large thin-walled cells; stores food
Xylem (a) Spring xylem (b) Summer xylem	Thick layer of cells next to pith; widest layer of cells in stem; transports water and supports stem Portion of xylem with large cells; produced in spring Portion of xylem with small cells; produced in summer
Vascular cambium	Single layer of cells at top edge of xylem; produces new xylem and phloem cells
Phloem	Groups of thin-walled cells, inside cortex; transports food
Bast fibers	Groups of thick-walled cells; no hollow center visible; surrounds phloem; supports stem

Part C. Two-Year-Old Tree Stem

1. Compare Figure 7, showing a two-year-old tree stem, with the one-year-old stem in Figure 6. How many bands or sections of xylem are present in

   a. a one-year-old tree stem? _____

   b. a two-year-old tree stem? _____

2. Note that the difference in the size of spring and summer xylem cells forms a line separating one year of xylem from the other.

   a. Which xylem has larger cells? _____

   b. Why do you suppose that cell diameter of spring and summer xylem may differ? (HINT: A lot of water results in good growth of cells. A lack of water results in slower

   growth.) _____

   _____

   c. A new band of xylem tissue is formed each year. This band is called an annual ring. An annual ring includes both spring and summer xylem. Which tissue forms this band?

   _____

   d. Each new band of xylem forces the older band toward the stem's center. Which stem tissue is the oldest band of xylem found

   closest to? _____

3. Label Figure 7, showing a two-year-old stem cross section. Use the following labels: *first year xylem* (oldest), *second year xylem* (youngest).

Part D. Three-Year-Old Tree Stem

1. Compare Figure 8, showing a three-year-old tree stem, with the one- and two-year-old stems.

2. Label the tissues in Figure 8. Use the label lines along the right side of Figure 8 to label each band of xylem as *first year xylem* (oldest), *second year xylem,* and *third year xylem* (youngest).

3. Use the following terms to label the areas along the left side of Figure 8.

   bark—all tissue from cork through vascular cambium

   wood—all tissue from youngest xylem band through pith

   vascular ray—narrow, one-cell-thick tissue extending through the xylem.

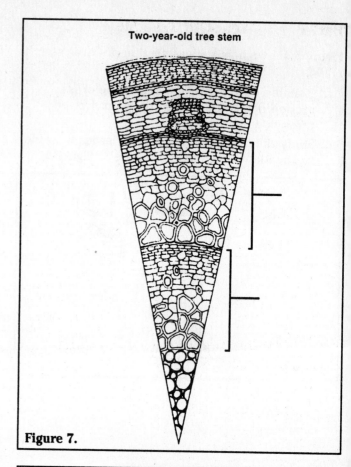

Figure 7.

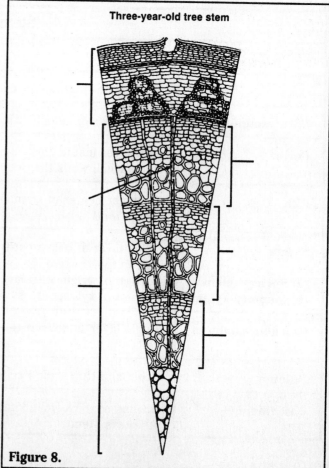

Figure 8.

DATA AND OBSERVATIONS

Table 3.

Parsnip Root Section Under Low-Power Magnification	
Region A	**Region B**

Table 4.

Root Regions and Their Functions			
Region	**Name**	**Description**	**Function**
A	Cortex	Widest area of root	
B	Central cylinder	Center area of root	

ANALYSIS

1. How is the shape and function of the central cylinder of a root similar to a water pipe or blood vessel?

_____

_____

2. How is the structure of root cortex cells adapted for their function? _____

_____

3. Use Tables 1 and 2 to complete the following chart. Write the names of root and stem tissues that carry out each function listed.

Function	Roots	Stems
Protection		
Food storage		
Transport food or water		
Absorb water from soil		
Produce new tissue		
Support stem		

4. **a.** Name the main cell types that make up the central cylinder of a root. _____

_____

b. What does each cell type transport? _____

_____

5. Skin is called a dermis. The prefix *epi-* means outside. *Endo-* means inside.

a. What are the functions of epidermis and endodermis? _____

_____

b. Are these cell layers properly named, based upon their location and function in roots? Explain.

_____

_____

6. How is the structure of thick bast fibers adapted for their function? _____

_____

7. A thin, waxy layer is present along the outside of a tree's cork. Explain how this layer helps cork

function. _____

_____

8. **a.** How many bands of xylem does a three-year-old tree stem have? _____

b. Does a new band of xylem form in a tree stem during each year of growth? _____

c. How can a tree's age be determined? _____

_____

9. Annual rings vary in thickness due to environmental factors. What kinds of environmental factors during

the year might influence growth? _____

_____

FURTHER EXPLORATIONS

1. Based upon Figure 8 of a three-year-old tree stem, draw and label
the cross section of a six-year-old tree stem, assuming similar
environmental conditions.
2. Suppose that during a ten-year period rainfall in a region increased
each spring by ten percent. During the next ten-year period, rainfall
decreased each spring by ten percent in the same region. What
effect do you think this weather pattern would have on the annual
rings of a 20-year-old tree? Construct a graph to illustrate your
answer.

How Do Gymnosperm Stomata Vary?

24–2
LAB

INVESTIGATION

▶ Gymnosperms are vascular plants that produce naked seeds. Conifers are a division of gymnosperms whose seeds develop in cones or, in a few cases, on branches. There are about 550 living species of conifers, including pines, spruces, firs, yews, redwoods, bald cypress, junipers, and cedars. Most conifers are evergreen trees or shrubs that form vast forests in the taiga biomes of the world, but some species are found in temperate zones in mixed forests and in warmer climates such as in the southeastern United States.

The leaves of most conifers are needle-like or scaly and vary in number and type depending on the species. As in the leaves of other types of plants, conifer needles have stomata and guard cells, which control the exchange of gases between the cells and the environment. The stomata also allow transpiration from the needles, which regulates water loss and water movement within the tree. The number, location, and distribution of guard cells and stomata vary from one type of gymnosperm to another.

In this lab you will investigate whether the number of stomata on gymnosperm leaves varies in different climates. Specifically, do they seem to be more numerous in climates where there is ample water and fewer in number in climates that are dry? Are they fewer in number at higher altitudes, where winds increase the potential for water loss? You will also investigate whether the position and distribution of stomata on the leaves vary between temperate, moist environments and those with harsh weather conditions.

OBJECTIVES

- Make a hypothesis that relates the number, location, and distribution of stomata on a gymnosperm leaf to the type of environment in which the species of gymnosperm lives.

- Calculate the number of stomata per millimeter of leaf surface using prepared slides of different species of gymnosperms.
- Relate the structure of the gymnosperm leaf to the survival of the plant.

MATERIALS

Prepared single-leaf slides of various species of gymnosperm, such as:
  hemlock leaf
  redwood leaf
  yew leaf
  juniper leaf

compound microscope
calculator
transparent metric ruler
field guide to conifers

PROCEDURE

1. Make a hypothesis that relates the number, location, and distribution of stomata on a gymnosperm leaf to the environment in which the gymnosperm lives.

2. In order to estimate the number of stomata in a millimeter of leaf surface, you must first calculate the diameter of the field of view of your microscope under low and high power. **(Note: See page 1136 of the *Skill Handbook* in your textbook for more information.)** Turn the nosepiece of your microscope until the low-power objective clicks into place. Put a transparent metric ruler on the stage of the microscope. Position the millimeter line at the edge of the field of view, as shown in the hypothetical diagram in Figure 1. Count the number of millimeters that span the field of view. Record your results in Table 1.

3. Record the magnification of the high power and low power objectives of your microscope in Table 1.

4. Divide the magnification of the high power objective by the magnification of the low power objective. Record your answer.

5. Now divide the diameter of the field of view under low power by your answer in step 4. The result will be the diameter of the field of view under high power. This is the measurement that you will use when estimating the number of stomata per millimeter on a gymnosperm leaf.

6. You are now ready to examine prepared slides of the leaves of various species of gymnosperms as seen in cross-section. Observe each slide at low power to locate the leaf and to focus the microscope.

7. Turn the magnification to high power and focus with the fine adjustment knob. Locate the stomata on each needle.

8. For each specimen, count the total number of stomata on each edge of the leaf. Record your answer in Table 2.

9. Now estimate the number of stomata one would find per millimeter along the *length* of each leaf. Record your answer in Table 2.

10. Describe the location of the stomata for each specimen and the way the stomata are distributed on the leaf.

11. Sketch each leaf on a separate sheet of paper. Note that each needle is enclosed by an epidermis covered by a cuticle. The mesophyll cell layers may be tightly packed or loosely packed. The stomata on some leaves may be more recessed than others. Some gymnosperms have canals that produce resin, which is used to protect the needle against insects. Also note the central vessel composed of xylem and phloem tissue. Relate these various features and structures of the leaf to the survival of a conifer in a given habitat.

12. Refer to a field guide for conifers, and research the habitats in which the specimens you studied can be found. Record this information in Table 2.

HYPOTHESIS

_____

_____

_____

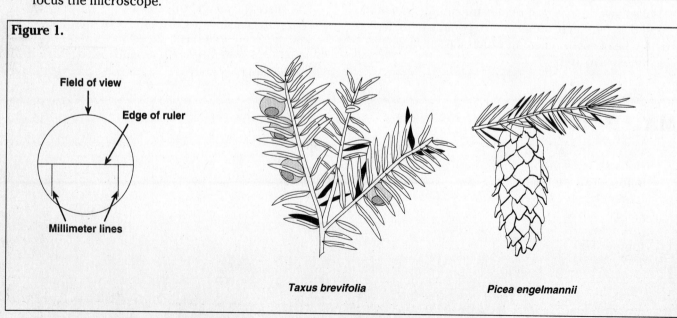

Figure 1.

Field of view

Edge of ruler

Millimeter lines

Taxus brevifolia   *Picea engelmannii*

DATA AND OBSERVATIONS

Table 1.

Diameter of the field of view under low power =	mm
Magnification of the high power objective =	
Magnification of the low power objective =	
High power objective magnification divided by the low power magnification =	
Diameter of the field of view under high power =	mm

Table 2.

Name	Total Stomata	Stomata per mm.	Location	Distribution	Habitat
Hemlock leaf (*Tsuga caroliniana*)					
Redwood leaf (*Metasequoia*)					
Yew leaf (*Taxus)*					
Spruce leaf (*Picea*)					

ANALYSIS

1. Suppose two species of gymnosperms live in the same habitat but do not share the same number of stomata per millimeter. What might be an explanation for this?

_____

_____

_____

_____

2. Explain the number of stomata on the gymnosperm leaf in terms of natural selection.

_____

_____

_____

_____

3. Describe any relationship you noticed between the shape, width, and internal anatomy of the leaves you observed and the environments in which the various species live.

_____

_____

_____

_____

CHECKING YOUR HYPOTHESIS

Was your **hypothesis** supported by your data? Why or why not? _____

_____

FURTHER EXPLORATIONS

1. When the gymnosperms in your area start producing pollen in the spring, examine the pollen from several species under the microscope. Relate the structure of the pollen to the method of gymnosperm pollination.

2. The red spruce in the Adirondack Mountains is particularly sensitive to acid rain at high elevations. Use resources in the library to determine why this conifer is more susceptible to acid rain than are other conifers.

INVESTIGATION

How Are Ferns Affected by Lack of Water?

25–1

LAB

▶ The life cycle of a fern involves the production of spores and the production of sperm and eggs. The fern life cycle is said to show alternation of generations. Some stages of this life cycle are more dependent on water than others. In this Investigation, you will discover which stages occur under wet conditions, and which ones occur under dry conditions.

OBJECTIVES

- Hypothesize how different stages of the life cycle of a fern are affected by water.
- Observe the release of spores from a fern plant.
- Observe the germination of fern spores.
- Observe the release of sperm from a fern prothallus.

MATERIALS

mature fern plant
scalpel
microscope slides (4)
coverslips (4)
droppers (2)

glycerin
microscope
wax marking pencil
toothpick
petroleum jelly

fern-spore culture
mature prothallus
pencil with eraser
water

PROCEDURE

Part A. Alternation of Generations

1. Study the life cycle of the fern in Figure 1. Notice that the cycle includes two different plants: the large, familiar fern plant and a smaller plant called a prothallus. Locate these stages in the diagram.

2. Notice that spores are produced by the leafy fern plant.

3. Find the germinating spores in the diagram. A spore germinates to form a group of threadlike structures that develop into a tiny, heart-shaped plant called a prothallus.

4. Notice that eggs and sperm are produced by the prothallus.

5. Make a **hypothesis** that states if water is needed for each of these three stages of the life cycle: release of spores, germination of spores, and release of sperm. Write your hypothesis in the space provided.

2. With a scalpel, carefully scrape one of these dots onto a microscope slide. This structure contains several sporangia. **CAUTION:** *Use the scalpel with care. The tip and edge are sharp.*

3. Add a drop of glycerin and a coverslip to the slide. Glycerin will draw the water out of the sporangia.

4. Tap the coverslip with the eraser end of a pencil. This will help release the spores.

5. Observe the slide under low power of a microscope. When you find a sporangium, center it in the field, and change to high power.

6. Observe the sporangium for several minutes. Draw the sporangium before and after any change. Draw your observations of a sporangium in the space marked "Sporangium" in Data and Observations.

Part B. Release of Spores

1. Obtain a fern plant. Look for brown dots on the back of the fronds of the fern.

Part C. Germination of Spores

1. Use a wax marking pencil to write your initials on two microscope slides.

Figure 1.

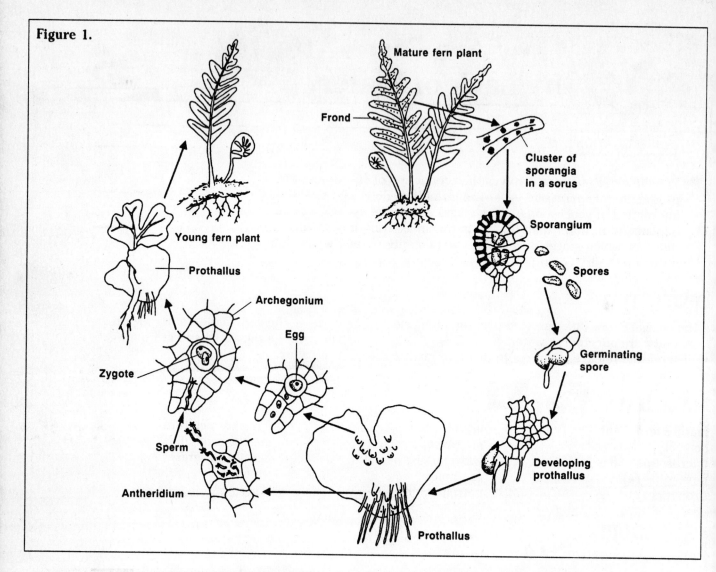

2. Use a toothpick to make a ring of petroleum jelly about the size of a dime on one slide. Place a drop of fern-spore culture inside the ring, and add a coverslip.

3. Observe the slide under low power of a microscope. Make a drawing of your observation in the space marked "Spore culture sealed slide" in Table 2.

4. Make a second wet mount of fern-spore culture, but do not use petroleum jelly to seal the coverslip to the slide.

5. Observe this slide under low power. Make a drawing of your observation in the space marked "Spore culture unsealed slide" in Table 2. Set the slide aside.

6. One or two days later, observe both slides again. Look for changes in the spores. Draw your observations in the appropriate spaces in Table 2.

Part D. Release of Sperm

1. Obtain a mature prothallus. Make a wet mount of the prothallus. Add a coverslip.

2. Observe the slide under low power. Near the notch in the heart-shaped prothallus, locate the structures that produce the eggs, called archegonia.

3. Near the pointed base of the prothallus, locate the structures that produce the sperm, called antheridia. Use the eraser end of a pencil to press gently on the coverslip. Watch closely to see if sperm are released. If sperm are released, observe their movement. Record your observations in Table 1.

4. Set the slide aside for several minutes, and allow it to dry out. Repeat step 3.

HYPOTHESIS

_____

_____

DATA AND OBSERVATIONS

Table 1.

Observations	Wet prothallus	Dry prothallus
Release of sperm		
Movement of sperm		

Sporangium

Table 2.

Spore culture sealed slide	Spore culture unsealed slide
After 1–2 days	After 1–2 days

ANALYSIS

1. Does adding glycerin to a slide simulate moist conditions or dry conditions? Explain your answer.

   _____

   _____

2. Describe what happened to the sporangium in glycerin. _____

   _____

   _____

   _____

3. Under what conditions are spores released? _____

   _____

4. How did the slides of spore cultures differ after a day or two? How did this difference affect the spores?

   _____

   _____

   _____

   _____

5. Under which conditions were sperm released by the prothallus? _____

   _____

CHECKING YOUR HYPOTHESIS

Was your **hypothesis** supported by your data? Why or why not? _____

_____

FURTHER INVESTIGATIONS

1. Investigate the effect of the lack of water on a fern plant. Obtain two fern plants of the same species and size. Plant them, using the same amount of soil for each. Keep the plants under the same conditions, varying only the amount of water they receive. Record the amount of water each plant receives. Also record the growth and development of each plant.

2. Investigate the conditions under which germinating spores grow best. Use some of the fern-spore culture from this Investigation. Determine the condition you would like to investigate, such as moisture, light, or growth medium. Design and set up an experiment to determine which conditions are best for the development of germinating spores.

What Is the Effect of Light Intensity on Transpiration?

26-1

LAB

INVESTIGATION

▶ In order to carry out photosynthesis, a plant must exchange gases with the atmosphere through the stomata. One consequence of having open stomata is the loss of water vapor. Most of the water taken up by a plant's roots passes through the plant's vascular system and evaporates from the surfaces of the leaves during transpiration. Although a plant benefits from this gas exchange because photosynthesis can occur, thereby producing energy, the plant loses valuable water vapor to the atmosphere. Many different environmental factors, including temperature, humidity, light quality and intensity, and wind velocity, can influence a plant's rate of transpiration.

OBJECTIVES

- Prepare a setup that will test the effect of light intensity on a plant's transpiration rate.
- Make a hypothesis to describe the effect of light intensity on the rate of transpiration.

- Observe the effect of high and low light intensity on transpiration rate.
- Compare the rates of transpiration in a plant exposed to two different light intensities.

MATERIALS

stop watch
22-cm Pasteur pipette
permanent marker
pan, approx. 30 × 50 × 20 cm,
    with water approx. 10 cm deep
pruning shears
petroleum jelly

metric ruler
10 cm rubber surgical tubing, 5 mm bore
incandescent lamp
25- and 100-watt bulbs
dicot branch with leaves
thermal mitt
laboratory apron

PROCEDURE

1. Put a light coating of petroleum jelly around the wide end of the Pasteur pipette and around the dicot branch, 3 cm from the cut end. Be careful not to get petroleum jelly on the end of the branch. See Figure 1 for help.
2. With the permanent marker, make two marks on the pipette, the first mark 1 cm from the drip tip and the second mark 9 cm from the first mark. See Figure 2. **CAUTION:** *The tip of the pipette is fragile. Do not press down on it.*
3. Slip the rubber tubing over the wide end of the pipette. The fit should be snug with no air leaks.

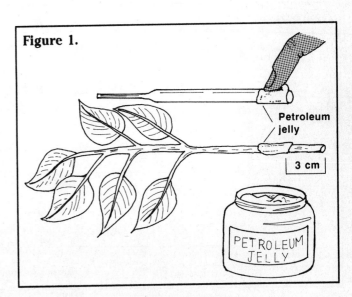

Figure 1.

Petroleum jelly

3 cm

PETROLEUM JELLY

Figure 2.

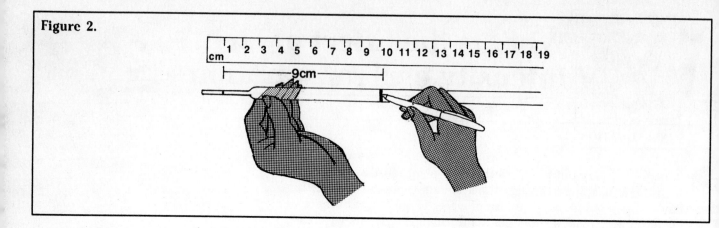

4. Immerse the pipette and tubing in the pan of water so that the insides of both parts are completely filled with water.

5. Immerse the cut end of the dicot branch in the water. Use the pruning shears to cut off an additional 2 cm of stem under water. **CAUTION:** *Pruning shears are very sharp.* Slip the branch into the free end of the rubber tubing while keeping pipette, tubing, and branch under water. There should be no air bubbles in the setup. Compare your setup with that in Figure 3.

6. Test your setup for air bubbles by raising the drip tip of the pipette. Air bubbles except those immediately next to the branch will rise to the pipette tip. If there are air bubbles, disassemble and repeat steps 1 through 5.

As the branch transpires, the water in the pipette will recede from the tip. You will be measuring the time it takes for the branch to transpire the same amount of water under two different light intensities.

7. Make a **hypothesis** to describe the effect of light intensity on transpiration. Write your hypothesis in the space provided.

8. Lay your setup on the counter. Position an incandescent lamp containing a 25-watt bulb so that the bulb is 15 cm above the leaves of the branch. Allow the plant to adjust under the light until the water in the pipette recedes to the 1-cm mark.

9. When the water reaches the 1-cm mark, begin timing the retreat by starting the stop watch. When the water reaches the 9-cm mark, stop the stop watch and record in the table the time it took for the water to move the 8 cm.

10. Refill the pipette by immersing the pipette tip in the pan of water. Gently squeeze the rubber tubing to expel the water and air. Release the pressure on the tube and refill the pipette with water.

11. Repeat steps 6, and 8 through 10 two more times. Record your results in the table under Trials 2 and 3.

Figure 3.

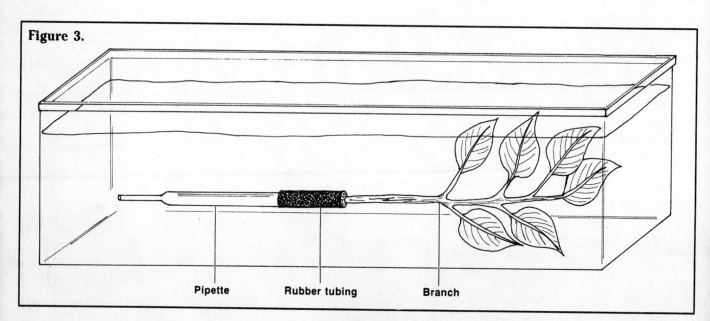

Pipette   Rubber tubing   Branch

12. Change to the 100-watt bulb by disconnecting the lamp, then unscrewing the bulb while wearing a thermal mitt. **CAUTION:** *The 25-watt bulb will be hot.* Repeat steps 6, and 8 through 10 three times. Record your results in the table.
13. Calculate the average elapsed time for each light intensity and record these numbers in the table.

HYPOTHESIS

_____

_____

_____

DATA AND OBSERVATIONS

Table 1.

Light intensity	Elapsed time (in seconds)			
	Trial 1	Trial 2	Trial 3	Average
Low (25-watt bulb)				
High (100-watt bulb)				

ANALYSIS

1. What does the movement of water in the tube suggest about what is happening to water in the plant?

_____

_____

2. Compare the time it took for water to move the 8 cm in the tube at the two light intensities. At which intensity was the rate of movement greater? _____

_____

_____

3. Would you expect the trend you observed to continue if the light intensity were increased to that of a 200-watt bulb? Explain. _____

_____

_____

4. Why does a change in light intensity change the transpiration rate? _____

_____

5. How would you expect the transpiration rate of a branch with twice as many leaves as your branch to compare with the rate you measured in this Investigation? Why? _____

_____

_____

CHECKING YOUR HYPOTHESIS

Was your **hypothesis** supported by your data? Why or why not? _____

_____

FURTHER INVESTIGATIONS

1. Plan and conduct an experiment to test the effect on the transpiration rate of coating the upper and lower leaf surfaces with petroleum jelly.

2. Devise other experiments and hypotheses that test for the effects of wind, heat, humidity, or other environmental factors on the transpiration rate of plants.

Do Dormant and Germinating Seeds Respire? 27–1

LAB

▶ Seeds that are purchased for planting are viable. Viable seeds are alive but show no signs of any life processes. Scientists refer to seeds in this condition as being dormant. When dormant seeds are soaked in water, they respond by showing evidence of life processes. Seeds that show evidence of life processes are called germinating seeds. One of the life processes carried out by seeds is respiration. During respiration, a seed takes in oxygen from the air and gives off carbon dioxide. The volume of oxygen taken in and carbon dioxide given off is about equal.

OBJECTIVES

- Prepare respiration chambers to measure the amount of oxygen used by dormant and germinating seeds.
- Compare the respiration rate of dormant and germinating seeds.

- Predict the volume of oxygen used by seeds in several experimental conditions.
- Hypothesize how much oxygen is used by dormant and germinating seeds.

MATERIALS

test tubes (3)
bean seeds, soaked (5)
bean seeds, dry (5)
absorbent cotton

metric ruler
soda lime
small beaker

water, colored
measuring half-teaspoon
masking tape

marking pen
rubber band
laboratory apron

PROCEDURE

1. Place five dry, dormant bean seeds into a test tube.
2. Add a small cotton plug just above the seeds. Pack the cotton just tightly enough so that the seeds cannot fall out when the tube is turned upside down.
3. Add a half-teaspoon of soda lime (a carbon dioxide gas absorber) to the test tube. **CAUTION:** *Do not handle soda lime with bare hands.*
4. Insert a second cotton plug over the soda lime as shown in Figure 1.
5. Use masking tape and a pen to label the test tube "dormant." Place the label near the middle of the test tube.
6. Prepare a second test tube in a similar manner. However, put five soaked, germinating seeds in this test tube. Label this test tube "germinating."

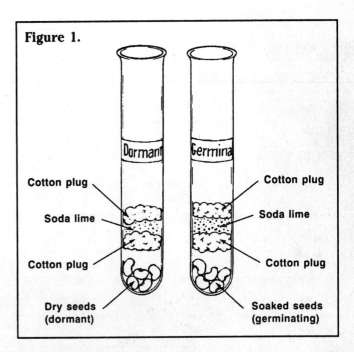

Figure 1.

Dormant | Germina

Cotton plug
Soda lime
Cotton plug
Dry seeds (dormant)

Cotton plug
Soda lime
Cotton plug
Soaked seeds (germinating)

7. Prepare a third test tube without seeds. Add a cotton plug, one half-teaspoon of soda lime, and a second cotton plug just above the soda lime.

8. Place a rubber band around all three test tubes to prevent their tipping over. Turn all the test tubes upside down and place them into a small beaker containing 30 mL of colored water.

9. With a metric ruler, measure in millimeters the height to which the colored water rises in each test tube (Figure 2). Measure the water level in the test tube, not in the beaker.

10. Record in Table 1 the height to which the colored water rises in each test tube. Record the height as 0 mm if water does not move into a test tube.

11. Do not disturb the test tubes for at least 24 hours.

12. Make a **hypothesis** to explain which one of the test tubes will show the greatest difference in water height. Write your hypothesis in the space provided.

13. After 24 hours, measure and record the height of water inside each test tube. DO NOT move the beaker or remove the test tubes from the beaker until you have measured the height of water within each test tube.

14. Calculate the difference in water height from start to end of the Investigation by subtracting

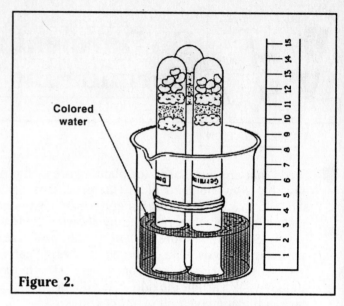

Colored water

Figure 2.

the height of water in each tube at the start from the height of water in each tube after 24 hours. Record this amount in the last column of Table 1.

HYPOTHESIS

_____

_____

DATA AND OBSERVATIONS

Table 1.

Height of Water in Test Tubes			
Test tube contents	Height of water at start (mm)	Height of water after 24 hours (mm)	Difference in water height (mm)
Dormant seeds			
Germinating seeds			
No seeds			

ANALYSIS

1. **a.** What do you think is trapped within each test tube turned upside down in the water? _____

   b. What prevents water from moving into each test tube at the start of the experiment? _____

   _____

2. If air within a test tube is used, water will rise within the test tube to replace the missing air.

   a. Which test tube shows the most air used in 24 hours? _____

152

b. What is the evidence? _____

c. What gas in the air is being used? _____

3. What life process is responsible for the use of this gas? _____

4. a. Which kind of seed, dormant or germinating, carries out this process at a faster rate? _____

    b. What is the evidence? _____

    _____

5. What does the test tube without seeds show? _____

_____

6. Soda lime absorbs carbon dioxide from air.

    a. Recalling that seeds use oxygen and release carbon dioxide, predict the level of water 24 hours later

    in all test tubes if soda lime is not used. _____

    b. Explain. _____

    _____

7. A seed contains a food supply. This food is used during respiration.

    a. Which kind of seed, dormant or germinating, carries out respiration at a slower rate? _____

    b. What evidence do you have to support your answer? _____

    c. Which type of seed, dormant or germinating, would use food at a slower rate? _____

    d. Explain why using food slowly may help plants survive. _____

    _____

    _____

8. Suppose you place 20 germinating bean seeds into one test tube and add soda lime. Into a second test tube, you place 10 germinating bean seeds and add soda lime. You then invert both test tubes into a beaker of water. Complete Table 2.

Table 2.

Number of seeds	Expected height of water 24 hours later
20 germinating seeds	
10 germinating seeds	

9. Give reasons for the measurements in Table 2. _____

_____

10. Suppose you place 20 germinating bean seeds into one test tube with soda lime. Into a second test tube you place 20 germinating bean seeds without soda lime. You then invert both test tubes into a beaker of water. Complete Table 3.

Table 3.

Treatment	Expected height of water 24 hours later
20 germinating seeds with soda lime	
20 germinating seeds without soda lime	

11. Give reasons for the measurements in Table 3. _____

_____

12. Suppose you boil some germinating seeds for 30 minutes and place these boiled seeds into a test tube with soda lime. You then prepare a second test tube with unboiled germinating seeds and soda lime and invert both test tubes into water. Complete Table 4.

Table 4.

Seed treatment	Expected height of water 24 hours later
Boiled germinating seeds with soda lime	
Unboiled germinating seeds with soda lime	

13. Give reasons for the measurements in Table 4. _____

_____

CHECKING YOUR HYPOTHESIS

Was your **hypothesis** supported by the data you recorded? Why or why not? _____

_____

FURTHER INVESTIGATIONS

1. Try the same investigation using other kinds of seeds, such as grass, radish, or corn. Do you suppose that you will get similar results?
2. Do differences in water temperature affect the respiration of dormant and germinating seeds? Design an experiment that will show the effect of hot water (32°C), warm water (22°C), cool water (12°C), and cold water (2°C) on the respiration of dormant and germinating seeds.

How Do Hormones Affect Plant Growth?

27–2

LAB

INVESTIGATION

▶ Living systems produce many chemicals that influence or regulate specific cell processes in organisms. One type of chemical is called a hormone. Plant hormones include auxins and gibberellins. These hormones control plant processes such as stem cell elongation. They also direct roots to grow downward in the direction of the force of gravity and stems to grow upward away from the force of gravity.

Auxins are present in young, newly forming roots and are distributed unevenly within the root. Cells along the top of the root have less auxin than cells in the lower portion because the force of gravity pulls the auxins downward. This causes the cells to grow faster along the top of the root than those along the bottom. As a result, the new young root curves downward in the direction of the force of gravity.

Young stem cells respond to these auxin amounts in a way directly opposite to that of roots. Stems also have more auxin along the bottom surface due to the force of gravity and less auxin along the top surface. Their top surface grows more slowly, while the bottom surface grows more quickly. This causes the stem tissue to turn upward away from the force of gravity.

OBJECTIVES

- Record the positions and directions of growth of new roots and stems for four corn seeds over a minimum 48-hour period.
- Make a hypothesis that explains how the direction a seed faces influences the growth of the stem and roots.

- Compare the direction of growth shown by young corn roots and stems when positioned in different orientations to the force of gravity.

MATERIALS

bowl or shallow dish
corrugated cardboard
stapler
metric ruler

paper towels
corn seeds, soaked (4)
plastic bag

straight pins (4)
masking tape
water

PROCEDURE

1. Cut a piece of corrugated cardboard that measures 26 cm × 13 cm. Fold the cardboard in half so you have two 13 cm × 13 cm sections. See Figure 1A.

2. Staple several thicknesses of paper toweling to the outside surface of one side of the cardboard. See Figure 1B.

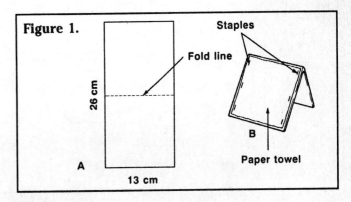

Figure 1.

26 cm

Fold line

Staples

Paper towel

B

A

13 cm

Figure 2.

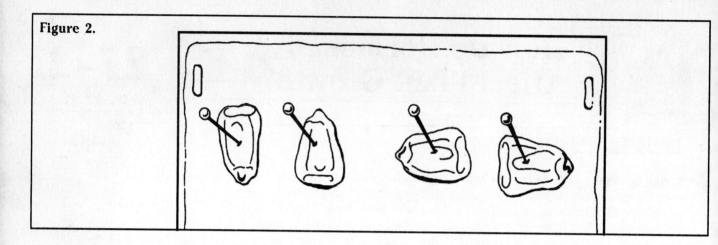

3. With straight pins, attach four soaked corn seeds to the top of the paper-toweling side of your cardboard. Position the seeds as in Figure 2.

4. Stand the cardboard in a shallow dish as shown in Figure 3. Pour in enough water to cover the bottom of the dish.

5. Slip a plastic bag over the top to reduce drying out of the seeds. Add a label with your name and the date to the outside of the plastic bag.

6. Make a **hypothesis** that explains the relationship of the position and direction of a seed to the growth of its roots and stem. Write your hypothesis in the space provided.

7. Place your seeds in an area designated by your teacher. Observe the seeds after 48 or 72 hours. Examine them for the directions of root and stem growth.

8. Record the exact positions and directions of growth of new roots and stems for each seed by drawing these structures on the seeds shown in Data and Observations. Label the root and stem of each plant.

    NOTE: *Roots originate from the pointed end of the seed. Stems originate from an area directly*

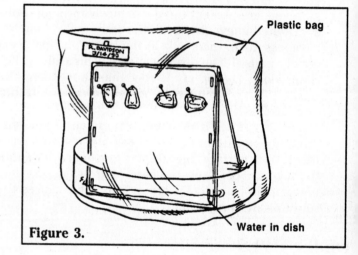

Plastic bag

Water in dish

Figure 3.

above where the root emerges. Stems may already appear green. Roots may already show some branching.

HYPOTHESIS

_____

_____

DATA AND OBSERVATIONS

ANALYSIS

1. What is the direction of root growth from the seed with the pointed end facing:

   a. down? _____

   b. up? _____

   c. left? _____

   d. right? _____

2. a. What influence does the direction in which a seed is pointing have on the direction of root growth?

   _____

   _____

   b. Do roots seem to grow toward or away from the force of gravity? _____

   _____

3. What is the direction of stem growth from the seed with the pointed end facing:

   a. down? _____

   b. up? _____

   c. left? _____

   d. right? _____

4. a. What influence does the direction in which a seed is pointing have on the direction of stem growth?

   _____

   _____

   b. Do stems grow in the direction toward or away from the force of gravity? _____

   _____

5. What affects the way roots and stems grow? _____

   _____

   _____

   _____

6. Do root and stem cells respond to the presence of auxins in a similar or different manner? Explain.

   _____

   _____

   _____

7. When you garden, must you always be careful about the direction a seed faces when planted

in the ground? Explain. _____

_____

_____

CHECKING YOUR HYPOTHESIS

Was your **hypothesis** supported by your data? Why or why not? _____

_____

FURTHER INVESTIGATIONS

1. Test for the influence of light on the direction of stem and root growth. Design an experiment to determine how stem and root growth respond to light.
2. Plan and demonstrate an investigation of the influence of magnetism on the direction of plant stem and root growth. Determine whether magnetism affects the gravitational pull on the roots and stems based upon the direction a seed faces.

EXPLORATION

Symmetry

28–1
LAB

► Animals vary in the patterns of embryonic development, type of support, mechanisms for food capture, type of nervous response, structure of digestive and transport systems, methods of excreting wastes, and methods of locomotion. Animals also exhibit differences in symmetry, or the arrangement of their body parts. Organisms that lack a regular arrangement of parts are categorized as asymmetric. Animals with parts that radiate from a central point or from a central axis have radial symmetry. And animals that can be divided into approximately mirror-image halves along a central plane are bilaterally symmetrical. Animals with bilateral symmetry have definite front and back (anterior and posterior) ends and a concentration of nervous tissue in the anterior region, often with external sense organs. The symmetry of an animal affects the way it moves and obtains its food. Some scientists think that symmetry even affects mate choice and serves as an indicator of an organism's health.

OBJECTIVES

• Identify the symmetry of a variety of animals.

• Relate symmetry to food capture and locomotion.

MATERIALS

museum mount of a bird
preserved or dried specimens of
    a sand dollar, sea urchin,
    brittle star, jellyfish, butterfly
    or moth, sea anemone, and
    squid
live crayfish, goldfish, lizards,
    grasshoppers
living cultures of
    Daphnia
    Planaria

Hydra
    rotifers
yeast boiled in Congo red
grass
beef liver
living flies
lettuce
guinea pig pellets
fish food
aquarium or fish bowl
terrarium

spring, pond, or aquarium water
watchglass
compound microscope
pipette
hand lens or dissecting
    microscope
depression slide
coverslip
glass jars, large (2)

PROCEDURE

Part A. Identifying Symmetry

1. Study the animals drawn in Figure 1. Identify the type of symmetry that characterizes each animal's body plan and record it in Table 1. Draw a line through the center of each animal to help you make the identifications.

2. Examine each of the preserved or dried specimens and record their symmetry in Table 2. Can the organisms be divided along any plane into roughly equal halves? If so, classify the organism as radially symmetrical. Is the organism irregular in shape? Then its body plan exhibits asymmetry. Can the organism be divided along its

length to form mirror-image halves? Does the anterior end look different from the posterior? Does the dorsal surface look different from the ventral surface? Then the organism is bilaterally symmetrical.

Part B. Symmetry and Behavior

In this part of your exploration, you will be working with a variety of living organisms to identify their symmetry and observe how it affects the way they move and feed. Use your observations and information from your text to complete Table 3. Follow your teacher's instruction regarding the care and handling of the organisms.

1. *Planaria*: Place a few drops of water on a watchglass. Use a pipette to remove a planarian from the culture jar and place it in the water. Introduce a small piece of liver, and observe the movement and feeding behavior of the planarian by using a hand lens or dissecting microscope. Record your observations in Table 3.

2. **Rotifer:** Using a pipette, place a drop of the rotifer culture on a depression slide. Add a drop of the stained yeast to the slide and place a cover slip over it. Use a compound microscope to examine the rotifers' method of obtaining and ingesting the yeast cells. Record your observations.

Figure 1.

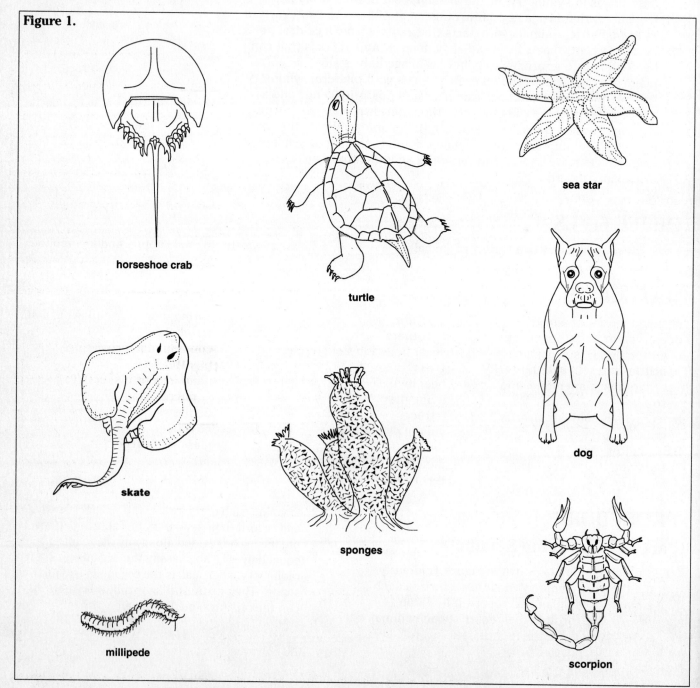

horseshoe crab

turtle

sea star

skate

sponges

dog

millipede

scorpion

3. *Hydra*: Place some water on the watchglass. Use the pipette to add one or more *Hydra*, and then place the watchglass under the dissecting microscope. Release about 0.5 mL of the *Daphnia* culture near the *Hydra*. Observe the *Hydra* under the dissecting microscope as they capture and eat the *Daphnia*. Record your observations in Table 3.

4. **Crayfish:** Place a crayfish in an aquarium or fishbowl and fill it with water. Drop a few pellets of guinea pig food a short distance in front of the crayfish. If it is hungry, it should respond quickly to the food. Record its movements and feeding behavior in Table 3.

5. **Goldfish:** Place a goldfish in a large jar of aquarium water. Sprinkle some flakes of food on the surface of the water. Observe and record the behavior of the fish.

6. **Lizard:** The lizard can be observed in a terrarium that has a screened top. Living flies are the food source. Do not distract the lizard after introducing the flies into the terrarium. Watch patiently for food capture and consumption.

7. **Grasshopper:** Observe the grasshopper's movements in a large glass jar. Slowly place a small piece of lettuce in the jar and observe how the insect responds. Avoid distracting the grasshopper with your own movements.

DATA AND OBSERVATIONS

Table 1.

Organism	Symmetry
horseshoe crab	
sea star	
turtle	
skate	
dog	
sponges	
millipede	
scorpion	

Table 2.

Organism	Symmetry
sand dollar	
sea urchin	
brittle star	
jellyfish	
sea anemone	
butterfly/moth	
squid	

Table 3.

Organism	Symmetry	Description of Movement and Eating Behavior
Planaria		
Rotifer		
Hydra		
Crayfish		
Goldfish		
Lizard		
Grasshopper		

ANALYSIS

1. What do your observations suggest about the relationship between movement and symmetry?

_____

_____

_____

2. What do your observations suggest about the relationship between food capture and symmetry?

_____

_____

_____

3. What are the advantages and disadvantages of radial symmetry?

_____

_____

_____

4. What appears to be the evolutionary trend in symmetry? Why do you think this is so?

_____

_____

_____

FURTHER EXPLORATIONS

1. Design an investigation that would compare the internal and external symmetry of an animal.
2. Using the library and other textbooks as a resource, show the correlation between early embryonic development and the symmetry of at least three organisms.

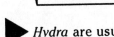
EXPLORATION

Hydra Behavior

▶ *Hydra* are usually sessile organisms. They capture prey with their nematocysts, located on their tentacles. The nerve net signals the tentacles to push the prey through the mouth and into the gastrovascular cavity. In this Exploration, you will observe the triggering of nematocysts and determine the kinds of stimuli that cause the triggering of nematocysts.

OBJECTIVES

- Prepare a wet mount of a hydra for microscopic observation.
- Observe the triggering of nematocysts caused by dilute acid.

- Observe the reactions of *Hydra* to food and non food stimuli.

MATERIALS

droppers (3)
Hydra culture
depression slides (2)

stereomicroscope
dilute acetic acid
small beaker

aquarium water
dissecting probe
Daphnia culture

PROCEDURE

1. Using a dropper, transfer a hydra and a few drops of culture water to a depression slide. Allow time for the hydra to relax and extend its tentacles. **CAUTION:** *Hydra are living animals. Handle them with care.*

2. Observe the hydra under low and then high power of the stereomicroscope. See Figure 1.

3. With the hydra in view, carefully add one drop of dilute acetic acid to the depression slide. **CAUTION:** *If acetic acid is spilled, rinse with water and call your teacher immediately.* The

acetic acid will cause the nematocysts to discharge. If they do not discharge immediately, carefully add acetic acid one drop at a time until they do.

4. After observing the triggering of nematocysts, remove the hydra to a small beaker of aquarium water by dipping the slide into the beaker.

5. Using a dropper, transfer a few hydra and a few drops of culture water to the second depression slide. Allow the hydra to relax and extend its tentacles.

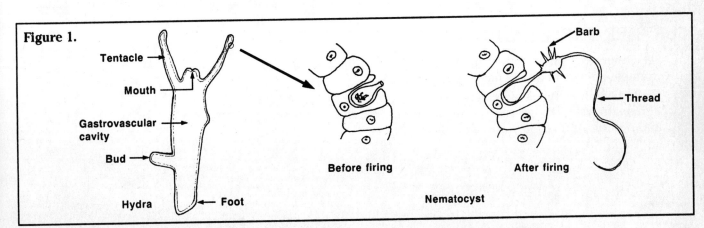

Figure 1.

Tentacle
Mouth
Gastrovascular cavity
Bud
Hydra — Foot

Before firing

After firing   Barb   Thread

Nematocyst

6. With a hydra in view under the high power of the stereomicroscope, carefully shake the depression slide. In Table 1, record the reaction of the hydra. With the dissecting probe, carefully touch the body and a tentacle of the hydra. Record the reaction of the hydra.

7. Allow the hydra to relax. Then, use a dropper to transfer a few *Daphnia* to the depression slide. In Table 1, record the reaction of the hydra when a *Daphnia* comes close to it.

DATA AND OBSERVATIONS

Table 1.

Reactions of *Hydra*	
Stimulus	**Response**
Shaking of slide	
Touching with probe	
Presence of *Daphnia*	

ANALYSIS

1. Why do you think the nematocysts trigger when a hydra is exposed to acetic acid? _____
_____

2. How does the reaction of *Hydra* to *Daphnia* differ from the reaction to being shaken or touched? _____
_____

3. Describe the feeding process of *Hydra*. _____
_____
_____

4. *Hydra* cannot pursue their prey. What adaptation makes this unnecessary? _____
_____

5. Of what survival value is *Hydra's* quick response to touch? _____
_____

FURTHER EXPLORATIONS

1. Set up an experiment to test the types of food *Hydra* will eat. Use examples of both aquatic plants and animals.
2. *Hydra* have four types of nematocysts. Research the different types of nematocysts. Design an experiment to observe the four types and determine how they are used.

Earthworm Dissection

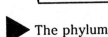

EXPLORATION

30-1
LAB

▶ The phylum Annelida includes earthworms, leeches, and segmented marine worms. The earthworm is a burrowing animal that literally eats its way through the soil. As the soil passes through the earthworm's digestive tract, decaying plant material is digested.

OBJECTIVES

- Dissect and identify internal and external features of the earthworm.
- Draw and label a diagram of the external anatomy of the earthworm.
- Label a diagram of the internal anatomy of the earthworm.
- Describe the major features of the earthworm phylum.

MATERIALS

colored pencils (4)
dissecting pins (10–15)
dissecting pan
hand lens or stereomicroscope

single-edged razor blade or scalpel
earthworm (preserved
   or freshly anesthetized)
laboratory apron

PROCEDURE

Part A. External Anatomy

1. Put on the laboratory apron. Place an earthworm in a dissecting pan. If you are using a preserved earthworm, first rinse it well with water.
2. Examine the ringlike **segments** that make up the length of the earthworm's body.
3. Identify the **anterior** (head) and **posterior** ends. At the anterior end is a small lobe on the **ventral** (lower) surface that is used for burrowing. Just behind the lobe is the **mouth**, which opens into a long, tubelike digestive system that ends at the anus. Note that the **dorsal** (upper) surface of the worm is a slightly different color from the ventral surface.
4. Rub the ventral surface of the earthworm in both directions with your finger. The bristles you feel are called **setae**. The setae are used in locomotion. Use a stereomicroscope or a hand lens to observe and count the setae.
5. Examine the prominent bandlike structure called the **clitellum**. It is located one-third of the way from the anterior end of the worm's body. This structure is important in reproduction.

6. Make a diagram of the ventral side of the earthworm in the space provided in Data and Observations. Label a segment, the mouth, setae, and clitellum.

Part B. Internal Anatomy

1. Stretch out the earthworm in the dissecting pan so that the dorsal surface faces up. Place a pin through each end of the worm to hold it in place.
2. Make a shallow cut lengthwise along the dorsal surface with a razor blade or scalpel. **NOTE:** An earthworm's skin is very thin. Make a very shallow cut. Cut just through the body wall and to the right or left of the dorsal blood vessel that runs down the "back." **CAUTION:** *The blade will be very sharp. Cut away from your fingers.*
3. Spread the edges of the worm's body carefully, and pin the skin and muscle to the dissecting pan. Place the pins at an angle as shown in Figure 1.

Figure 1.

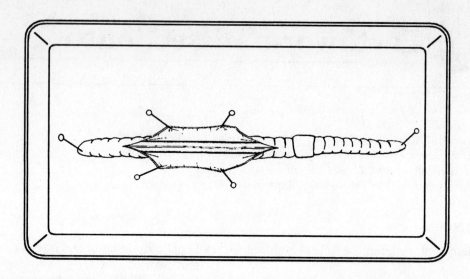

4. Examine the space between the body wall and the internal organs. This cavity is called the **coelom** and is divided into compartments by **septa**. The cavity is filled with fluid in the living earthworm and gives support to the body.

5. Locate and identify in the earthworm the following anatomical structures that are printed in bold print. Label them on Figure 2 under Data and Observations.

Digestive System

The **pharynx** is just posterior to the mouth. The pharynx is an enlarged, muscular structure that helps ingest food. The more narrow esophagus leads from the pharynx to the thin-walled **crop**, an enlarged storage sac. Food passes from the crop to the muscular **gizzard**. The earthworm swallows small stones that help grind food in the gizzard. Ground up food passes to the **intestine** where chemical digestion and absorption occurs. The solid wastes pass out through the anus.

Circulatory System

The earthworm has a closed circulatory system. Five dark vessels called **aortic arches**, or "**hearts**," surround the esophagus. These five vessels pump blood posteriorly through the **ventral blood vessel** located under the digestive system and anteriorly through the **dorsal blood vessel**.

Nervous System

The small white **brain** is located anterior and dorsal to the pharynx in the area of the third segment. A **nerve ring** leaves the brain and connects to the white **ventral nerve cord** that runs the length of the worm. From this nerve cord arise smaller nerves that go out to various structures in each segment of the worm's body.

Reproductive System

Each earthworm has both male and female reproductive organs and is said to be a *hermaphrodite*. Earthworms, though hermaphroditic, mate with other earthworms. They don't fertilize their own eggs. The **seminal vesicles** are the three large white structures on each side of the esophagus. These vesicles hold a worm's own sperm. In mature specimens, the lobed, sperm-producing **testes** are in segments ten and eleven below the **seminal vesicles.** The smaller round white structures anterior to the testes are the **seminal receptacles**. They receive and store sperm from another earthworm. The egg-producing **ovaries** are in segment 13, posterior to the testes. They are very small and may be difficult to find. After exchanging sperm, each earthworm's clitellum secretes a ring of mucus into which eggs and sperm are deposited.

6. Color the structures of Figure 2 according to the following instructions.

> digestive system – yellow
> circulatory system – red
> nervous system – blue
> reproductive system – green

7. Dispose of your earthworm according to your teacher's instructions.

DATA AND OBSERVATIONS

In the space provided, draw and label the external view of the earthworm.

Figure 2.

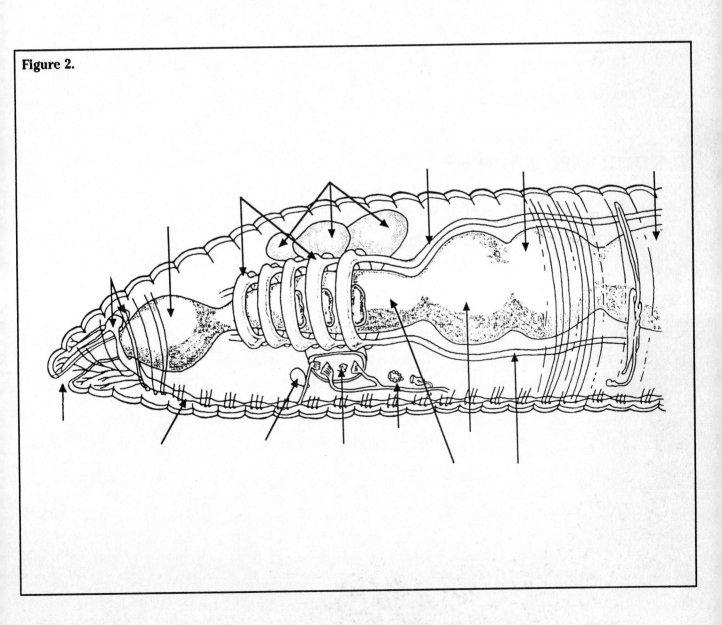

ANALYSIS

1. Which surface of the earthworm is darker in color—dorsal or ventral? _____

2. How many setae can you find per segment? _____

   In which direction do they point? _____

   What is the function of setae? _____

   _____

3. List the sequence of structures that food passes through after being taken in by the earthworm. _____

   _____

4. The earthworm does not undergo self-fertilization. What is the advantage (if any) to being

   hermaphroditic? _____

   _____

5. List three features of the earthworm that distinguish it from members of Phylum Platyhelminthes and

   Phylum Nematoda. _____

   _____

FURTHER EXPLORATIONS

1. Observe movement in live earthworms. Determine how the
   movement is similar to peristaltic movement of intestines.
2. Obtain specimens of each class of Annelida. Draw and label their
   external features. Compare their similarities and differences.

How Do Snails Respond to Stimuli?

30-2

LAB

INVESTIGATION

▶ The organization of an animal's nervous system often reflects its patterns of behavior. Animals that are very active usually show a greater degree of brain development and better-developed sense organs. Though not a rapid mover, the snail is an active animal and its nervous system is well developed. The snail has sense organs in its mantle cavity and at the ends of its tentacles. The way it responds to environmental stimuli may mean the difference between survival and death.

OBJECTIVES

- Observe a snail's response to gravity, touch, and light.
- Test a snail's response to a chemical substance and to substances common in its environment.
- Make a hypothesis that describes a snail's response to an acid substance.
- Measure the rate at which a snail moves.

MATERIALS

white paper
stereomicroscope or hand lens
lettuce
pencil
glass jar (1 pint) or 250-mL beaker
paper towels

live snails (5)
moist soil
camel hair brush
vinegar or lemon juice
small plastic tray with lid,
   approx. 15 cm × 20 cm × 5 cm

cardboard, approx. half
  the size of the tray
metric ruler
stopwatch or clock

PROCEDURE

Part A. Geotaxis and Touch

Geotaxis is the response of an animal to the force of gravity. The response may be either positive (toward gravity) or negative (away from gravity).

1. Place five snails on a wet paper towel on a flat surface.
2. Locate the head regions and tentacles of the snails. Using a hand lens or a stereomicroscope, locate the eyes on the ends of the tentacles as shown in Figure 1.
3. Place two or three snails in the bottom of a clean jar or beaker. Place the jar on its side and observe snail movement. Return the jar to an upright position and place the snails on the bottom again. Place the jar on its side and observe. Record in Table 1 how many snails moved toward the force of gravity and how many moved away from the force of gravity.

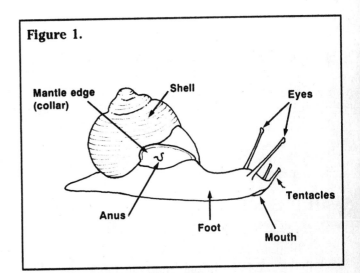

Figure 1.

Mantle edge (collar) Shell Eyes Anus Foot Mouth Tentacles

4. Very gently tickle the tentacles with the paintbrush and observe the behavior of the snails. Record your observations in Table 1.

Part B. Phototaxis

Phototaxis is the movement of an animal in response to light. The phototactic response can be either positive (toward light) or negative (away from light).

1. Place a piece of paper towel in the bottom of a small plastic tray.
2. Divide the tray into two halves by marking a line on the paper towel with a pencil as shown in Figure 2.
3. Moisten the paper towel with water and place the five snails on the line.
4. Place a piece of cardboard on the top of one-half of the tray. This will provide a darkened area for the snails.
5. Observe the snails for 15 minutes.
6. Record in Table 1 how many snails showed a positive phototactic response and how many snails showed a negative response.

Part C. Chemotaxis

Chemotaxis is the movement an animal makes in response to a chemical stimulus. Chemotactic movements can be either positive (toward the chemical) or negative (away from the chemical).

1. Remove the moist paper towel from the plastic tray.
2. Divide the plastic tray into halves with a pencil line.
3. Place a piece of lettuce in one half of the tray as shown in Figure 3.
4. Place the five snails on the center line and cover the tray with its lid. Observe the activity of the snails for 15 minutes. Record your observations in Table 1.
5. Repeat steps 3 and 4, substituting moist soil for the lettuce.
6. Repeat step 3, substituting a paper towel soaked in vinegar or lemon juice for the lettuce.
7. Make a **hypothesis** that describes how the snails will react to this chemical substance. Write your hypothesis in the space provided.
8. Repeat step 4.

Part D. Snail's Pace

1. Draw a short line on a piece of white paper.
2. Moisten the paper with water.
3. Place the front of one snail on the line and allow the snail to move for 60 seconds.
4. Measure the distance (in mm) traveled by the snail. Record the distance in Table 2 for Trial 1.

Figure 2.

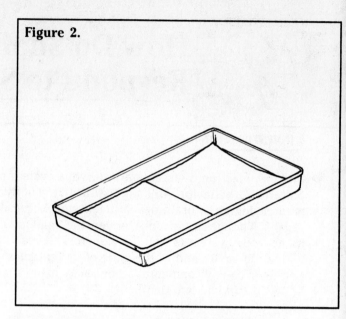

Figure 3.

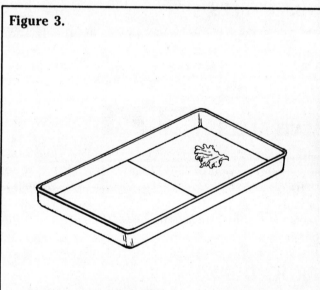

5. Repeat steps 3 and 4 three more times. Calculate the average speed first in mm per 60 seconds (mm/60 s) and then in mm per second (mm/s). Record your results next to Table 2.
6. Compare the speed of your snail with those of others in your class.
7. Return the snails to their proper place, as directed by your teacher, at the conclusion of this Investigation.

HYPOTHESIS

_____

_____

DATA AND OBSERVATIONS

Table 1.

Responses of Snails to Stimuli	
Test	**Response**
Geotaxis	
Touch	
Phototaxis	
Chemotaxis	
Lettuce	
Soil	
Vinegar or lemon juice	

Table 2.

A Snail's Pace
mm/60 s
Trial 1   _____
Trial 2   _____
Trial 3   _____
Trial 4   _____

average speed = _____ mm/60 s

average speed = _____ mm/s

ANALYSIS

1. **a.** Was the snails' behavior toward or away from the force of gravity? _____

   b. Is this positive or negative geotaxis? _____

2. **a.** How did the snails react when given the choice between light and dark conditions? _____

   _____

   b. How might this type of behavior be beneficial for a snail? _____

   _____

3. How did the snails react when touched on the tentacle? _____

4. **a.** How did the snails react when exposed to moist soil and lettuce? _____

   b. Explain how this behavior could be beneficial. _____

   _____

5. a. Did the snail exhibit a positive or a negative response to the vinegar or lemon juice? _____

   b. How might this type of behavior be beneficial for a snail? _____

_____

CHECKING YOUR HYPOTHESIS

Was your **hypothesis** supported by your data? Why or why not? _____

_____

FURTHER INVESTIGATIONS

1. Design an experiment to test a squid's response to the same stimuli. Compare the squid's responses with those of the snail. Account for any differences in response by comparing the environments of the animals.

2. Squid have chromatophores in their skin, which contain various colored pigments. Chromatophores are under nervous and hormonal control and can create a complex pattern of colors and color flashes by turning on and off. The color patterns created are used in communication. Research this subject in the library and prepare a report.

Squid Dissection

▶ Mollusks are soft-bodied animals and include snails, clams, octopus, squid, and slugs. They are bilaterally symmetrical and have well-developed digestive, circulatory, excretory, and respiratory systems. Squid are characterized by a large prominent head with conspicuous eyes and a mouth surrounded by ten tentacles. The class Cephalopoda, which includes the squid and the octopus, is considered to be the most complex and highly developed group of mollusks.

OBJECTIVES

- Dissect and identify the organs and major organ systems of the squid.
- Describe the major features of the squid phylum.

- Determine the function of various squid features.

MATERIALS

squid (fresh or thawed)
scissors
dissecting pins (5–10)

dissecting pan
stereomicroscope or hand lens
laboratory apron

PROCEDURE

Part A. Body Organization

1. Put on the laboratory apron. Place the squid in the dissecting pan. Note that there is no external shell and that the major part of the body is enclosed by the soft, muscular **mantle**. There are ten conspicuous arms or **tentacles**, derived from the mollusk foot.

2. Arrange your specimen so that the dorsal end points away from you. See Figure 1. Turn the animal so that the siphon faces you. The eyes should be on the right and left sides of the body.

3. Slit open the mantle cavity by inserting the tip of the scissors under the mantle at the siphon and cutting to the apex. Cut with care so that you do not disturb the internal organs.

4. Pin down the mantle to the pan, slanting the pins at an angle away from the specimen as shown in Figure 2.

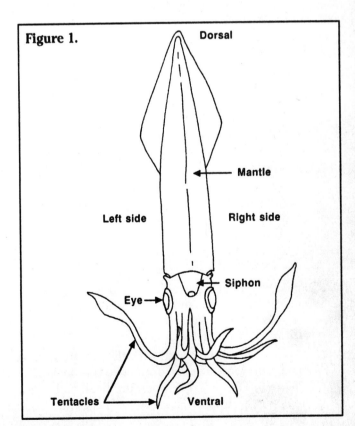

Figure 1.

Dorsal

Mantle

Left side

Right side

Siphon

Eye

Tentacles

Ventral

Figure 2.

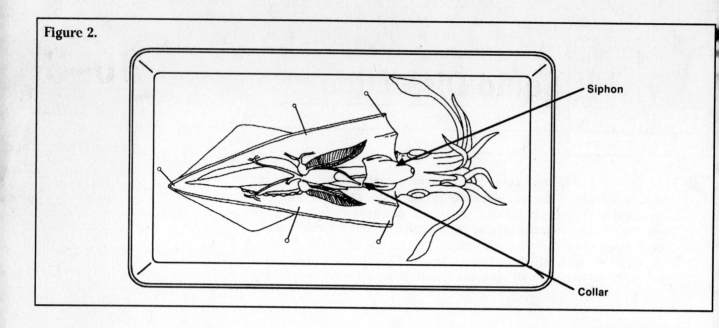

Siphon

Collar

Part B. Mantle Cavity and Respiratory System

1. Examine the mantle cavity. The walls of the mantle cavity are very muscular. This cavity is involved in propelling a squid through the water. In the living squid, the mantle cavity expands by muscular action and fills with water. The **collar** locks tightly against the head, leaving the **siphon** as the only exit for the water. The mantle muscle then contracts and water is squeezed out through the siphon. This method of movement is referred to as jet propulsion.

2. Examine the siphon. The siphon is well equipped with muscles and can be pointed for directional jet propulsion. Note that the tip of the siphon has a muscular valve.

3. Find the two **gills** shown in Figure 3. These structures are oriented so that incoming water passes over them.

4. Locate and remove the **pen**, the vestigial internal shell. Grasp the tip of the pen and tug gently.

Part C. Feeding and Digestive Systems

1. Examine under a stereomicroscope or a hand lens the structure and organization of the **suckers**. The suckers, which are located on the tentacles, are used to hold onto prey.

2. Remove the siphon and, with the scissors, make an incision into the head. Expose the beak as shown in Figure 4. Pry open the beak and observe the tonguelike **radula**. Trace the

Figure 3.

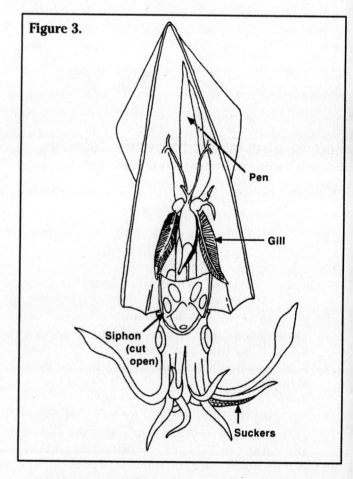

Pen

Gill

Siphon
(cut
open)

Suckers

esophagus, which is surrounded by the **liver**, to the thick-walled **stomach**. The stomach emerges to form the **caecum**. Note the **pancreas**. The **intestine** runs from the stomach and terminates in the **rectum**. An **ink sac** arises from the intestine near the anus, which is located near the siphon. It is used for defense.

Figure 4.

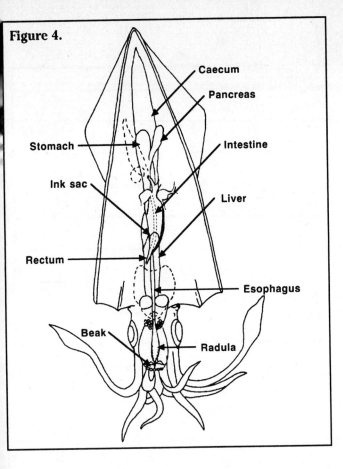

Figure 5.

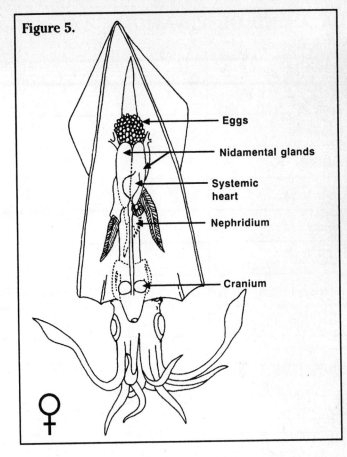

Part D. Circulatory, Excretory, Nervous, and Reproductive Systems

1. Locate the systemic **heart** as shown in Figure 5. This is a difficult structure to find because it is transparent.

2. Examine the **nephridium**, a kidneylike excretory organ that removes liquid waste products from the blood.

3. Locate the white mass of the **cranium** above and between the eyes. This structure contains the squid's **brain**.

4. Locate the reproductive organs as shown in Figure 6. Determine the sex of your squid, but be sure to examine squid of both sexes. The male has **testes** that lie beneath the caecum. The female has two large **nidamental glands** that secrete a protective covering over the **eggs**. Eggs might or might not be present.

5. Remove the **eye** and cut it in half. Examine the transparent **lens** and the shiny black **retina** at the back of the eye.

6. Complete Table 1 in Data and Observations. Consult your textbook if necessary.

Figure 6.

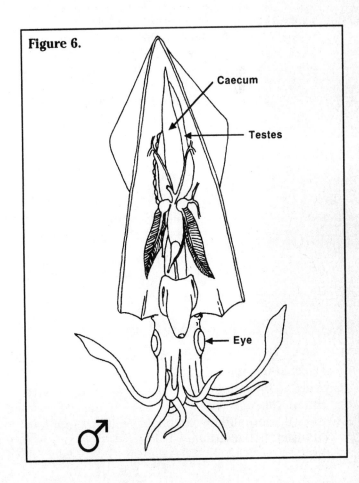

DATA AND OBSERVATIONS

Table 1.

Organ	Function
Mantle	
Siphon	
Gills	
Suckers	
Pen	
Ink sac	
Nephridium	
Nidamental gland	

ANALYSIS

1. Describe squid locomotion. _____

_____

_____

2. What anatomical features show that a squid is well adapted to a predatory existence? _____

_____

3. What could be the advantage of a reduced internal shell? _____

_____

4. Cephalopods are thought to have the most highly developed eyes in the invertebrate world. What

anatomical evidence did you see that indicates this? _____

_____

5. Which features of the squid are common to all members of the phylum Mollusca? _____

_____

FURTHER EXPLORATIONS

1. Write a report on the chambered nautilus. Learn how it controls its depth in the water.
2. Many squid contain bioluminescent bacteria in their skin. Try to culture *Photobacterium* from the surface of the squid's skin. Many college microbiology lab books contain directions on how to culture this bacterium.

How Does Temperature Affect Mealworm Metamorphosis? 31–1

LAB

INVESTIGATION

▶ Moths, butterflies, bees, and beetles are examples of insects that have four stages to their life cycles: egg, larva, pupa, and adult. This sequence of events is called complete metamorphosis. The female adult lays eggs, which hatch into larvae. Larvae eat an enormous amount of food and grow very rapidly. When a larva reaches a certain size, it enters an immobile stage called a pupa. Inside its protective covering, the pupa undergoes many changes controlled by hormones. When metamorphosis is complete, an adult emerges from the pupa. The mealworm, *Tenebrio,* is an excellent insect for the study of complete metamorphosis.

OBJECTIVES

- Observe the four stages of the life cycle of the mealworm, *Tenebrio.*
- Hypothesize how an increase in temperature will affect the rate of metamorphosis.
- Conduct an experiment to test the effect of temperature on the development and emergence of an adult mealworm from a pupa.

MATERIALS

samples of mealworms (egg, larva, pupa, adult)
mealworm pupae (of same age) (4)
wax marking pencil
stereomicroscope
chart of *Tenebrio* life cycle

plastic vials (4)
foam plugs (4)
incubator (at 30°C)
thermometer

PROCEDURE

1. Examine samples of the four stages of mealworms under the stereomicroscope. Relate them to the chart of the life cycle of *Tenebrio.*
2. Make a **hypothesis** that predicts how an increase in temperature will affect the length of time it takes a mealworm pupa to become an adult. Write your hypothesis in the space provided.
3. With your marking pencil, label the four plastic vials Room Temp. A, Room Temp. B, 30°C A, and 30°C B. These labels indicate the temperature at which the pupae will be stored. Label them also with your name (or group name) and the date.
4. Place one pupa in each of the four vials and stopper with foam rubber plugs. The foam plugs will allow the insects to breathe.

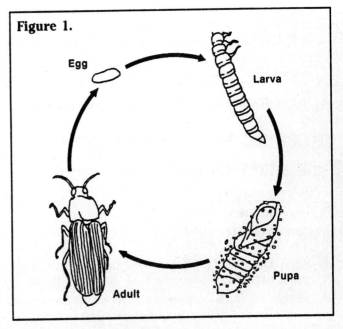

Figure 1.

Egg

Larva

Pupa

Adult

5. Store your vials at their proper temperatures with those of the rest of the class. Record the starting date in Table 1. Record the actual room temperature in Table 2.

6. Check your vials daily for the presence of adult mealworms. When you observe an adult in a vial, record in Table 1 the number of days needed for metamorphosis.

7. Dispose of the adults after they emerge and clean your vials following the instructions of your teacher.

8. Wait for metamorphosis to be complete for mealworms of the entire class, compile the class data, and complete Table 2. Calculate the average time for emergence by dividing the total number of days by the total number of pupae.

HYPOTHESIS

_____

_____

DATA AND OBSERVATIONS

Table 1.

Tenebrio Metamorphosis		
Temperature	Starting date	Length of time for emergence (Days)
Room temp. A		
Room temp. B		
30°C A		
30°C B		

Table 2.

Calculations			
Temperature	Total number of days for entire class	Total number of pupae for entire class	Average time for emergence
Room temp. (_____°C)			
30°C			

ANALYSIS

1. How did an increase in temperature affect the time needed for metamorphosis? _____

2. Why do you think a temperature increase causes this effect? _____

_____

3. Why might the class averages be a more accurate measurement of the time for metamorphosis than your

data alone? _____

4. Sequence the stages of a Tenebrio life cycle. _____

CHECKING YOUR HYPOTHESIS

Was your **hypothesis** supported by your data? Why or why not? _____

_____

FURTHER INVESTIGATIONS

1. Repeat the Investigation with other insects, such as Drosophila, to see if temperature affects their metamorphosis.

2. Design an experiment to test the effect of temperature on other stages in the life cycle of Tenebrio.

Identifying Insects

31–2

LAB

EXPLORATION

▶ Insects make up a large class in the phylum Arthropoda. Their external anatomical features have been used to classify insects into orders. Insects in the same order share many features and generally look similar to one another. For example, bees and wasps belong to the order Hymenoptera. Insects in different orders might resemble those of other orders at first glance, but a careful look at external characteristics makes it possible to identify them correctly. Biological keys make it easier to use external characteristics to identify the different kinds of insects.

OBJECTIVES

• Become familiar with various insect anatomical characteristics.
• Become familiar with using a biological key.

• Use the characteristics of insects and a biological key to identify insects to their order.

MATERIALS

white adhesive labels
hand lens or stereomicroscope
insect collection from Exploration 1–1

PROCEDURE

1. Use the procedures learned in Investigation 20–1, "Can a Key Be Used to Identify Organisms?", to identify each one of your fifteen insects using the key included with this Exploration under Data and Observations. For assistance with mouth parts, refer to Figure 1. Use the hand lens or a stereomicroscope to help you see the insects' structures.

2. Prepare a label showing the order of each insect you identify.

3. Place the label on the polystyrene board beneath each keyed insect.

4. Turn in your keyed insect collection.

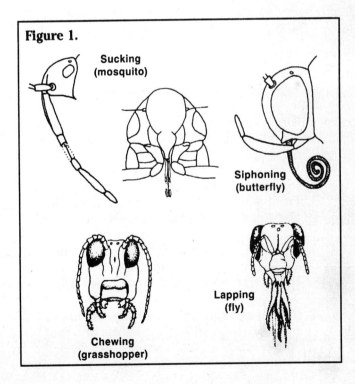

Figure 1.

Sucking (mosquito)

Siphoning (butterfly)

Chewing (grasshopper)

Lapping (fly)

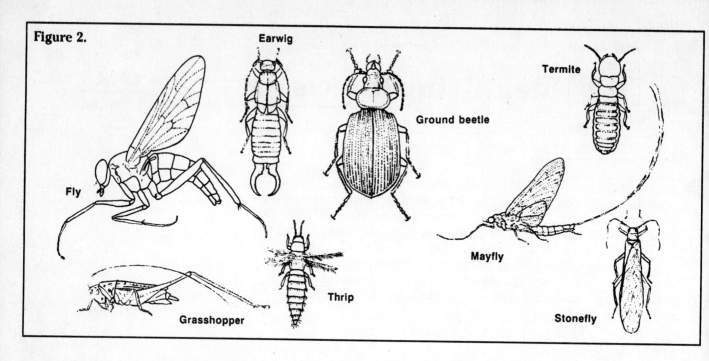

Figure 2.

Earwig

Ground beetle

Termite

Fly

Mayfly

Grasshopper

Thrip

Stonefly

DATA AND OBSERVATIONS

Key to Insect Orders

1. A. Insects with wings.. Go to 2
   B. Insects without wings or only rudimentary wings .... Go to 16

2. A. Insects with only one pair of thin, usually transparent wings; second pair
   replaced with short, pinlike structures (flies, mosquitoes) .... Order Diptera
   B. Insects with two pairs of wings .... Go to 3

3. A. Two pairs of wings not alike in structure (not equal in thickness or
   transparency) (beetles, true bugs, grasshoppers) .... Go to 4
   B. Two pairs of wings of similar structure (bees, butterflies, dragonflies) .... Go to 7

4. A. First pair of wings horny and meeting in a straight line down the back (beetles,
   earwigs).. Go to 5
   B. First pair of wings not as in 4A .... Go to 6

5. A. Tip of abdomen with a prominent pair of pincers (earwigs)................... Order Dermaptera
   B. Tip of abdomen without pincers (beetles) .... Order Coleoptera

6. A. Front wings leathery at base, membranous and overlapping at tip; mouthparts
   for sucking (true bugs) .... Order Hemiptera
   B. Front wings leathery with veins; hind wings folded lengthwise; mouthparts for
   chewing (crickets, katydids, grasshoppers) .... Order Orthoptera

7. A. Wings wholly or for the most part covered with scales; mouthparts formed for
   siphoning (moths, butterflies) .... Order Lepidoptera
   B. Wings transparent or thinly clothed with hairs (bees, mayflies, dragonflies) .... Go to 8

8. A. Mouthparts for sucking attached to hind part of lower surface of head; wings
   when at rest held like the halves of a roof (cicadas, leafhoppers, treehoppers,
   aphids) .... Order Homoptera
   B. Mouthparts not as in 8A .... Go to 9

9. A. Body usually brown and slender; insect mothlike, with long slim antennae; no
mouthparts evident except for a pair of slender palpi; wings frequently hairy,
usually broadest beyond the middle; wings held like halves of a roof over
abdomen (caddisflies) .... Order Trichoptera
   B. Body and other characteristics not as in 9A .... Go to 10

10. A. Wings with few or no cross veins (bees, thrips) .... Go to 11
   B. Wings with many cross veins (dragonflies, lacewings) .... Go to 12

11. A. Front wings the larger pair; hind wings frequently hooked to front wings;
mouthparts for lapping, chewing, or sucking (bees, wasps) .... Order Hymenoptera
   B. Front wings the same size; wings very narrow, bristly hairs at margin (thrips) ... Order Thysanoptera

12. A. Front pair of wings much larger than hind pair; wings held vertically above
body; long, fragile-jointed tails behind (mayflies) .... Order Ephemeroptera
   B. Front pair of wings and other characteristics not as in 12A .... Go to 13

13. A. Antennae short and inconspicuous; long slender insects with long narrow
wings (damselflies, dragonflies) .... Order Odonata
   B. Antennae longer and conspicuous .... Go to 14

14. A. Abdomen usually with two short appendages; back wings much broader than
front wings folded lengthwise (stoneflies) .... Order Plecoptera
   B. Abdomen and characteristics not as in 14A .... Go to 15

15. A. Wings equal in size with many distinct veins; tarsi five-jointed
(antlions, lacewings) .... Order Neuroptera
   B. Wings equal in size with indistinct veins; tarsi four-jointed; thorax in front of
wings very short (termites) .... Order Isoptera

16. A. Insects narrow-waisted, antlike (ants) .... Order Hymenoptera
   B. Insects not narrow-waisted but still antlike .... Go to 17

17. A. Bodies antlike but with wide waists, not flattened, light colored (termites) .... Order Isoptera
   B. Bodies not as in 17A .... Go to 18

18. A. Insects small and plump, soft-bodied with small heads, two short tubes
extending from back of abdomen; sucking mouthparts (aphids) .... Order Homoptera
   B. Insects not as in 18A .... Go to 19

19. A. Body of insect small and narrow, flattened on the side; sucking mouthparts;
hind legs for jumping; five tarsal segments (fleas)...................... Order Siphonaptera
   B. Body of insect not as in 19A .... Go to 20

20. A. Body covered thickly with scales; mouthparts for sucking or are absent
(butterflies, moths) .... Order Lepidoptera
   B. Body of insect not as in 20A .... Go to 21

21. A. Insect very delicate with chewing mouthparts and long, jointed, threadlike
tails and antennae (bristletails, firebrats, silverfish) .... Order Thysanura
   B. Insect not as in 21A .... Go to 22

22. A. Insect delicate with chewing mouthparts; abdomen of six segments; underside
of abdomen frequently has a long, usually double appendage used for leaping
(springtails) .... Order Collembola
   B. Insect not as in 22A .... Go to 23

181

23. A. Mouthparts for chewing .... Go to 24

B. Mouthparts for sucking .... Go to 25

24. A. Antennae threadlike; face directed forward and downward (crickets, roaches, grasshoppers, walkingsticks) .... Order Orthoptera

B. Antennae beadlike, clublike or comblike, without prominent forceps at tip of abdomen (beetles) .... Order Coleoptera

25. A. Insect apparently legless, frequently covered with a waxy scale; insect usually tightly attached to plant leaves or stems (scale insects)................... Order Homoptera

B. Insect with well developed legs with a sucking beak that arises at front of head and held between the legs (true bugs) .... Order Hemiptera

ANALYSIS

1. The first separation in the key involved the presence or absence of wings. What is another feature that

could have been used to separate the insects into two main groups? _____

_____

2. Must animals that belong to the same order also belong to the same class? Explain. _____

_____

3. Choose two of your insects that are similar to one another. What features make them appear to be

similar? _____ What characteristics make it possible to

distinguish them? _____

4. Would color be a good characteristic to use to classify insects and include in a key? Explain. (Examine

other students' collections and compare the same kind of insects.) _____

_____

_____

5. List three main characteristics that are used repeatedly in the key to identify the insects. _____

_____

_____

FURTHER EXPLORATIONS

1. Prepare a classroom display of the various types of insect mouthparts by drawing and labeling various parts on poster board.

2. Consult the library or ask your teacher for biological keys that will allow you to key some of your insects to the family, genus, and species levels. Keying below the level of family usually requires a great deal of care and precision.

Comparing Arthropods

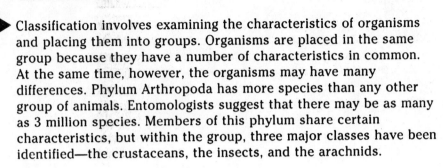

31–3
LAB

▶ Classification involves examining the characteristics of organisms and placing them into groups. Organisms are placed in the same group because they have a number of characteristics in common. At the same time, however, the organisms may have many differences. Phylum Arthropoda has more species than any other group of animals. Entomologists suggest that there may be as many as 3 million species. Members of this phylum share certain characteristics, but within the group, three major classes have been identified—the crustaceans, the insects, and the arachnids.

OBJECTIVES

- Examine the characteristics of a spider, a crayfish, and a grasshopper.
- Determine which characteristics are phylum characteristics.
- Determine which characteristics are class characteristics.
- Construct a dichotomous taxonomic key.

MATERIALS

preserved spider
preserved crayfish

preserved grasshopper
dissecting probes (2)

dissecting pan
stereomicroscope or hand lens

PROCEDURE

Part A. Observation

1. Place a preserved spider, crayfish, and grasshopper in a dissecting pan.
2. Examine the following features of the three arthropods. Record your observations in Table 1.

Segmented Body

Examine both the dorsal (back or top) and ventral (underside) surfaces of the specimens' bodies. Do the animals appear to be segmented in any body region?

Type of Skeleton

Touch and lightly squeeze the animal. If the outside is hard, the animal has an external skeleton. If soft, the animal has an internal skeleton. Is the skeleton external or internal?

Types of Appendages

An appendage is any limblike structure that projects from a body or organ. See if the segments that make up the appendages will bend. If appendages bend

between segments, they are jointed. Are the appendages "jointed?"

Modifications of Appendages

Are the appendages structurally and functionally different? List some of the functions of the various appendages. Use your text to help you.

Wings

Does the animal have wings?

Number of Body Regions

Examine the dorsal and ventral surfaces. Look for the head, thorax, and abdomen. How many body regions does each of the specimens have?

Fusion of Body Regions

Are the head and thorax fused together to form a cephalothorax region? Look for a groove or depression between the two. This groove suggests that they were once separate regions.

Figure 1.

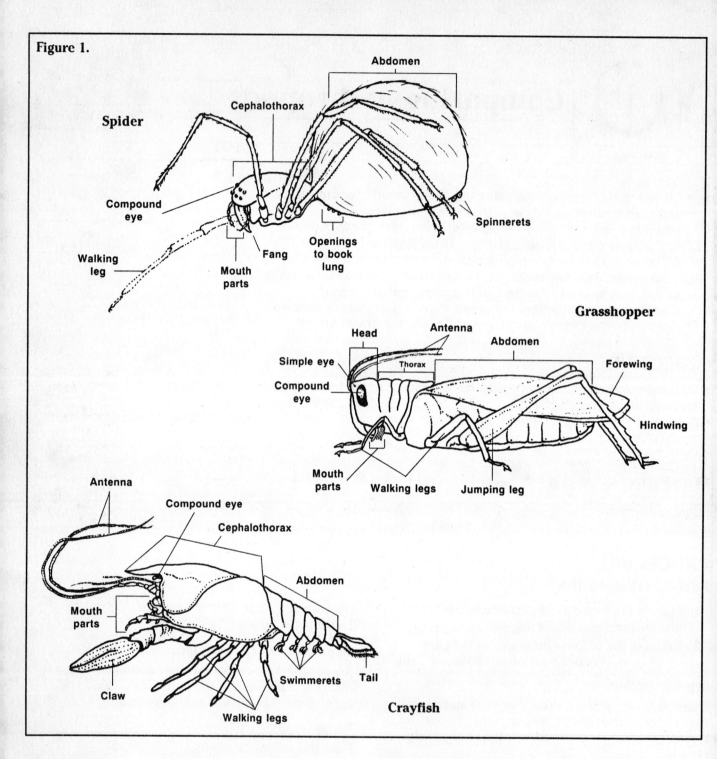

Number of Eyes

How many simple eyes and compound eyes does the animal have? (You will need to examine the specimens with a hand lens or stereomicroscope.)

Number of Antennae Pairs

Does the animal have antennae present? If so, how many pairs of antennae does the animal have?

Number of Leg Pairs

How many pairs of legs does each animal have? (Count the large claw structures on the crayfish as legs. The small leg-like appendages on the crayfish abdomen are not legs. They are swimmerets and function in reproduction.)

Part B. Making a Key

1. Prepare a taxonomic key that can be used to identify organisms that belong to these three classes. For assistance, consult with your teacher or refer to *Organizing Information* in the *Skill Handbook,* pages 1143–1148. Write your key in the space provided in Data and Observations.

DATA AND OBSERVATIONS

Table 1.

Characteristic	Crustacea (crayfish)	Insecta (grasshopper)	Arachnida (spider)
Segmented body?			
Type of skeleton?			
Type of appendages?			
Appendages structurally different?			
Functions of appendages?			
Wings present?			
Number of body regions?			
Body regions fused?			
Number of simple eyes?			
Number of compound eyes?			
Number of antenna pairs?			
Number of leg pairs?			

Characteristics of Arthropods

Student Key:

ANALYSIS

1. Which of the characteristics are the same in all three animals? _____

   _____

2. In which characteristics do the three animals appear to differ? _____

   _____

3. At which level in the classification hierarchy (phylum or class) do members of a group have more

   characteristics in common? _____

   _____

4. The classification systems used today attempt to reflect genetic relationships. Why can you say that
   members of a group are more closely related to one another than they are to members of another group?

   _____

   _____

   _____

   _____

FURTHER EXPLORATIONS

1. Examine centipedes or millipedes, or photographs of these
   arthropods. What phylum and class characteristics are shared with
   the other classes of arthropods? What characteristics are different?
2. Investigate how to build a spider-web box. Build a spider box and
   observe web-building behavior.

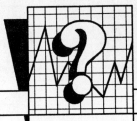

Do Starfish Respond to Gravity?

32–1
LAB

INVESTIGATION

▶ Starfish respond to stimuli by means of their tube feet. Investigators have shown that starfish move by pushing off with their tube feet as well as by pulling with them. Thus, the starfish moves, or walks, in response to external stimuli such as the force of gravity. Geotaxis, the response to the force of gravity, can be either positive (in the same direction as the force of gravity) or negative (in the opposite direction as the force of gravity).

OBJECTIVES

- Observe the external structures of a living starfish.
- Observe the walking behavior of a living starfish.
- Construct a hypothesis that describes a starfish's reaction to gravity.
- Determine whether or not a starfish responds to the force of gravity.

MATERIALS

living starfish
seawater

colored celluloid
glass container for starfish

glass plate
drawing paper

PROCEDURE

1. Carefully place a living starfish, dorsal surface up, in a glass container of seawater.

2. Make drawings on a sheet of drawing paper of the dorsal and ventral surfaces of your starfish. Refer to the figures in Exploration 32–2. Correctly draw and label the following.

   dorsal surface — central disc (central body)
   rays (five arms)
   madreporite plate (round structure)

   ventral surface — mouth
   ambulacral grooves (5)
   tube feet

The tube feet are connected internally to a canal system that in turn connects to the madreporite plate. Water filters into the canal system through the madreporite plate. Within the canals, water is under pressure and helps to extend and retract the tube feet. The extending and retracting of the tube feet is used by the starfish for locomotion.

3. Gently remove the starfish from the container. Place the piece of celluloid in the water at the bottom of the dish. Place the starfish on top of the piece of celluloid. See Figure 1.

4. Observe the stepping action of the tube feet on the celluloid sheet by looking at the underside of the container while another student holds it up. Observe the movement of the celluloid sheet. After completing your observations, carefully remove the celluloid from the starfish. Write a description of the movement in Data and Observations.

5. Make **hypotheses** that describe how a starfish will react to (1) being turned over, and (2) being placed on an inclined surface. Operate under the assumption that the reactions will be responses to the force of gravity. Write your hypotheses in the space provided.

Figure 1.

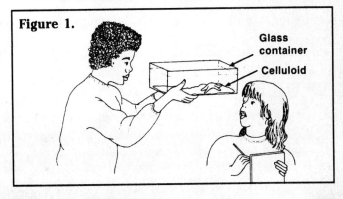

Glass container
Celluloid

6. Turn the starfish over in its container so that the ventral surface faces up. Note the reaction of the starfish. Record your observations in Data and Observations.

7. Lean one edge of the glass plate on the bottom of the container. Allow the starfish to crawl onto the plate. When it is well onto the plate, tilt the plate at a 45 degree angle. See Figure 2.

8. Observe how the starfish reacts to being on an inclined surface.

9. After several minutes, shift the glass plate so that the edge that was once down is now up, as in Figure 2. Observe the reaction of the starfish. Record your observations in Data and Observations.

10. After making your observations, return the living starfish to its aquarium.

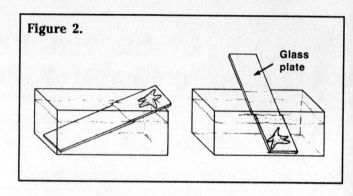

Figure 2.

Glass plate

HYPOTHESES

_____

_____

DATA AND OBSERVATIONS

1. Describe what happened to the celluloid as the starfish moved. _____

_____

2. Describe the movements of the starfish after it was turned over. _____

_____

3. Did one ray of the starfish lead the way? _____

4. How many rays were attached to the dish before the body of the animal turned over? _____

5. How did the starfish react to finding itself on an inclined surface? _____

6. How did the starfish react to the reversal of "up" and "down"? _____

ANALYSIS

1. Did the piece of celluloid move in relation to the starfish's body? _____ In what direction did

   the starfish move in relation to the direction of movement of the celluloid? _____

2. Does a starfish show positive or negative geotactic behavior? _____

CHECKING YOUR HYPOTHESES

Were your **hypotheses** supported by your data? Why or why not? _____

_____

FURTHER INVESTIGATIONS

1. Test the righting response of a sea cucumber or a sea urchin.

2. Examine the movements of a starfish in reaction to a clam. Starfish that are hungry will eat a live clam. It may take several hours for a starfish to get the clam open and extend its stomach into the clam.

Starfish Dissection

32-2
LAB

Phylum Echinodermata is a group of marine animals that includes starfish. "Echinos" means spiny, and "derma" means skin. Starfish are the most common representative animals of this phylum. Both internally and externally, the anatomy of starfish in the genus *Asterias* is characterized by five arms and a five-part anatomical arrangement.

OBJECTIVES

• Dissect a preserved starfish.
• Identify and label the major external and internal structures of a starfish.

• Describe how the water vascular system enables a starfish to move.

MATERIALS

scissors
preserved starfish
dissecting tray

dissecting probe
laboratory apron

PROCEDURE

Part A. External Anatomy

1. Place your starfish in the dissecting tray so that the dorsal (top) surface faces upward as shown in Figure 1.
2. Examine the animal's dorsal surface. Locate the **central disc** and the five arms or **rays** that extend from the central disc. Find these structures on Figure 1.

3. Note the many **spines** scattered over the surface of the arms and the central disc. These spines are attached to the plates of the starfish skeleton just under the skin. These plates are called ossicles. Find the spines on Figure 1.
4. Examine the demonstration slide that your teacher has set up on the stereomicroscope. Among the spines are small pincer-like pedicellarias and gills. The **pedicellarias** keep the surface of the skin free of debris, while the **gills** are involved in respiration. Find these structures on Figure 2.

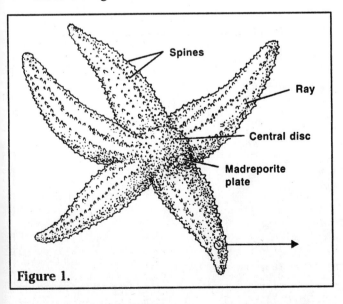

Figure 1.

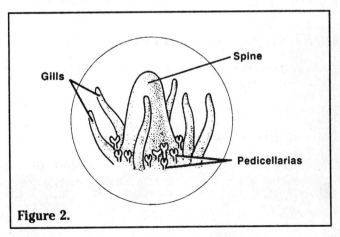

Figure 2.

5. Turn the starfish over and locate the mouth. Examine the five **ambulacral grooves** that extend from the **mouth** along the middle of each ray. Numerous **tube feet** used for locomotion and absorption of dissolved oxygen are present along the grooves. Find these structures on Figure 3.

Part B. Internal Anatomy

Digestive System

1. Once again, place the starfish in the dissecting tray so that the dorsal surface faces upward.

2. Locate the **madreporite plate**. It is a round, sievelike structure on the dorsal surface of the starfish, and it looks almost like a wart. Use Figure 1 as a reference and find the madreporite plate.

3. Carefully cut a ring around the madreporite plate.

4. Use your scissors to cut off the tip of any arm except for the two arms next to the madreporite plate.

5. Starting at the end of the arm with its tip cut off, use your scissors to remove the remaining skin/skeleton from the central disc and from the top of the arm. This will expose the starfish's internal organs. Start your cut at the end of the cut arm.

6. Examine the **digestive gland**, a large olive-green gland with two branches that fills most of the arm.

7. Draw and label this digestive gland on the outline of the starfish in Figure 6.

8. Turn your starfish over and locate the **mouth**. The mouth is attached to the pouchlike **stomach**, which can be seen through the opening you have cut in the dorsal surface.

9. Draw and label the mouth and stomach on the starfish in Figure 6.

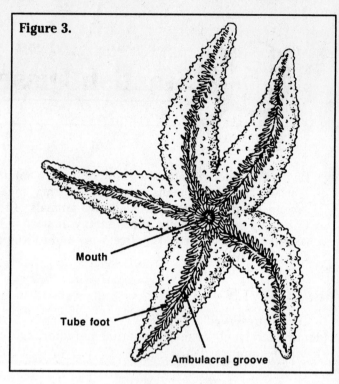
Figure 3.

Mouth

Tube foot

Ambulacral groove

10. Locate the short tube that connects the stomach to the digestive glands.

Reproductive Organs

1. Remove the entire digestive gland from the dissected arm of your starfish.

2. Locate the pale, lumpy organs under the digestive gland near the central disc as shown in Figure 4. These are the reproductive organs. They are called **gonads**. Starfish have separate sexes. During spawning, the gonads are very large, but in preserved specimens, they are usually quite small. The male and female gonads look very much alike in preserved specimens. In living starfish, the testes are gray and the ovaries are orange.

3. Draw and label the gonads in Figure 6.

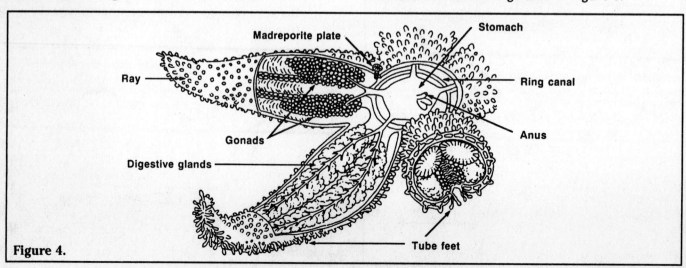

Madreporite plate

Stomach

Ray

Ring canal

Anus

Gonads

Digestive glands

Tube feet

Figure 4.

Water Vascular System

1. Carefully remove the reproductive organs and the remaining parts of the digestive system (including the stomach). This will expose the **water vascular system**. Be careful not to damage the madreporite plate.

2. Study your starfish. Find on Figure 5 each of the structures of the water vascular system that are in boldfaced print in the following paragraph.

Water enters the system through the **madreporite plate**. The madreporite plate is connected to the circular **ring canal** by the **stone canal**. The water is then distributed to the **radial canals** that are in each of the rays. These canals deliver water to the **tube feet**. The tube feet contain **ampullae** (bulblike structures). By alternating between contracting and expanding the ampullae, the starfish is able to move. As the ampullae contract, they force water into the tube feet, and the tube feet lengthen. The starfish places the lengthened tube feet in the direction it is going. Then the ampullae relax, and expand. When this happens, water leaves the tube feet, thus shortening each tube foot and creating suction at its end.

3. Trace the pathway that seawater takes from the madreporite plate to the tube feet. Mark this pathway on Figure 5 with short dashed lines.

4. After completing the dissection, dispose of your starfish according to your teacher's instructions.

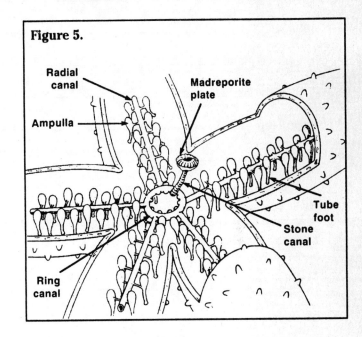

Figure 5.

Radial canal

Madreporite plate

Ampulla

Tube foot

Stone canal

Ring canal

DATA AND OBSERVATIONS

Figure 6.

ANALYSIS

1. List the functions of the:

   madreporite plate. _____

   spines. _____

   gills. _____

   digestive glands. _____

   pedicellarias. _____

   tube feet. _____

   gonads. _____

   radial canal. _____

2. Describe how the water vascular system helps a starfish move around. _____

   _____

   _____

   _____

3. What external and internal features indicate that the starfish is organized in a pattern of fives? _____

   _____

   _____

FURTHER EXPLORATIONS

1. Demonstrate how a plastic dropper can be used as a model to show the action of tube feet and ampullae as the dropper is filled and emptied of water.

2. Examine the anatomy of the other groups of echinoderms (sea cucumbers, crinoids, sea urchins, sand dollars) and note the similarities and differences with starfish anatomy.

Making an Echinoderm Key

32-3
LAB

▶ Taxonomy is the area of biology that includes classifying, naming, and identifying organisms. One of the most useful devices for identifying organisms is a taxonomic key. A key is a step-by-step listing of characteristics that eventually leads to the identity of an organism. Keys are available for practically every group of animals and plants.

OBJECTIVES

• Construct a key that can be used to identify various echinoderms.

• Utilize the key to identify five echinoderms.

MATERIALS

Specimens, models, plastimounts, or photographs of the following:

starfish   brittle star
sea cucumber   sea lily
sea urchin or sand dollar   (one of each per group of students)

PROCEDURE

1. Study each of the specimens available and observe the structural features of each.

2. Prepare a list of similarities and differences of the five specimens.

3. Divide the specimens into two groups based upon one characteristic that clearly separates each group. This characteristic must be present in only one group for the key to work.

4. Use this characteristic to fill in the first two lines of your key. Write the two alternative conditions for this one characteristic on the lines marked 1A and 1B in Data and Observations. Add directions by writing "Go to 2" on the first line. Directions for line 1B will be filled in later. (For extra help, refer to *Organizing Information* in the *Skill Handbook* of your textbook.) As an example, a group of animals can be separated by the number of their legs. 1A might read "Has 4 legs . . . go to 2" and 1B might read "Has 2 legs."

5. Allow your teacher to check this part of your key before you continue.

6. Continue to work with the group identified by the characteristic in 1A and repeat steps 3 and 4. Place the next characteristic difference on

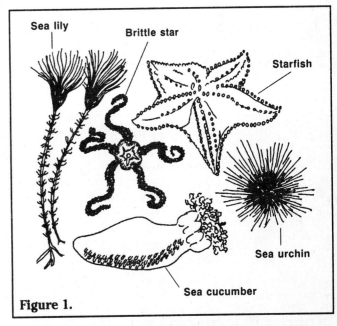

Figure 1.

the next lines under 2A and 2B. If a line of your key leads to only one organism, write the name of the organism at the far right side of the line. Repeat this procedure, constructing your key, until all organisms from the 1A group have been keyed out and identified to species.

7. Repeat steps 3 through 6 with the group from 1B. Place these lines of your key below those lines from step 6. Fill in the directions on line 1B from step 4.

8. Allow your teacher to check your completed key.

9. Identify your organisms using your newly constructed key.

10. Exchange your key with that of another student and use his or her key to identify your specimens. Make a note of the characteristics that the other student used to construct the key. Retrieve your key and make any changes in it that will make it easier to use.

DATA AND OBSERVATIONS

1A _____ **Go to 2**

1B _____

2A _____

2B _____

_____

_____

_____

_____

_____

_____

_____

_____

ANALYSIS

1. List three characteristics that were used in your key. _____

_____

2. What features, if any, did you use in your key that were used by other students? _____

_____

3. Why shouldn't color or food habits be used in making a key? _____

_____

FURTHER EXPLORATIONS

1. Use the key you have constructed to try to identify other echinoderm specimens that are available. Specimens you might see on trips to the ocean or an aquarium would be good choices.

2. Make a key for another phylum such as mollusks, spiders, or protozoans.

Frog Dissection

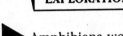
EXPLORATION

▶ Amphibians were probably the first animals to possess four limbs, and they demonstrate a greater complexity over fishes. Most of the changes made life on land possible. By examining the anatomy of amphibians, one can see many structures that are basic to both amphibians and all the more advanced vertebrates.

OBJECTIVES

- Identify the external organs of a preserved frog.
- Perform procedures to identify the internal organs of a frog.

- Draw and label various features of a frog.
- Give a function of each major organ of a frog.

MATERIALS

frog (preserved)
scissors
dissecting pins (6–10)
dissecting pan

stereomicroscope
dissecting probe
forceps
microscope slides (2)

coverslip
microscope
laboratory apron

PROCEDURE

Part A. External Anatomy

Refer to Figure 1 for this part of the Procedure.

1. Put on a laboratory apron. Rinse a preserved frog well with water. Place it ventral surface down in your dissecting pan.

2. Note the arrangement of the spots and the coloration of the frog. The color of the frog is caused by scattered granules in the epidermis and chromatophore cells in the dermis. Chromatophore cells are cells that contain pigments.

3. Remove a 1 cm × 1 cm section of the dorsal skin containing one of the frog's spots. Make a wet mount of this piece of skin.

4. Place the wet mount on the microscope under low and then high power. Chromatophores are usually star shaped. Dispersal of the pigment into the rays makes the skin darker. When the pigment is concentrated in the center of the chromatophore, the skin is lighter.

5. Make a drawing of the chromatophore in the space provided in Data and Observations.

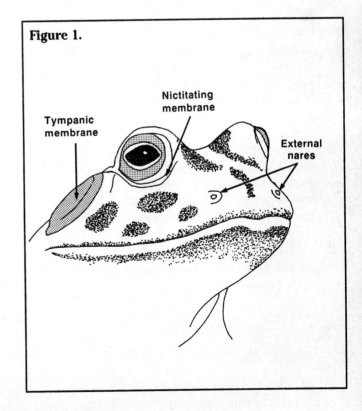

Figure 1.

Tympanic membrane

Nictitating membrane

External nares

6. Locate the thin membrane that covers the eye from below. This is the **nictitating membrane**. This membrane covers the eye when the frog is under water.

7. Notice the large **tympanic membranes** behind the eyes. These membranes function as eardrums to receive sound waves.

8. Examine the forelegs and hindlegs of the frog, noting the number of toes. If your frog is a male, it will have roughened pads near the thumbs. These are used to hold the female during mating.

9. Answer questions 1, 2, 3, and 4 in the Analysis.

10. Find in Figure 1 the structures that are in bold print in Part A.

Part B. Oral Cavity

Refer to Figure 2 for this part of the Procedure.

1. Open the mouth by cutting the jaws with the scissors. Locate the **maxillary teeth** around the edge of the jaw. They hold food but are not used for chewing.

2. Locate the slit-like **glottis** at the back of the throat. This opening leads to the respiratory system. Above the glottis is the **opening to the esophagus**.

3. Find the **eustachian tubes** at the posterior corners of the upper jaw. Probe with your dissecting probe to find out where they lead. These tubes equalize pressure within the ear.

4. Locate in a male frog the opening that leads to the **vocal sacs** at the widest corner of the lower jaw. They amplify the male's mating call.

5. Notice the shape of the **tongue** and where it is attached. It can be flipped forward to catch prey.

6. Locate the **nostril openings** in the roof of the mouth. Between the nostril openings are two **vomerine teeth**. Feel these teeth and the maxillary teeth with your fingers.

7. Locate in Figure 1 the **external nares** on the dorsal surface of the frog's head.

8. Answer questions 5, 6, and 7 in the Analysis.

9. Find in Figure 2 the structures that are in bold print in Part B.

Part C. Digestive System

Refer to Figures 3 and 4 for this part of the Procedure.

1. Place your frog dorsal surface down in the dissecting pan. Open the frog by cutting the skin around the abdomen in the manner shown in Figure 3.

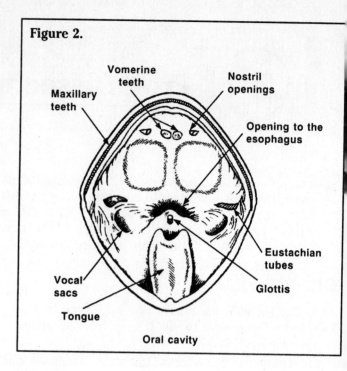

Figure 2.

Oral cavity

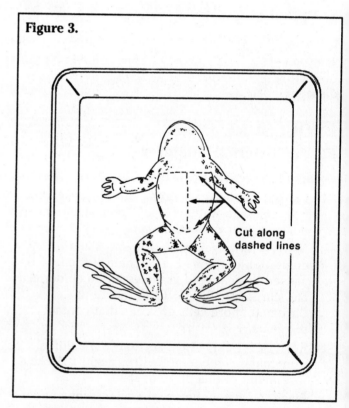

Figure 3.

Cut along dashed lines

2. Insert the point of your scissors just through the muscle above the anal opening and make a cut extending to the lower jaw. Cut sideways at both the forelegs and hindlegs as shown in Figure 3. Pin the muscles down to the dissecting pan. **NOTE:** If your frog is female and contains black eggs, they must be removed carefully before the internal organs can be observed.

3. Locate the **esophagus** as shown in Figure 4. Pass a probe into the **stomach**. Note that the lower end of the stomach is constricted. This constriction is the **pyloric sphincter**. It regulates the amount of food that enters the small intestine.

4. Cut open the stomach and observe its lining. If food is present in the stomach, can it be identified? Usually insect body parts will be present.

5. Follow the digestive tract beyond the pyloric sphincter to the coiled **small intestine**. The first portion that usually runs parallel to the stomach is the **duodenum**.

6. Cut open the lower part of the small intestine. Place a piece of the small intestine on a clean glass slide with the inside surface up. Observe the surface of the small intestine under a stereomicroscope. Note the villi, the many small folds in the lining that increase the absorption of nutrients into the bloodstream.

7. Follow the digestive tract below the small intestine where it widens into the large intestine or **colon**. The colon ends in the **rectum**, which in turn opens into the **cloaca**. The cloaca opens to the outside of the frog. The digestive, reproductive, and excretory systems all open into the cloaca.

8. Note the large brown **liver**. Lift the lobes of the liver to locate the green **gallbladder**. The gallbladder stores bile that is secreted by the liver. Bile aids in the digestion of fats.

9. Locate the **pancreas**, a soft, irregular pinkish organ that produces digestive enzymes, found lying in a membrane between the stomach and duodenum.

10. Answer questions 8, 9, 10, and 11 in the Analysis.

11. Find in Figure 4 the structures that are in bold print in Part C.

Part D. Respiratory and Circulatory Systems

Refer to Figure 4 for this part of the Procedure.

Air is drawn into the mouth by expansion of the throat. The external nares close, then the throat muscles contract and air is forced into the lungs through the glottis. Air is expelled as the nares remain closed, the throat expands, and air enters the mouth again from the lungs. The glottis closes, the nares open, and the throat contracts, forcing the air out through the nares. This method of breathing differs from that of more advanced vertebrates because the frog lacks a diaphragm.

1. Probe the glottis to see where it leads. Locate the **trachea**, the passageway between the glottis and the lungs.

2. Locate the pinkish-gray **lungs**.

3. Notice the three-chambered **heart** between the lungs and posterior to the trachea. The pointed **ventricle** is lighter in color than the rest of the heart. The two thin-walled **atria** are darker colored.

4. Lift the stomach and find the **spleen**, a round red organ. The spleen filters the blood, taking out improperly functioning red blood cells.

5. Answer question 12 in the Analysis.

6. Find in Figure 4 the structures that are in bold print in Part D.

Part E. Excretory and Reproductive Systems

Refer to Figure 4 for this part of the Procedure.

1. Examine the **kidneys** that lie against the dorsal body wall in the posterior region of the body cavity. Each kidney has a yellow stripe, known as the **adrenal body** that secretes hormones. The kidneys filter the blood and urine that drains into the **urinary bladder**, a thin-walled bag that attaches to the cloaca.

2. Locate in a female frog, two lobed, grayish **ovaries** that lie close to the kidneys. In a mature female, the two ovaries might be filled with black and white eggs.

3. Locate in a male frog, the white **testes** that can be found close to the kidneys. Look at the reproductive organs of both sexes.

4. Examine the yellow, finger-like **fat bodies** attached near the kidneys. Compare their size with these in a frog of the opposite sex. The fat bodies provide nourishment for the gametes.

5. Answer question 13 in the Analysis.

6. Find in Figure 4 the structures that are in bold print in Part E.

Part F. The Brain

Refer to Figures 5 and 6 for this part of the Procedure.

1. Turn over your frog so that the dorsal side once again faces up.

2. Insert the point of your scissors through the skin at the base of the head and remove the skin from the head area.

3. Bend the frog to determine the approximate region of the "neck."

Figure 4.

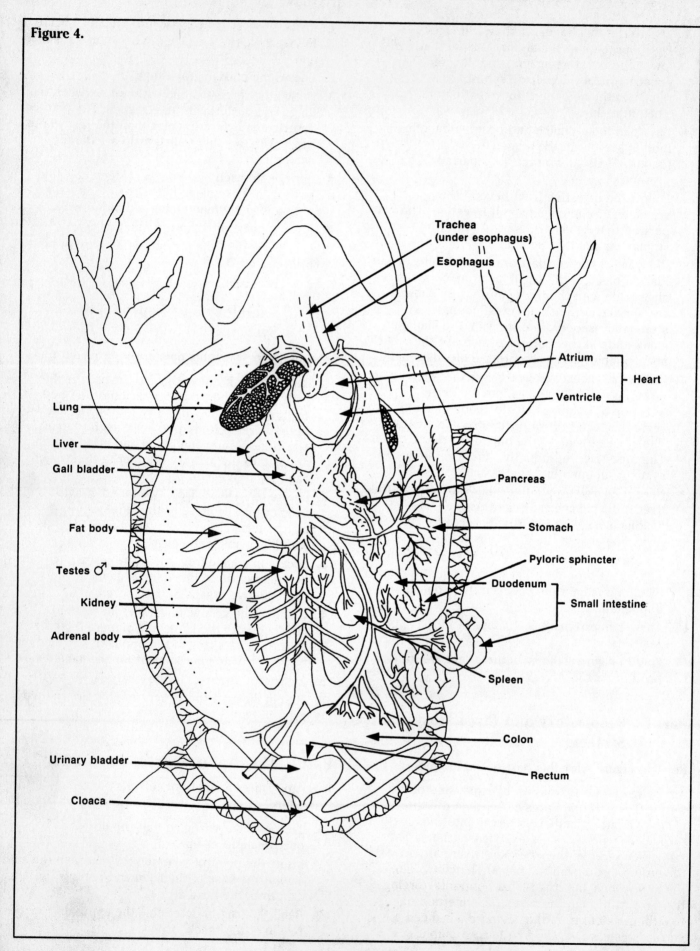

Trachea (under esophagus)

Esophagus

Atrium

Ventricle

Heart

Lung

Liver

Gall bladder

Pancreas

Stomach

Fat body

Pyloric sphincter

Testes ♂

Duodenum

Small intestine

Kidney

Adrenal body

Spleen

Colon

Urinary bladder

Rectum

Cloaca

Figure 5.

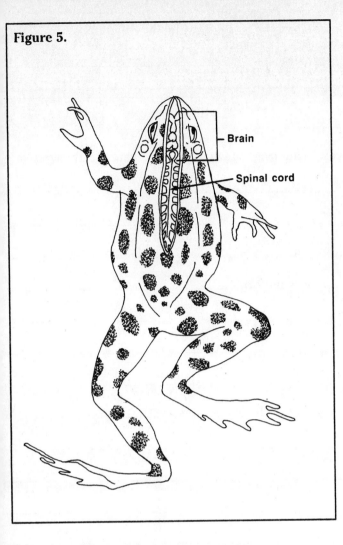

Brain

Spinal cord

Figure 6.

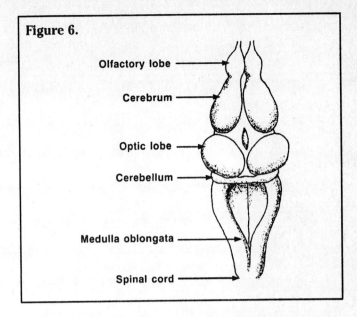

Olfactory lobe

Cerebrum

Optic lobe

Cerebellum

Medulla oblongata

Spinal cord

4. Insert your scissors and clip across the upper spinal cord in the region of the neck,
5. Locate the white **spinal cord** enclosed within the vertebrae.
6. Use your forceps to remove the bone above the spinal cord, working forward until you have reached the nostril area. You will be exposing the brain, as shown in Figure 6.
7. Locate the **olfactory lobes**, **cerebrum**, **optic lobe**, **cerebellum**, and **medulla oblongata** of the brain and the spinal cord, using Figure 6 as a guide.

DATA AND OBSERVATIONS

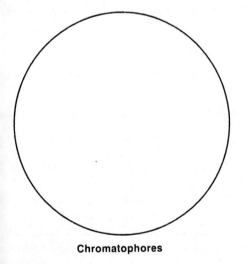

Chromatophores

ANALYSIS

1. What is the function of the nictitating membrane? _____

_____

2. **a.** How are the eyes positioned in the frog's head in comparison to the positioning of your eyes? _____

_____

b. How is this positioning of the eyes an adaptive advantage for a frog? _____

_____

3. Compare the colors of the dorsal and ventral surfaces of the frog. Of what adaptive value to the frog is

each of these colorations to its survival? _____

_____

_____

4. Is the dispersal of the pigment in all of the chromatophores uniform? _____

5. How is the tongue attached? _____

6. Where does the glottis lead? _____

7. To what structures do the eustachian tubes lead? _____

8. How does the lining of the stomach compare with the lining of the small intestine? _____

_____

9. What is the function of the villi? _____

10. How many lobes does the frog's liver have? _____

11. What is the function of the pyloric sphincter? _____

_____

12. Sequence the passage of air into and out of a frog. _____

_____

13. Are the fat bodies larger in male or female frogs? _____ Why is this so? _____

_____

14. The largest parts of the frog's brain are the olfactory lobes and optic lobes, the centers of smell and

vision. How is this adaptation an advantage for the frog's lifestyle? _____

_____

FURTHER EXPLORATIONS

1. Find out what amphibians can be found in the area in which you live.

2. Observe the behavior of a living frog in an aquarium. Observe how it floats, swims, and uses its eyes, as well as other behaviors.

Capillary Circulation in Fish

33–2 LAB

EXPLORATION

▶ A circulatory system is a transport system that contains a pumplike organ called a heart. The heart is a muscle that forces a liquid tissue called blood through a series of vessels. All vertebrates have closed circulatory systems. Blood leaves the heart in large vessels called arteries. Arteries branch into smaller arterioles as they enter the tissues. Arterioles branch into smaller capillaries. Capillary walls are only one cell thick. Arteriole and artery walls are composed of several layers of cells. In the capillaries, branching allows the blood to come in contact with cells or layers of cells. Gases and other necessary materials are exchanged. Capillaries then join together to form larger vessels called venules. Venules join together to form larger veins. Veins return deoxygenated blood to the heart.

OBJECTIVES

- Determine the diameter of a field of view.
- Locate and identify an arteriole, a capillary, and a venule in the tail of a goldfish.
- Observe and describe the direction of the blood flow in the tail.
- Measure and record data for the diameter of each blood vessel.

MATERIALS

absorbent cotton, 5 cm square (2)
aquarium water
petri dish half

goldfish in aquarium
aquarium net
glass slides (2)

microscope
dropper
transparent plastic metric ruler

PROCEDURE

Part A. Diameter of Field of View

1. Place the millimeter section of a transparent plastic ruler over the center of the stage opening of the microscope.
2. Using low power, locate the measured lines of the ruler in the center of the field of view. Move the ruler until one of the millimeter lines is visible at one edge of the field of view. See Figure 1.
3. Estimate in millimeters the diameter of the field of view on low power. Calculate the diameter in micrometers (1 micrometer = 0.001 mm).
4. Record the diameter in Table 1.

Part B. Capillary Circulation

1. Saturate two pieces of absorbent cotton 5 cm² with aquarium water.
2. Place one piece on the bottom of a petri dish with a small amount of aquarium water.

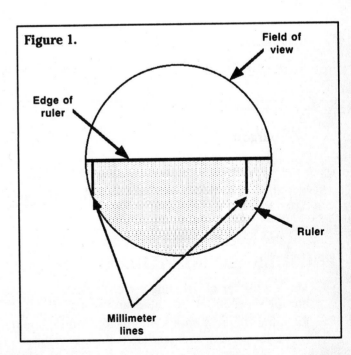

Figure 1.

Field of view

Edge of ruler

Ruler

Millimeter lines

3. Place the head of a goldfish on the wet cotton. Place the second piece of cotton over the head and gills of the fish.
4. Place a clean microscope slide under the tail of the fish and another slide on top of the tail (Figure 2). This will hold the tail in place.
5. Add fresh aquarium water periodically in order to add oxygen. **CAUTION:** *Work quickly. Be sure the cotton stays wet to provide oxygen for the fish.*
6. Place the petri dish on the stage of your microscope. Using low power only, focus on a section of the tail that shows blood flowing.
7. Locate and identify an arteriole, a capillary, and a venule.
8. Determine the direction of blood flow.
9. Calculate the diameter of each vessel in micrometers by estimating how many vessels could be placed side by side across the field of

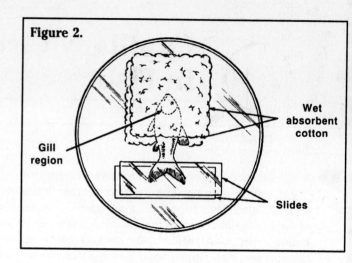

Figure 2.

view and dividing the field of view diameter by the number of vessels. Record the diameters in Table 1.
10. Return the goldfish to the aquarium.

DATA AND OBSERVATIONS

Table 1.

	Field of view	Arteriole	Capillary	Venule
Diameter in micrometers				

ANALYSIS

1. In which of the three vessels does the blood throb? Why? _____

_____

2. In which of the three vessels does the blood flow at a uniform rate? Why? _____

_____

3. Do arterioles carry blood to or from the tail? Give reasons for your answer. _____

_____

4. Which type of blood vessel has the smallest diameter? Why? _____

_____

5. Which type of blood vessel has the largest diameter? Why? _____

_____

FURTHER EXPLORATIONS

1. Make a poster diagraming a closed circulatory system of a fish and an open system of a crustacean.
2. Examine *Daphnia* with a microscope and compare its open circulatory system with the goldfish's closed system.

EXPLORATION

Examining Bird Feet

34–1
LAB

▶ You may have seen a sandpiper hurrying along the shore away from an oncoming wave or a sparrow perching on a telephone wire. Birds' feet are adapted for a variety of functions. Birds run, hop, walk, scratch, perch, attack, defend, take off for flight, preen, swim, or obtain food with their feet. Birds' feet have claws and are composed primarily of bone, tendon, and tough scaly skin. Most birds have four toes arranged in a variety of ways on the foot. The toes have various sizes and shapes and are adapted to particular lifestyles. By examining a bird's foot, you may be able to tell something about where the bird lives, how it gets food, what it eats, or how it defends itself.

OBJECTIVES

- Identify adaptations of birds' feet that make them suited for particular habitats and lifestyles.
- Compare and contrast the feet of various species of birds.

- Relate the characteristics of birds' feet to their functions.

MATERIALS

field guide to birds

PROCEDURE

1. Examine Figure 1 and note the various parts of a bird's foot.
2. Examine the diagrams of birds' feet in Table 1. Fill in the "Structural Adaptations" column. Describe the features of each foot type that make it suited for its particular function. Look for the following traits.

   a. number of toes
   b. length of toes
   c. position of toes: all facing forward or three in front; curled or straight
   d. width of toes
   e. presence of webbing between toes
   f. presence of claws
   g. length and thickness of claws
   h. shape of claws
   i. covering on foot consisting of scales, feathers, and so on.

3. Look at the drawings of different birds in Table 2. Compare their feet to those pictured in Table 1.

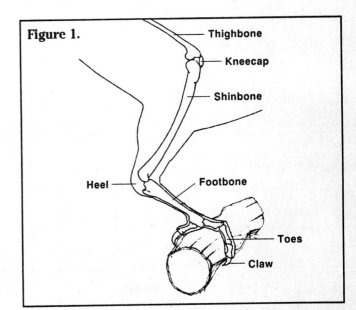

Figure 1.

Thighbone
Kneecap
Shinbone
Heel
Footbone
Toes
Claw

4. Write in the appropriate column of Table 2 the function of these birds' feet. Also write in Table 2 what habitat you think these birds might live in. Consult a field guide to birds to help you.

DATA AND OBSERVATIONS

Table 1.

Adaptations of Birds' Feet			
Type of feet	Function	Habitat	Structural adaptations
A.	Walking, scratching, maintaining body heat	On the ground in prairies	
B.	Running	Hard flat ground	
C.	Swimming	Lakes, ponds, streams	
D.	Grasping prey	Meadows with streams	
E.	Perching	Meadows and woodlands	
F.	Climbing	Bark of trees	

Table 1. (continued)

Type of feet	Function	Habitat	Structural adaptations
G.	Wading	Ponds with soft, muddy bottoms	
H.	Clinging	Vertical surfaces of cliffs	
I.	Full-time flight	In the air	
J.	Swimming, pushing (for a tobogganing-type sliding movement on breast and belly)	Water, ice, and snow	
K.	Walking on water plants	Shorelines of lakes	
L.	Defense	Various habitats	

Table 2.

Birds and Their Feet

Bird	Closeup view of foot	Feet functions	Habitat
Penguin			
Woodpecker			
Osprey			
Heron			
Ruffed Grouse			
Killdeer			

Table 2. (continued)

Bird	Closeup view of foot	Feet functions	Habitat
Hummingbird	4X actual size		
Cliff swallow			
Jacana			
Robin			
Mallard duck			
Pheasant			

ANALYSIS

1. Birds that run on the ground have toes that all point forward. Why might this be an advantage to a

   running bird? _____

   _____

   _____

2. How does webbing on the foot of a swimming bird help it to swim? _____

   _____

   _____

3. What is the adaptive advantage of having very long toes for birds that walk on water plants? _____

   _____

   _____

4. In the wild, birds perch on branches of varying size and hardness. As a result, their claws wear naturally as they grow and their feet are well exercised. What might happen to a pet bird that sits on one plastic

   perch in its cage all year? _____

   _____

   _____

5. Many zoos are using rubberized netting or artificial carpets in aviaries instead of wire mesh on the floor.

   How might this be beneficial to the birds? _____

   _____

   _____

FURTHER EXPLORATIONS

1. Look at pictures of birds' beaks to see how they are adapted to particular lifestyles.
2. Go on a bird-watching trip. Identify the birds you see and hear. Use binoculars to observe the type of feet the birds have and how they use them. Classify their feet types using Table 1.

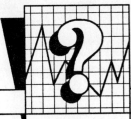

How Do Densities of Bird and Mammal Bones Compare? 34-2

LAB

INVESTIGATION

▶ Birds have skeletons similar to the skeletons of mammals. Both birds and mammals have skulls, backbones, ribs, pectoral and pelvic girdles, and limbs. The skeletons of birds and mammals must provide varying amounts of support, strength, and flexibility to the body of the animal depending on its lifestyle. In addition, you would expect birds' skeletons to be adapted for flight. Flying is easier if the skeleton is lightweight. Reduced skeletal mass results if birds have fewer bones. Birds have fewer vertebrae in parts of the back and fewer bones at the ends of the limbs. Skeletal mass also will be reduced if bone density is decreased. Density is a measure of how much mass there is in a given space.

OBJECTIVES

- Hypothesize which bones will be more dense—bird or mammal.
- Calculate the densities of various bird and mammal bones.

- Develop reasons that will explain the results of your measurements.

MATERIALS

clean, dry mammal bones: pieces of ribs
  (two of beef, two of pork)
500-mL graduated cylinder
dissecting probe
pencil

clean dry bird bones: leg bone and wing bone
  (one each of turkey and duck)
100-mL graduated cylinder
balance

PROCEDURE

Density is the ratio of mass per unit volume. This relationship between mass and volume can be expressed by the formula

$$\text{density} = \frac{\text{mass}}{\text{volume}}$$

1. Label the bones with a pencil using the following abbreviations:

   BR1 = beef rib   TL = turkey leg
   BR2 = beef rib   TW = turkey wing
   PR1 = pork rib   DL = duck leg
   PR2 = pork rib   DW = duck wing

2. Make a **hypothesis** that predicts which bones will be the least dense. Write your hypothesis in the space provided.

3. Find the masses of all eight bones, using the balance. Record these data in Table 1.

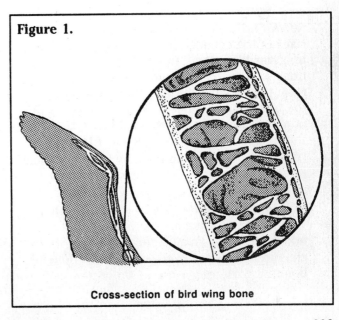

Figure 1.

Cross-section of bird wing bone

Figure 2.

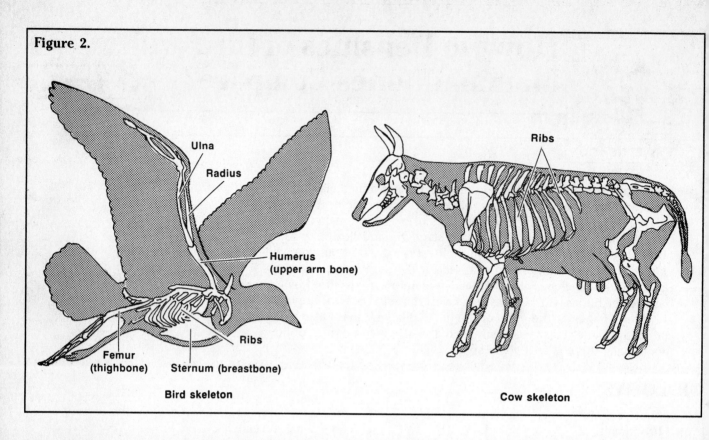

Bird skeleton

Cow skeleton

4. Fill the graduated cylinders 3/4 full of water. Adjust the volume so the bottom of the meniscus rests at a line. (For help in measuring the volume, see Appendix B on page 275.)

5. Choose one of the bones. Select the smallest graduated cylinder into which the bone will fit. Measure the volume of water in the cylinder. Record the volume in Table 1.

6. Place the bone carefully into the water in the graduated cylinder. Be careful not to splash any of the water out of the cylinder. If the bone floats above the water, push it under the water with a dissecting probe as shown in Figure 3.

7. Record in Table 1 the level to which the water rises in the cylinder when the bone is immersed. To determine the volume of the bone, subtract the initial volume of the water from the volume to which the water rises with the bone immersed. Record the volume of the bone in cubic centimeters (cc) in Table 1. Note that one mL of water is equivalent to one cc.

8. Divide the mass of the bone by its volume and record this number as its density in Table 1.

9. Repeat steps 5 through 8 for each of the remaining bones. Be sure to remeasure the volume of water before you immerse each bone, because a small amount of water will cling to and be removed with the previous bone.

Figure 3.

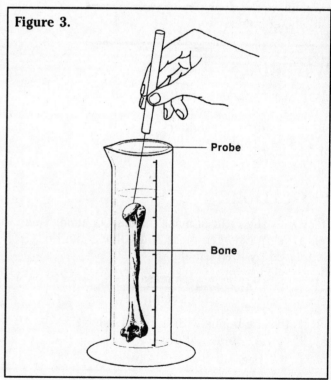

HYPOTHESIS

_____

_____

DATA AND OBSERVATIONS

Table 1.

Bone	Mass (grams)	Volume of water in cylinder (mL)	Volume to which water rises (mL)	Volume of bone (cc)	Density
			Bone Data		
BR1					
BR2					
PR1					
PR2					
TL					
TW					
DL					
DW					

ANALYSIS

1. What is the average density of the pork bones? _____ of the beef bones? _____

2. What is the density of turkey leg bones? _____ of duck leg bones? _____

3. What is the density of the turkey wing bones? _____ of duck wing bones? _____

4. Why is the density of pig and cattle bones, and that of turkey and duck leg bones, similar? _____
_____

5. Why are a duck's wing bones less dense that a turkey's wing bones? _____
_____

6. Of what adaptive value is it for wing bones to be less dense than all of the other bones you measured?
_____
_____

7. Do bones that have a density of less than 1 g/cc sink or float? _____ Why? _____
_____

8. Besides lightweight skeletons, what other adaptations do birds have for flight? _____

_____

_____

9. What types of mammals would benefit by having bones less dense than those of a pig or a cow? _____

_____

_____

_____

CHECKING YOUR HYPOTHESIS

Was your **hypothesis** supported by your data? Why or why not? _____

_____

FURTHER INVESTIGATIONS

1. Using the methods in this Investigation, calculate the densities of other bones of other mammals and birds or other vertebrates such as fish, reptiles, and amphibians.
2. Test the strength of bones from mammals and birds by dropping increasing sizes of known masses on them until they are damaged or broken. Wear safety goggles while doing this.

Mammal Teeth

EXPLORATION

▶ A beaver with sharp chisel-like teeth gnaws the bark of young aspen trees. A cow collects grasses by holding the plants between the lower teeth and a pad in its upper jaw and tearing off the blades of grass with a jerk of its head. A rat is captured by a weasel with a piercing bite to the back of the rat's neck. A whale strains plankton from sea water with fibrous plates that contain frayed hairlike strands. Mammals have teeth adapted for obtaining specific types of food. By examining a mammal's teeth, you can learn a great deal about its feeding habits and lifestyle.

OBJECTIVES

- Observe adaptations of mammals' teeth for eating particular types of food.
- Recognize the relationships of mammal feeding habits to body features other than teeth.
- Predict the types of teeth possessed by mammals adapted to specific habitats and behaviors.

MATERIALS

PROCEDURE

Part A. Mammal Teeth

1. Examine the following figures of mammal teeth. Compare the descriptions with their matching drawings.

   Figure 1. Evolutionary changes of the cusps, which are the grinding surfaces of teeth, reflect changes in the way teeth are used.

   A. **Mammal-like reptile** – Fossil evidence shows that the reptilian ancestors of mammals had teeth that probably interlocked when the jaw closed. The teeth may have been adapted for piercing the hard coverings of insects.

   B. **Early mammal** – Some early mammals may have had triangular teeth that could rub against one another, the beginning of a grinding ability.

   C, D. **Later mammals** – The change from piercing types of teeth to teeth with larger areas of touching surfaces enabled mammals to cut and grind their food.

Figure 1.

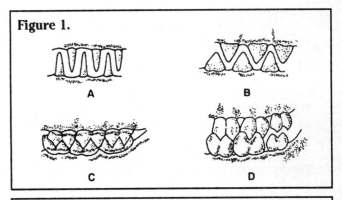

A   B

C   D

Figure 2.

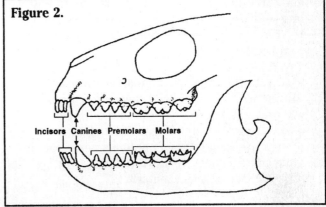

Incisors Canines Premolars Molars

Figure 2. Mammal tooth types – Most mammals have a variety of types of teeth. Incisors gnaw or cut, canines stab and tear, premolars crush and shear, molars grind.

Figure 3. Insectivores – In insectivores, such as shrews and moles, the incisors are simple and peglike. The canines are usually the same shape as the incisors. The molars and premolars have sharp points suited for shearing and cutting through hard insect coverings.

Figure 4. Bats are varied in their food preferences.

A. Fruit-eating bats – Teeth are specialized for crushing small, hard fruits.

B. Nectar-feeding bats – Teeth are very small.

C. Insect-feeding bats – Teeth are all very sharp and can easily capture and pierce insects with hard coverings.

Figure 5. Anteaters – Anteaters have no teeth or only a few tiny teeth. They use their tongues to catch ants and termites. They expose the insects by pulling apart the insect nests with their powerful front arms.

Figure 6. Rodents – Rodents, such as the beaver, are equipped with large incisors adapted to gnawing trees. The teeth continue to grow throughout the rodent's life. The sharp chisel-like edge is maintained by one side wearing away more quickly than the other. The molars grind very tough bark.

Figure 7. Carnivores – Carnivores, such as the coyote, have long canines, shearing premolars, and crushing molars. These teeth enable them to capture and kill their prey.

Figure 8. Omnivores – Omnivores, such as the black bear, often have large canines for tearing and large molars for grinding. Premolars and incisors are not extremely large and sharp. In the case of this type of bear, they are greatly reduced.

Figure 9. Ungulates – Ungulates, such as the pronghorn antelope, are browsers or grazers. Incisors are adapted for nipping off grasses and shrubs. Canines are absent. Premolars and molars are long in young animals and short in older animals that have worn them away by grinding coarse plant material.

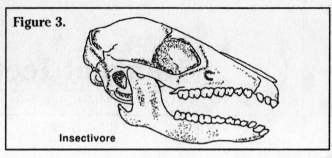

Figure 3.

Insectivore

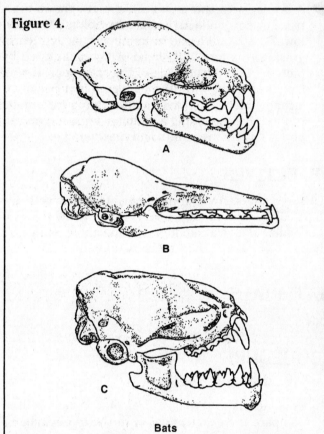

Figure 4.

A

B

C

Bats

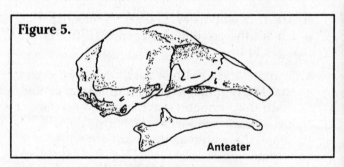

Figure 5.

Anteater

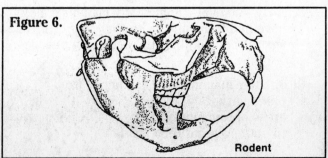

Figure 6.

Rodent

Figure 7.

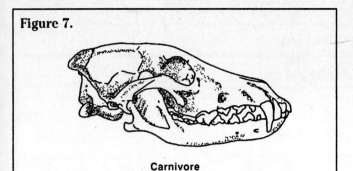

Carnivore

Figure 8.

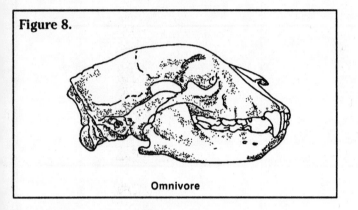

Omnivore

Figure 9.

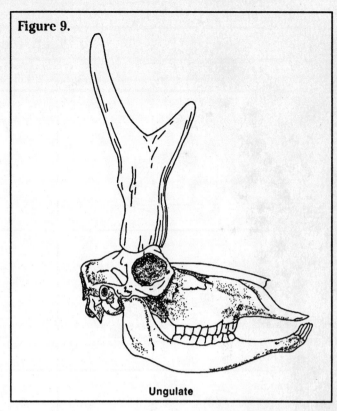

Ungulate

Part B. Skull Identification

1. Examine the mammal skulls in Table 1 and compare their teeth with the teeth in Part A.
2. Identify the type of animal represented by each skull and write its name next to the skull.
3. Write in the characteristics used for your identification.

DATA AND OBSERVATIONS

Table 1.

Skull Identification		
Mammal skull	**Type of mammal**	**Characteristics**
1.		
2.		

Table 1. (continued)

Skull Identification		
Mammal skull	Type of mammal	Characteristics
3.		
4.		
5.		
6.		
7.		
8.		

Table 1. (continued)

Skull Identification		
Mammal skull	Type of mammal	Characteristics
9.		
10.		
11.		
12.		

ANALYSIS

1. Based on your observations of the skulls in Table 1, tell whether each of these 12 mammals eats plants, insects, meat, meat and insects, or a variety of foods.

1. _____   4. _____   7. _____   10. _____

2. _____   5. _____   8. _____   11. _____

3. _____   6. _____   9. _____   12. _____

2. Based on your answers to question 1, hypothesize whether each mammal moves fast or slowly.

_____

_____

3. Compare the differences between the teeth of omnivores, carnivores, and herbivores.

_____

_____

4. Figure 10 is the skull of a vampire bat. These bats make an incision in the skin of a cow or horse and lap the blood with their tongues. Explain how the teeth of a vampire bat are adapted to this behavior.

_____

_____

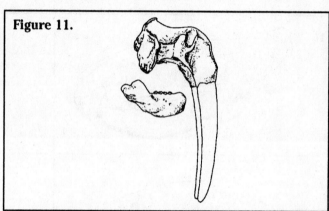

Figure 10.

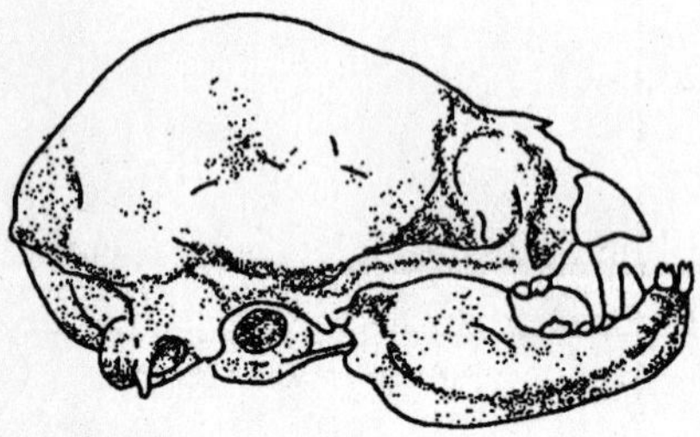

Figure 11.

5. Figure 11 is the skull of a walrus. Walruses feed on mollusks that they rake from the sea floor. Explain

how the teeth of a walrus are adapted to this lifestyle. _____

_____

6. Describe and diagram the teeth of a mammal that lives in a tropical climate and forages on the floor of

the forest for worms. _____

_____

7. If a mammal had to feed on only mosses and lichens, what would its teeth be like? _____

_____

8. If a sheep were a predator, how would its teeth be different? _____

_____

FURTHER EXPLORATIONS

1. Take a field trip to a museum of natural history and examine the teeth of as many mounted mammals as you can. Make lists of those you think are herbivores, omnivores, and carnivores.

2. Have a dentist save some human teeth for you. Soak them overnight in a 10% solution of household bleach in order to disinfect them. Examine the differences between molars, premolars, canines, and incisors. Wear safety glasses and crack the teeth open with a hammer. Observe the parts of the teeth: enamel, dentine, and pulp cavity.

How Does Insulation Affect Thermal Homeostasis?

35–2

LAB

INVESTIGATION

▶ Most mammals maintain a constant body temperature regardless of the temperature of the environment. They produce heat through metabolic processes and maintain that heat in a variety of ways. The rate at which a body loses heat is proportional to the difference between a body's temperature and that of the environment. Therefore, reducing heat loss is essential for mammals that live in environments colder than their bodies. Some mammals control heat loss by limiting the blood supply to the surface of the body. Other mammals produce insulating tissues like fur or blubber that reduce heat loss. For example, the arctic fox has a wooly underfur and long guard hairs that enable it to exist comfortably at temperatures as low as -40°C.

OBJECTIVES

- Construct hypotheses to predict how two different bodies will cool under different conditions.
- Determine the effect of insulation on the cooling of a warm body at room temperature.
- Determine the effect of insulation on the cooling of a warm body under cold conditions.
- Compare cooling rates of warm bodies at room temperature to the cooling rates of warm bodies at cold temperatures.

MATERIALS

5-L pail
100-mL graduated cylinders (4)
newspaper (20 pages)
rubber bands (8)
thermometer, 30 cm long

food storage size plastic bags (4)
colored pencils (4)
graph paper
hot tap water (45–55°C)
ice cubes (24 standard cubes)

stirring rod
masking tape
clock or stopwatch

PROCEDURE

1. Make a **hypothesis** that predicts the effect of insulation on the cooling rate of a warm body. Write your hypothesis in the space provided.

2. Make a second **hypothesis** that describes the effect of surrounding environmental temperature on the cooling rate of a warm body. Write your hypothesis in the space provided.

3. Measure the air temperature of the room and record it in Table 1.

4. Fold 10 pages of newspaper in half crosswise as shown in Figure 1. Fold another 10 pages of newspaper in the same manner.

5. Place 24 ice cubes in your pail and add cold tap water. The water level should be about half the height of the graduated cylinders. Stir gently

Figure 1.

with a stirring rod until the temperature throughout the pail is the same. Measure the temperature of the ice water in the pail and record it in Table 1.

6. Wrap two of the 100-mL graduated cylinders tightly with the folded newspapers as shown in Figure 2. Keep the fold at the base of the cylinder as you wrap. Fasten each wrapped cylinder with two rubber bands. These will be the two insulated cylinders.

7. Place all four cylinders into separate plastic bags. Fasten each bag around the top of the cylinder with a rubber band as in Figure 3. The cylinders will remain open at the top to allow for insertion of a stirring rod and a thermometer.

8. Add 100 mL of hot tap water (45–55°C) to each of the four graduated cylinders.

9. Measure the temperature of the water in each of the four graduated cylinders. Record the temperatures in Table 1.

10. Label your four cylinders with your name, then place one wrapped and one unwrapped graduated cylinder into the pail of ice water as in Figure 4. Leave the other two cylinders at room temperature.

You now have four graduated cylinders containing hot water. These cylinders represent warm-blooded animals. Two animals (cylinders) are placed in cold surroundings (ice water pail), but one animal has body insulation (paper wrap) and one does not. Two animals are placed in room temperature surroundings, but one animal has body insulation (paper wrap) and one does not.

11. After 5 minutes, measure the water temperature in each of the graduated cylinders and record the data in Table 1. Stir the water in each cylinder with the stirring rod for 15 seconds before inserting the thermometer and reading the temperature.

12. Repeat step 11 every 5 minutes until 30 minutes have elapsed. Record your temperatures each time in Table 1.

13. Make a graph of the data in Table 1 by plotting temperature against time. Time should be plotted on the horizontal axis and temperature on the vertical axis. Plot the data for each cylinder using a different colored pencil.

14. Complete the additional calculations requested in Data and Observations.

Figure 2.

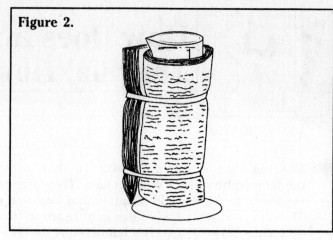

Figure 3.

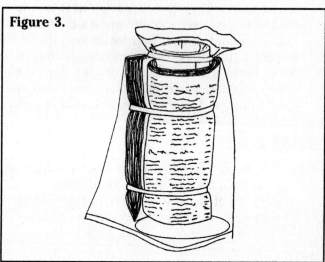

Figure 4.

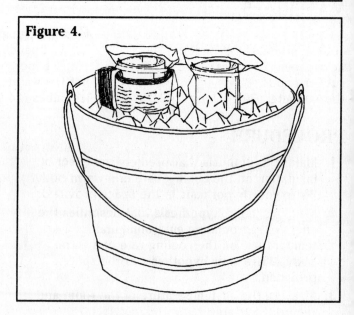

HYPOTHESIS 1

_____

_____

HYPOTHESIS 2

_____

_____

DATA AND OBSERVATIONS

Table 1.

Time	Room temperature _____ °C		Low temperature _____ °C	
	Uninsulated	Insulated	Uninsulated	Insulated
0 minutes (start)				
5 minutes				
10 minutes				
15 minutes				
20 minutes				
25 minutes				
30 minutes				

Temperature (°C) of Water in Graduated Cylinders

1. Calculate the difference between room temperature and the temperature in the uninsulated graduated cylinder at the beginning of the experiment. _____

2. Calculate the difference between the temperature of the ice water and the temperature in the uninsulated graduated cylinder at the beginning of the experiment. _____

Make the following calculations and write your answers in Table 2.

For each cylinder, calculate the temperature difference between starting and ending temperatures.

For each cylinder, calculate the percentage decrease in temperature.

$$\% \text{ decrease in temperature} = \frac{\text{temperature difference}}{\text{temperature at start}}$$

Table 2.

Graduated cylinder	Temperature difference	% decrease in temperature
Uninsulated at room temp.		
Uninsulated at low temp.		
Insulated at room temp.		
Insulated at low temp.		

ANALYSIS

1. Examine the lines of your graph.

   a. Which graduated cylinder lost heat the fastest? _____

   b. Which graduated cylinder lost heat most slowly? _____

   c. What effect did insulation have on the graduated cylinder at room temperature? _____

   _____

   d. What effect did insulation have on the graduated cylinder at low temperature? _____

   _____

2. For which of the graduated cylinders was the temperature difference between the contents of the

   cylinder and the environment greatest at the beginning of the experiment? _____

   _____

3. What effect does environmental temperature appear to have on a body's rate of cooling? _____

   _____

4. How do the percentages of temperature decrease compare between insulated and uninsulated cylinders

   a. at room temperature? _____

   b. at low temperature? _____

5. Hypothesize how insulation works to maintain heat in a warm body. _____

   _____

6. What other adaptations, in addition to insulation, do mammals use to maintain their body heat? _____

   _____

CHECKING YOUR HYPOTHESES

Were your **hypotheses** supported by your data? Why or why not? _____

_____

_____

_____

FURTHER INVESTIGATIONS

1. Conduct this same experiment to compare newspaper with other
   insulating materials such as cotton or wool socks.
2. Conduct the same experiment with the same amount of water in
   cylinders or beakers that have less surface area than the graduated
   cylinders used in this Investigation. Compare your results.

How Is Response Related to Nervous System Complexity? 36-1

LAB

▶ The movement of an organism toward or away from an environmental stimulus is called a taxis. Such an ability might mean the difference between life and death for an organism. Organisms that can respond instinctively to stimuli do not have to learn the correct response. All the behavior necessary for survival is programmed into their genetic material. Each pairing of stimulus and response is an innate behavior that contributes to the survival of the organism. Responses of invertebrates to the force of gravity (geotaxis), chemicals (chemotaxis), and light (phototaxis) can be compared.

OBJECTIVES

• Hypothesize the effect of nervous system complexity on responses to the same stimuli.

• Compare the responses of different invertebrates to the same stimuli.

MATERIALS

droppers (2)
flashlight
coarse salt
black construction paper
plastic thermometer tube
  with caps

duct tape
test-tube rack
wax marking pencil
ruler
stopwatch
hand lens

living cultures of
  Planaria
  vinegar eels (*Turbatrix aceti*)
  Daphnia
filter paper
laboratory apron

PROCEDURE

The three species of invertebrates in this Investigation have different levels of nervous system development. *Planaria* has the least amount of development, followed by the vinegar eel. *Daphnia* has the largest nervous system.

1. Make a **hypothesis** that predicts how a species' response to a stimulus relates to the complexity of its nervous system. Write your hypothesis in the space provided.

2. Place a cap tightly on one end of a thermometer tube. Fill the tube with water to test for leakage. If the cap leaks, tape it securely with a piece of duct tape as in Figure 1 so that it is watertight. Pour out the water. Use the ruler and marking pencil to divide the tube into thirds as shown in Figure 1.

3. Using a clean dropper, fill the tube with the *Planaria* culture medium. Be sure the tube contains at least three organisms.

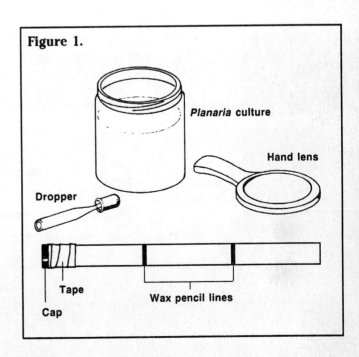

Figure 1.

Planaria culture

Hand lens

Dropper

Tape

Cap

Wax pencil lines

4. Place the second cap on the top of your tube and place the tube in the test-tube rack as shown in Figure 2. Immediately start your stopwatch and measure the amount of time it takes for you to see that the planarians are reacting to the force of gravity. Stop your stopwatch when you detect an overall trend in movement in the tube. Record this time in Table 1 in the section labeled **Geotaxis**.

5. After 5 minutes, observe the planarians and look for their general direction of movement. Count the number of planarians in the top third of the tube and the number in the bottom third of the tube. Do not count the planarians in the middle third. Use a hand lens to help you see the planarians. Record these numbers in the **Geotaxis** section in Table 1.

6. Place the tube containing the planarians in a horizontal position. Tape the second end with tape if it leaks. Cover one end with a piece of black construction paper and arrange a flashlight so that it clearly illuminates the other end of the tube as shown in Figure 3. Immediately start your stopwatch and measure the amount of time it takes for the planarians to react to light. Stop your stopwatch when you detect an overall trend in movement in the tube. Record this time in Table 1 in the **Phototaxis** section.

7. After 5 minutes, observe the planarians and note the general direction of movement. Remove the construction paper and flashlight. Count the number of planarians in the lighted third of the tube and the number in the dark third. Do not count planarians in the middle third of the tube. Record these numbers in the **Phototaxis** section in Table 1.

8. Place a few crystals of coarse salt in a small piece of filter paper and fold into a wad. Open the top of the thermometer tube and put the wad of paper into the cap as shown in Figure 4a. The paper wad should be large enough to stay stuck in the top cap. Put the cap back onto

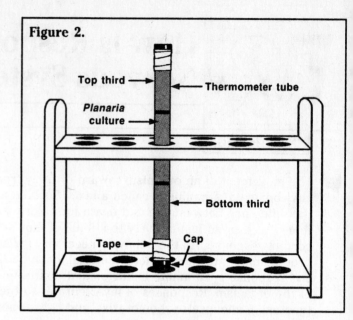

Figure 2.

Top third — — Thermometer tube

Planaria culture

Bottom third

Tape — Cap

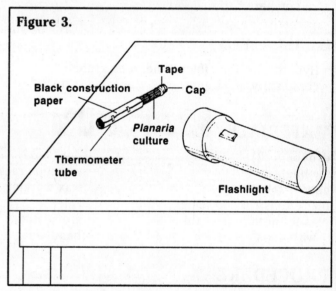

Figure 3.

Tape

Black construction paper — Cap

Planaria culture

Thermometer tube

Flashlight

the tube and carefully place the tube in a horizontal position as in Figure 4b. The salt will dissolve slowly in the liquid medium. Allow the tube to rest for 2 minutes.

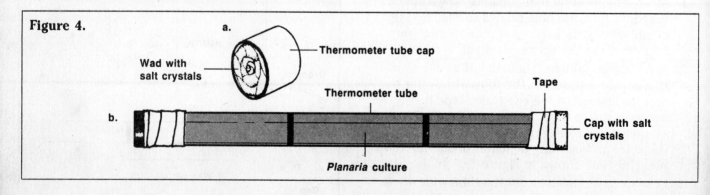

Figure 4.

a.

Wad with salt crystals — Thermometer tube cap

Thermometer tube

Tape

b.

Cap with salt crystals

Planaria culture

9. Start you stopwatch and measure the reaction time as described in steps 4 and 6. Record the time in the **Chemotaxis** section in Table 1.

10. After 5 minutes, observe the planarians and note their general direction of movement. Count the number of planarians in the third of the tube nearest the salt and the number in the third furthest away from the salt. Record these numbers in the **Chemotaxis** section in Table 1.

11. Complete Table 1 for *Planaria* by entering its responses to the stimuli. Follow the directions below Table 1.

12. Dispose of the planarians in your tube according to your teacher's directions. Clean and rinse the tube.

13. Repeat steps 2 through 12 for the other two organisms.

HYPOTHESIS

_____

_____

DATA AND OBSERVATIONS

Table 1.

Species' Response to Stimuli					
Response		*Planaria*	Vinegar eel	*Daphnia*
Geotaxis	Reaction time (min)			
	Number at top			
	Number at bottom			
	Response (+, –, 0)			
Phototaxis	Reaction time (min)			
	Number at light end			
	Number at dark end			
	Response (+, –, 0)			
Chemotaxis	Reaction time (min)			
	Number near salt			
	Number away from salt			
	Response (+, –, 0)			

If number close to the stimulus is much greater than number away from the stimulus, the response is strongly positive (+).

If number close to the stimulus is much less than number away from the stimulus, the response is strongly negative (–).

If number close to the stimulus is about equal to number away from the stimulus, the response is weak or absent (0).

ANALYSIS

1. Which species exhibited:

   positive geotaxis? _____ negative geotaxis? _____

   positive phototaxis? _____ negative phototaxis? _____

   positive chemotaxis? _____ negative chemotaxis? _____

2. In which species did the individual organisms exhibit the most diverse response to:

   the force of gravity? _____

   light? _____

   salt? _____

3. Is diversity of response among the individuals of a species a sign of a greater or less complex nervous

   system? _____

   Explain? _____

   _____

4. **a.** Which species had the fastest reaction time? _____

   b. Why would this automatic, rapid response to a stimulus be an adaptive advantage in a simple

   organism? _____

   _____

5. What might be some reasons for a species not demonstrating a strongly positive or negative response?

   _____

   _____

CHECKING YOUR HYPOTHESIS

Was your **hypothesis** supported by your data? Why or why not? _____

_____

FURTHER INVESTIGATIONS

1. Design an experiment to show whether or not invertebrates can learn. Design a simple maze and use rewards such as food to test a mealworm larva's ability to learn.

2. More complex organisms can experience innate responses similar to taxes, but often overcome them. For example, your natural response to a pin prick would be to move away, but many people give themselves injections for medical purposes. Investigate the parts of the mammalian brain, including those responsible for simple instinctive responses and those that allow higher learning.

Conditioning in Guinea Pigs

36–2
LAB

▶ Animals respond to stimuli in their surroundings. Usually each stimulus causes a specific response. An animal can be trained to give an established response to a new stimulus when the new stimulus is given at the same time as the original stimulus. This procedure is repeated until the animal learns to respond to the new stimulus when it is given alone. This kind of training is called conditioning, because the animal learns to respond to a new set of conditions. The response to new stimuli is called a conditioned response.

OBJECTIVES

- Observe feeding behavior in guinea pigs.
- Design and carry out an experiment to produce a conditioned response in guinea pigs.
- Determine the amount of time needed to produce a conditioned response in guinea pigs.

MATERIALS

guinea pig with cage and bedding
pellet-type food

lettuce, carrots, apples, or other fresh food
bell, whistle, or other source of sound

PROCEDURE

1. Obtain a guinea pig; be sure that it has an adequate cage and appropriate bedding materials.
2. Become familiar with the nutritional needs of the guinea pig. A guinea pig needs fresh water and between 40g and 70g of food daily. Table 1 lists some foods that are appropriate for a

guinea pig. If the guinea pig receives a high-protein, high carbohydrate food, less food is needed than if the diet is lower in proteins and carbohydrates. Although many of the dry foods contain less protein and carbohydrate, they are important in the animal's diet because they contain a variety of essential vitamins.

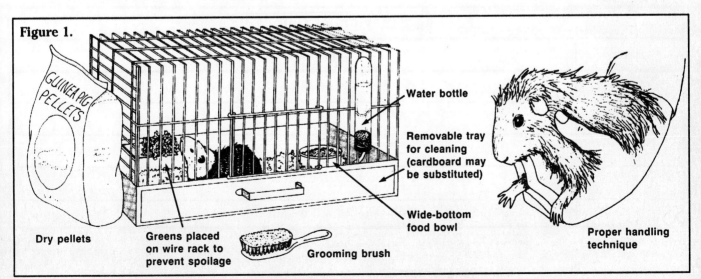

Figure 1.

Guinea pig pellets

Water bottle

Removable tray for cleaning (cardboard may be substituted)

Wide-bottom food bowl

Dry pellets

Greens placed on wire rack to prevent spoilage

Grooming brush

Proper handling technique

Table 1.

Food	Percent protein	Percent carbohydrate
Corn	9	65
Peanuts	20	15
Dry whole-wheat bread	8	55
Guinea pig pellets	20	48
Alfalfa hay	4	10
Lettuce	2	8
Carrots	1	8
Turnips	0.7	9
Apples	0.2	12
Pears	0.5	14

3. Vary the guinea pig's diet until you determine which combination of foods it prefers.

4. As you prepare the food, have a classmate observe the animal's behavior. Guinea pigs have very good hearing, and they may associate the sound of food preparation with being fed.

5. When you feed the guinea pig, watch its behavior closely. See if you can determine the point at which it recognizes that it is about to be fed. Record your observations in Table 2. Guinea pigs make a variety of sounds. Be sure to include these in the data table, too.

6. Observe the guinea pig when you approach the cage but do not feed it. Record this non-feeding behavior in Table 3.

7. After five days of observations, study Tables 2 and 3. Look for patterns of behavior that are associated only with feeding. These will be the behaviors that you will look for to determine when the guinea pig has been conditioned.

8. Just before you feed the guinea pig, provide a new stimulus. Make a loud hand clap or another sound. You may wish to try a visual stimulus of some kind. Make sure that the stimulus is one that does not occur at any other time.

9. Continue to condition the guinea pig to your new stimulus for one week. In Table 4, record the animal's behavior. As you do this, be aware of any other stimuli that might be associated with feeding. You will need to remove them before you can test the effects of conditioning.

10. After one week, give the new stimulus, but do not feed the guinea pig immediately. Look for behavior related to feeding. Then feed the animal. Record your observations in Table 4 next to Testing.

11. If the guinea pig did not show feeding behavior at the new stimulus, repeat the conditioning process for another week.

12. After a week, test the conditioning again.

13. Repeat steps 11 and 12 until the guinea pig responds to the new stimulus with behavior that is related to feeding.

DATA AND OBSERVATIONS

Table 2.

Feeding Behaviors of Guinea Pigs	
Date	Behavior observed

Table 3.

Non-Feeding Behavior of Guinea Pigs	
Date	Behavior observed

Table 4.

Conditioning of Guinea Pigs		
Date	Procedure	Behavior observed
	Conditioning	
	Conditioning	
	Conditioning	
	Conditioning	
	Conditioning	
	Conditioning	
	Conditioning	
	Testing	
	Conditioning	
	Conditioning	
	Conditioning	
	Conditioning	
	Conditioning	
	Conditioning	
	Conditioning	
	Testing	

ANALYSIS

1. What did you choose as a new stimulus? _____

   _____

2. Why did the new stimulus have to be one that would not occur at any other time? _____

   _____

3. Why did other stimuli related to feeding have to be removed before the conditioning could be tested?

   _____

4. How long did it take to condition the guinea pig to the new stimulus? _____

   _____

5. How did the guinea pig respond to the new stimulus? _____

   _____

6. What behavior was seen only when the guinea pig expected to be fed? _____

   _____

   _____

FURTHER EXPLORATIONS

1. Stop using the new stimulus for a week. Then test it again. Determine whether the conditioned response appears. Explain.
2. Research Pavlov's experiments in conditioning. Make particular note of the way he controlled extra stimuli in his subjects' environments. Write a report on your research.

The Skeletal System

37–1

LAB

EXPLORATION

▶ The skeletal system of most vertebrates is composed of bones like those in your skeleton. Bone-forming cells originate in cartilage and secrete bone tissue into the cartilage. Pathways for blood vessels and nerves form in the bone.

OBJECTIVES

- Observe a microscope slide of compact bone and identify the parts.
- Identify bones by their shapes.

- Observe the features of joints.

MATERIALS

prepared slide of compact bone
microscope
human skeleton or pictures of the skeleton

PROCEDURE

Part A. Bone Cells

1. Obtain a slide of compact bone and use the 10X objective to focus on the tissue.

2. Compact bone is made up of many Haversian systems, elongated structures that surround canals. In cross section, a Haversian system looks like a series of rings, each surrounding a canal and containing osteocytes, the bone cells. Look at a Haversian system shown in Figure 1 and locate a similar structure on your slide.

3. A Haversian canal is located in the center of each system. Locate these structures on your slide. Blood vessels and nerves are found in each Haversian canal.

4. Bone cells, or osteocytes, are embedded in mineral salts in layers around Haversian canals. Find these layers on your slide. Within each layer are small canals called canaliculi that carry fluids between the blood vessels and osteocytes.

5. In Data and Observations, draw the Haversian system as observed on your slide. Label a Haversian canal, osteocytes, and canaliculi.

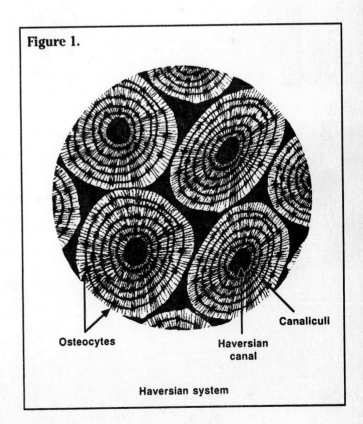

Figure 1.

Osteocytes

Canaliculi

Haversian canal

Haversian system

Part B. Recognizing Bones

1. Each bone has its own shape and function. Compare the bones shown in Figure 2 with a skeleton or a picture of a skeleton.

2. For each bone, determine where it is found in the human body. In Table 1, record the location, scientific name, and common name of each bone. **NOTE:** *Observe the ends of the bones to help you distinguish them.*

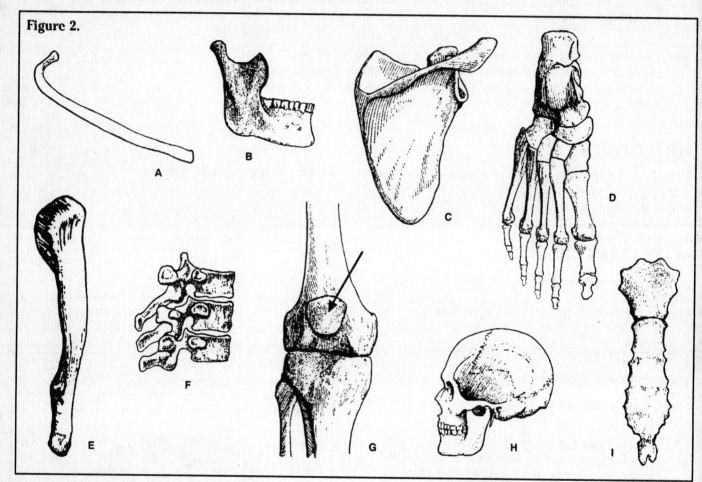

Figure 2.

Part C. Joints

1. Some examples of types of joints are fixed, gliding, hinge, pivot, and ball-and-socket. Three of these types of joints are shown in Figure 3. Joints are held together by tough strands of connective tissue called ligaments.

2. Study the examples of joints in Figure 3. Identify the type of joint shown in each drawing. In Table 2, record the type of joint shown and its location in the body.

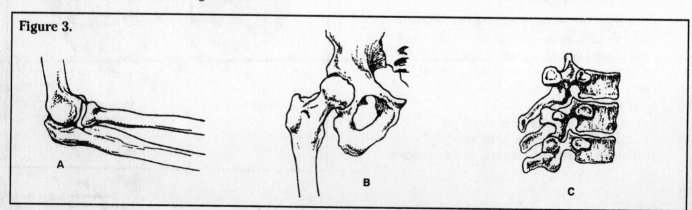

Figure 3.

232

DATA AND OBSERVATIONS

Table 1.

Bone	Common name of bone(s)	Scientific name of bone	Location of bone
A			
B			
C			
D			
E			
F			
G			
H			
I			

Table 2.

Joint	Type	Location in body
A		
B		
C		

ANALYSIS

1. In Part B, you examined various bones. What features did you use to distinguish the various bones from each other? _____

   _____

2. Which type of joint allows only for rotational movement around an axis? Give an example. _____

   _____

3. Which type of joint allows the bones to move back and forth in a single plane? Give an example. _____

   _____

4. Which type of joint allows the arm to move in a circle? Give an example. _____

   _____

5. What is the function of the:

   a. Haversian canal? _____

   b. canaliculi? _____

FURTHER EXPLORATIONS

1. If individual human bones or models of human bones are available, identify each bone by common name and by scientific name. Describe what features you used to identify the bones.

2. If a skeleton of another vertebrate, such as a cat, is available, identify the major bones of the skeleton. If such a skeleton is not available, do library research to find a drawing of a vertebrate skeleton. Make a table to list the names of bones found in this skeleton and in the human skeleton. In your table, list ways these bones are similar and ways they are different.

Endocrine Gland Studies

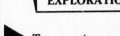

38-1
LAB

▶ Two systems regulate the body's functions. These systems are the nervous system and the endocrine system. The nervous system conducts impulses between the brain and spinal cord and the body. The endocrine system secretes chemicals from glands into the bloodstream. These chemicals, called hormones, carry messages to specific tissues or organs.

OBJECTIVES

• Observe the structure of the thyroid and adrenal glands.

• Analyze the relationship between the structure and the function of a gland.

MATERIALS

microscope
prepared slide of thyroid gland
prepared slide of adrenal gland

PROCEDURE
Part A. Thyroid Gland

1. Observe a slide of the thyroid gland with the 10X objective of a microscope.

2. Note that the gland consists of a large number of spherical sacs. These sacs are called follicles.

3. Switch to high power and observe an individual follicle.

4. Observe the central material in the follicle. The thyroid gland secretes a material called a colloid into the center of the follicle. The colloid is a stored form of thyroid hormone. The pituitary gland secretes a hormone that signals the thyroid follicle cells to transport the colloid molecules from the center of the follicle to the follicle cells. As the colloid crosses the membranes of the follicle cells, it is converted to thyroxine and then released into the bloodstream. Thyroxine increases the rate of most metabolic functions in body tissues.

5. Compare the follicle you observed with that in Figure 1. Notice the location of the follicle cells and the colloid.

6. Assuming your high-power field is 0.3 mm in diameter, estimate the diameter of each of 5 follicles. Record the diameters in Table 1. Calculate the average diameter of a follicle.

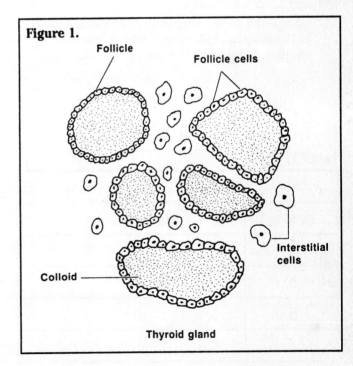

Figure 1.

7. Observe the cells between the thyroid follicles. These are interstitial cells. They secrete calcitonin, which regulates the calcium level in the blood.

8. Find the interstitial cells in Figure 1.

Part B. Adrenal Gland

1. Hold up the adrenal-gland slide. Note that there are two obvious layers of the adrenal gland. The outer layer is called the cortex. The inner layer is the medulla. The gland is covered by a fibrous connective tissue.

2. Locate the medulla under the 10X objective of your microscope. The medulla produces two hormones, epinephrine and norepinephrine. They help the body adjust to sudden stresses. They increase heart rate and the force of the heartbeat, and constrict blood vessels except those going to muscles. This causes an increase in blood pressure. These hormones also cause the liver to release stored sugar, which provides additional energy to the body under stress.

3. Observe the cortex of the adrenal gland with the 10X objective of your microscope.

4. The cortex is divided into three zones. Locate the innermost zone. This zone secretes small amounts of sex hormones.

5. Locate the large middle zone. The middle zone secretes glucocorticoids, a group of hormones concerned with metabolism and stress resistance. The middle zone also secretes cortisol, which increases carbohydrate, protein, and lipid metabolism.

6. Locate the outer zone of the adrenal gland. This area secretes aldosterone, which increases sodium and water retention by the kidneys.

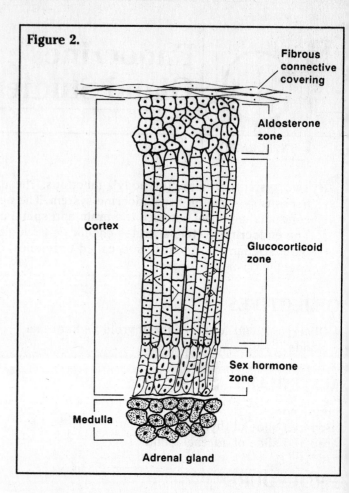

Figure 2.

Fibrous connective covering

Aldosterone zone

Cortex

Glucocorticoid zone

Sex hormone zone

Medulla

Adrenal gland

7. In Figure 2, notice the cortex, medulla, sex-hormone zone, glucocorticoid zone, aldosterone zone, and fibrous connective-tissue covering.

DATA AND OBSERVATIONS

Table 1.

Follicle Diameter	
Follicle	**Diameter (mm)**
1	
2	
3	
4	
5	
Average	

ANALYSIS

1. What is the function of the thyroid hormones? _____
_____

2. What is the function of calcitonin? _____
_____

3. Hyperthyroidism is a condition in which the thyroid secretes too much thyroid hormone. Predict what would happen to the amount of colloid in the center of the follicle if a person had hyperthyroidism. What

   would happen to the follicle cells? _____
_____
_____

4. Which part of Figure 3 shows the cells of a hyperthyroid follicle? _____

   Label it "hyperthyroid."

5. Hypothyroidism is a condition in which not enough thyroid hormone is produced. It can be caused by an iodine deficiency. Iodine is necessary for the synthesis of thyroid hormone. Without iodine, the follicle cells shrink and become inactive. Identify the part of Figure 3 that shows the cells of a hypothyroid

   follicle. _____

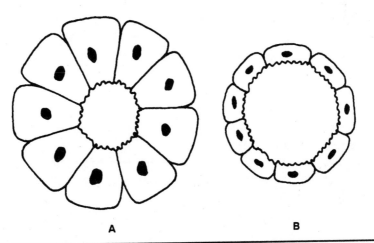

Figure 3.

A   B

Label it "hypothyroid."

6. What is the function of each of the following adrenal hormones?

   a. Epinephrine _____
_____

   b. Aldosterone _____
_____

c. Cortisol _____

_____

7. Epinephrine is also known as adrenalin. How could it benefit someone bleeding from an injury? _____

_____

8. How does epinephrine provide an adaptive advantage to an animal faced with danger? _____

_____

FURTHER EXPLORATIONS

1. View prepared slides of other hormone glands such as the parathyroid, pancreas, thymus, pineal, testis, ovary, and pituitary glands. Compare their structures with those of the glands observed in this Exploration.

2. Prepare a report relating gland structure to function. Include examples of structural and functional abnormalities.

EXPLORATION

Caloric Content of a Meal

38–2

LAB

▶ Food is the source of energy and building materials for the human body. This energy is used by cells to carry out respiration, protein synthesis, active transport, and other metabolic reactions as a person goes about his or her daily activities. When the foods a person eats provide more energy than is needed, the balance is converted and stored as fat. Conversely, when the body receives less energy than it needs, stored fat is burned.

OBJECTIVES

- Calculate the number of Calories and grams of carbohydrate, fat, and protein in two meals.
- Appraise the nutritional value of each meal.
- Plan a balanced, nutritional meal.

MATERIALS

food table (with Calories and grams of carbohydrate, fat, and protein listed)

PROCEDURE

1. Study Table 1. Assume a person eats three meals per day. Calculate the recommended intake *per meal* for all of the nutrients and Calories in Table 1. Record these in Table 1.
2. Examine the foods in the first column of Table 2. Two separate lunch plans are presented. Provide the information for the remaining columns by locating the proper food in the food table. If the food table does not present information for the same serving size, you will have to calculate the correct values. First, calculate the Calories and grams of

nutrients per ounce from the food table, and then multiply that number by the correct serving size.
3. Record your information, including the totals, in Table 2.
4. Compare the information for Meals 1 and 2 in Table 2 with the recommended intake per meal in Table 1.
5. Plan a meal that comes close to the intake recommendations of Table 1. Use the food table as a guide.

DATA AND OBSERVATIONS

Table 1.

Recommended Food Intake		
Nutrients	Amount per day	Amount per meal
Calories	2000–2500	
Carbohydrate (grams)	120	
Fat (grams)	60	
Protein (grams)	60	

239

Table 2.

Food	Serving size	Calories	Carbohydrate (grams)	Fat (grams)	Protein (grams)
Calories and Nutrients of Two Sample Meals					
Meal 1					
Spaghetti with meat sauce	6 oz.				
Green beans	4 oz.				
Garlic bread	2 slices				
Butter	1 Tbsp.				
Gelatin dessert	4 oz.				
Total					
Meal 2					
Hamburger bun	1				
Ground beef	4 oz.				
Cheese (American)	1 oz.				
Catsup	2 Tbsp.				
French fries	24				
Cola-type beverage	10 oz.				
Total					

ANALYSIS

1. Which of the two sample meals in Table 2 is higher in:

   Calories? _____ fat? _____ carbohydrate? _____ protein? _____

2. Which of the two sample meals is more nutritious? _____ Why? _____

3. Sample meal 1 is low in which important nutrient? _____ What kinds of foods could the

   other two meals of the day include to make up for this? _____

4. If you eat more Calories than your daily activities use, what will happen to your weight? _____

5. If you eat fewer Calories than you need, what will happen to your weight? _____

FURTHER EXPLORATIONS

1. Prepare a weekly meal plan that meets the recommended daily intake values of Calories, carbohydrates, fat, and protein.

2. Record the number of Calories and amounts of nutrients you had for breakfast and lunch. Determine how many Calories and how much of each nutrient you need for the evening meal. Plan this meal so that you will have received 100% of the RDAs for the day.

How Much Vitamin C Are You Getting?

INVESTIGATION

▶ Vitamin C is an essential vitamin for human health. Since the body cannot manufacture or store this chemical, vitamin C must be part of one's daily diet. It is found in potatoes as well as a variety of citrus fruits and can also be obtained from a vitamin supplement. A daily intake of 40 milligrams, or 0.040 grams, is considered the minimum daily requirement, as determined by the United States Food and Drug Administration.

The vitamin C concentration of various citrus juices can be determined by comparing them to a standard solution of known concentration. The microchemistry technique used to make the comparison is called a titration. In this experiment, titration is first used to measure the volume of iodine solution needed to react completely with a known volume of the standard solution. Once that volume of iodine solution is determined, it is then possible to titrate and analyze unknown concentrations of vitamin C. If, for example, it requires 10 drops of iodine solution to completely react with 5 milligrams of vitamin C, then an unknown solution requiring 20 drops of iodine solution must contain twice as much vitamin C as the standard solution.

The iodine solution is added to the standard solution one drop at a time. A chemical reaction occurs, during which the brown-red color of the iodine disappears. The end point of the reaction is reached when sufficient iodine is added to react with all the vitamin C. The end point can be detected by a starch indicator, which is added to the standard solution at the beginning of the experiment. The first drop of excess iodine causes the starch solution to turn blue-black, indicating that all of the vitamin C has been used up in the reaction.

OBJECTIVES

- Measure the volume of iodine solution needed to react with a standard concentration of vitamin C.
- Hypothesize which citrus juice sample contains the greatest concentration of vitamin C per serving.

- Measure the volume of iodine solution needed to react with unknown concentrations of vitamin C.
- Calculate the concentration of vitamin C in the citrus juice samples.

MATERIALS

standard vitamin C solution
(0.001 g/mL of water)
1% starch solution
distilled water
iodine solution
citrus juice samples (including fresh orange juice)

paper towels
plastic pipettes or microtip pipettes (3)
toothpicks
microplate, 96-well
microplate, 24-well
scissors or scalpel

tape
plastic straws (2)
wax marking pencil
laboratory apron
safety goggles

PROCEDURE

PART A. The Standard Solution

1. Construct a ring stand and pipette holder using a plastic drinking straw. With a scissors, cut the straw into two pieces on a slant, as shown in Figure 1. Insert the uncut end of one piece into a well on the edge of a 96-well microplate. This piece is a ring stand.

2. Cut off the slanted end of the other piece of straw. Insert one blade of the scissors into the straw and make a lengthwise cut to about the halfway point of the straw. Make a similar cut opposite and parallel to the first cut. Join the cuts by making a crosswise cut. Discard the piece of straw that results.

3. Cut the end of the flap of straw created in step 2 to make a point. Bend the flap and insert its point into the opening of the uncut portion of the straw, as shown. The loop you form in this way will serve as a pipette holder.

4. Hold the pipette holder with the loop horizontal and use a hole punch to make a hole at the end of the straw opposite to the loop. Slip the pipette holder onto the ring stand straw in the microplate.

5. Fill one of the pipettes with iodine solution. Insert the pipette into the ring stand and position the pipette directly over a well in the 24-well microplate (see Figure 1). This pipette will be used to titrate the iodine with the standard solution and then (in Part B) with the solutions of unknown vitamin C concentration. CAUTION: *If iodine is spilled, rinse with water and call your teacher immediately.*

6. Use a second pipette to place 10 drops of the standard vitamin C solution in the well of the 24-well microplate under the pipette.

Figure 1.

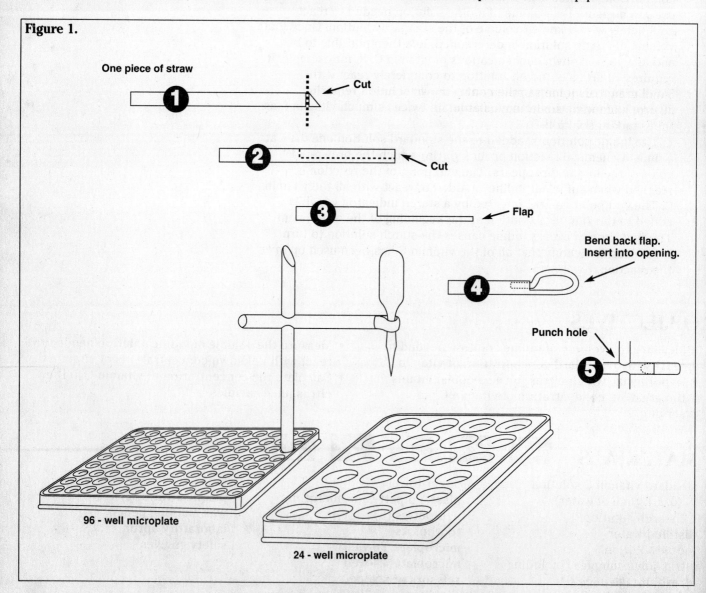

242

7. Use the third pipette to place 10 drops of water and 5 drops of starch solution in the well. Mix the contents of the well with a toothpick.

8. Now you are ready to begin the titration. Add the iodine solution to the well one drop at a time, stirring gently and counting each of the drops. When the brown-red color of the iodine begins to disappear, add the iodine solution more slowly. When the color has completely disappeared, the next drop or two may signal the end point of the chemical reaction. As soon as you see a blue-black tint, stop the titration and record the total number of drops of iodine you used in Table 1.

9. Discard the contents of the well according to your teacher's directions. Rinse the well with distilled water and dry it with a paper towel.

10. Repeat steps 5–9 two more times. Record each trial in Table 1 and average your results.

11. Rinse the stem and bulb of the pipette from the standard vitamin C in preparation for its use in Part B.

PART B. Unknown Solutions

1. Label the citrus juice sample containing fresh orange juice with the letter A. Label the other samples B and C or B-D.

2. Make a hypothesis to predict which juice sample has the highest vitamin C content. Write your hypothesis in the space provided.

3. If necessary, refill the pipette in the ring stand with iodine solution.

4. Add 10 drops of juice A to a well in the 24-well microplate.

5. Add ten drops of water and 5 drops of starch solution to the same well. Mix the contents of the well with a toothpick.

6. Add the iodine solution to the well drop by drop, stirring steadily and counting each of the drops.

7. Record in Table 1 the number of drops required to turn the solution blue-black.

8. Discard the contents of the well according to your teacher's directions. Rinse the well and dry it with a paper towel.

9. Repeat steps 3-7 two more times and average your results.

10. Repeat steps 3-9 for samples B-D.

HYPOTHESIS

_____

_____

DATA AND OBSERVATIONS

Table 1.

	Trial 1	Trial 2	Trial 3	Average
Drops Iodine Solution (Std. Vit. C)				
Drops Iodine Solution (Juice A)				
Drops Iodine Solution (Juice B)				
Drops Iodine Solution (Juice C)				
Drops Iodine Solution (Juice D)				

ANALYSIS

1. If your class has worked in teams, collect and share the data. Calculate the ratio of drops of iodine added to juices A–D to the drops of iodine added to the standard solution.

$$\text{Ratio 1} = \frac{\text{Avg. \# of Drops used with Juice A}}{\text{Avg. \# of Drops used with Standard}} = \underline{\hspace{2cm}}$$

2. What is the concentration of vitamin C in grams per mL of juices A–D?

    Concentration of vitamin C in juice sample A= \underline{\hspace{3cm}}

    Ratio 1 × 0.001 g/mL

3. How many grams of vitamin C would you get by drinking an average serving of each sample of juice? (The average serving of citrus juice is 150 mL.)

    \underline{\hspace{2cm}} g/mL (your answer from question 2)

    × 150 mL = \underline{\hspace{4cm}} g

4. Convert your answer to question 3 into milligrams.

    _____

5. What would have happened in your experiment if you forgot to add starch indicator solution?

    _____

    _____

6. What would be a way of calculating the amount of iodine solution used in each titration besides counting drops?

    _____

    _____

CHECKING YOUR HYPOTHESIS

Was your **hypothesis** supported by your classes data? Why or why not? \underline{\hspace{4cm}}

_____

_____

FURTHER INVESTIGATIONS

1. Design an experiment to assay the vitamin C content of potatoes.
2. Design an experiment to assay a multivitamin tablet for vitamin C content.

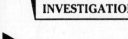

How Does Exercise Affect Heart Rate?

39–1
LAB

▶ The heart pumps blood to all the cells of the body. As you exercise, the muscle cells use more oxygen and food, which must be replenished. The cells also produce more wastes, which must be removed. The heart responds to the changing needs of body cells by pumping harder.

OBJECTIVES

- Determine your heart rate by taking your pulse.
- Hypothesize the effect of exercise on heart rate.

- Compare your heart rate during rest and during exercise.

MATERIALS

stopwatch or clock with second hand

PROCEDURE
Part A. Resting Pulse

1. Locate your pulse by placing your index and middle fingers on the carotid artery. This artery is located in your neck, under the corner of your jaw. Figure 1 shows the position your fingers should be in.
2. Press your fingers lightly against your neck to feel the pulse. Each pulse of blood is caused by one beat of the heart.
3. Work in pairs. Your partner will be the timekeeper. You will be the experimental subject.
4. Sit quietly for 2 minutes.
5. Count your pulse for 15 seconds, and record this number in Table 1.
6. Repeat step 5 twice more.
7. Calculate your heart rate per minute by multiplying each of the 15-second counts by 4.
8. Calculate your average resting heart rate per minute.
9. Make a **hypothesis** to explain how exercise will affect heart rate. Write your hypothesis in the space provided

Part B. Exercise

1. Follow your teacher's instructions for exercising. Remember that whatever exercise

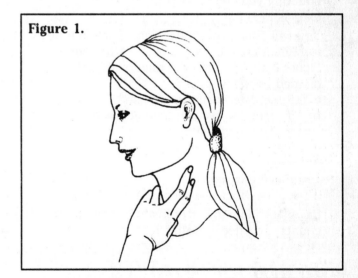

Figure 1.

you do should be done at a steady rate throughout all exercise periods.

2. You will exercise for 30 seconds of each minute for 10 minutes. In between exercise periods, you will count your pulse for 15 seconds and record this count in Table 2. Figure 2 shows that each minute should be divided as follows.
   a. 30 seconds—exercise
   b. 5 seconds—locate the pulse
   c. 15 seconds—count the pulse
   d. 10 seconds—record the count and prepare to exercise again

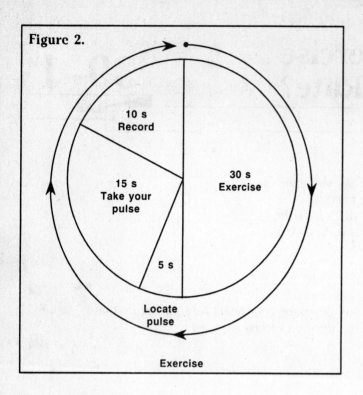

Figure 2.

- 10 s Record
- 15 s Take your pulse
- 5 s
- Locate pulse
- 30 s Exercise

Exercise

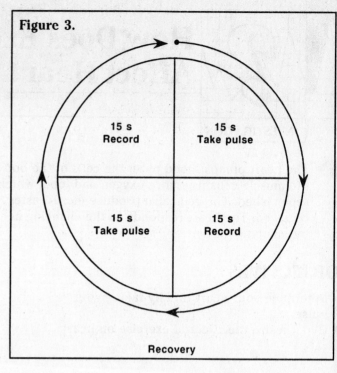

Figure 3.

- 15 s Record
- 15 s Take pulse
- 15 s Take pulse
- 15 s Record

Recovery

Part C. Recovery

1. Immediately after you finish the exercise period, sit down and begin the procedure for counting your pulse as you recover from exercising.
2. You will count your pulse twice each minute for 10 minutes. Each count lasts 15 seconds. Figure 3 shows that each minute should be divided as follows.
   a. 15 seconds—count the pulse
   b. 15 seconds—record the count and prepare for the next count
   c. 15 seconds—count the pulse
   d. 15 seconds—record the count and prepare for the next count
   All counts should be recorded in Table 2.
3. Calculate the number of beats per minute for all counts recorded in Table 2.

Part D. Graphing

1. Look at the grid provided in Data and Observations. The minutes are on the horizontal axis. The heart rates are on the verticle axis. Plot your average resting heart rate on the first vertical line.
2. Plot your heart rates during exercise for minutes 1 through 10.
3. Plot your recovery pulse rates for minutes 11 to 20. **NOTE:** *There are data for every thirty seconds in this time period. Be sure to plot the heart rate data against the correct times.*

HYPOTHESIS

_____

_____

DATA AND OBSERVATIONS

Table 1.

Condition	Trial	Pulse	
		Beats/15 seconds	Beats/minute
Rest	1		
	2		
	3		
	Average		

Table 2.

Condition	Minutes	Pulse	
		Beats/15 seconds	Beats/minute
Exercise	1		
	2		
	3		
	4		
	5		
	6		
	7		
	8		
	9		
	10		
Recovery	11:00		
	11:30		
	12:00		
	12:30		
	13:00		
	13:30		
	14:00		
	14:30		
	15:00		
	15:30		
	16:00		
	16:30		
	17:00		
	17:30		
	18:00		
	18:30		
	19:00		
	19:30		
	20:00		

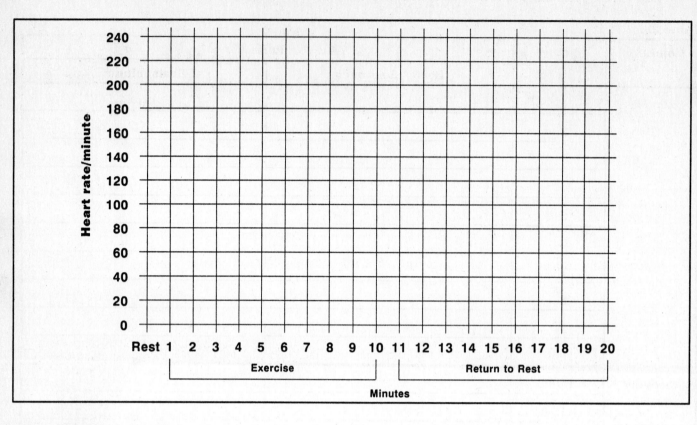

ANALYSIS

1. What was your average resting heart rate? _____

2. What was your highest heart rate? _____ When did this rate occur? _____

3. Did you return to your resting rate during Part C? _____ If so, how many minutes did it take? _____

4. Why do some classmates return to their resting rates more rapidly than others? _____

   _____

5. Describe the shape of your graph. _____

   _____

6. What factors, other than exercise, would increase heart rate? _____

   _____

CHECKING YOUR HYPOTHESIS

Was your **hypothesis** supported by your data? Why or why not? _____

_____

FURTHER INVESTIGATIONS

1. Compare the average heart rate of the males with that of the females in your class.
2. Test the effect of lying down, standing up, and sitting on your resting heart rate.

248

What Are the Locations of Taste and Smell Receptors? 40–1

LAB

▶ The senses of taste and smell are involved in chemoreception. Chemoreceptors for smell are stimulated by compounds in the air. Chemoreceptors for taste respond to compounds in solution. There are taste chemoreceptors for four different flavors: bitter, salty, sweet, and sour. Although these receptors are found all over the surface of the tongue, each type is more heavily concentrated in a particular area.

OBJECTIVES

- Hypothesize about the locations of different taste receptors on the tongue.
- Diagram the distribution of chemoreceptors on the tongue.

- Use correct lab techniques to dilute solutions.
- Determine the effect of the sense of smell on the ability to identify tastes.

MATERIALS

paper cups (14)
permanent marking pen
sweet solution
sour solution
bitter solution

salty solution
cotton swabs (42)
paper towel
paper bag
25-mL graduated cylinder

stirring rod
toothpicks (18)
potato, apple, and onion
  (6 pieces of each)
plastic wrap

PROCEDURE

Part A. Chemoreceptors on the Tongue

1. Work in teams of two. Each team should obtain six paper cups. Use a marking pen to label four of them with the four tastes: sweet, sour, bitter, salty. Label the fifth and sixth cups "water" and the name of a team member.

2. Measure about 20 mL of each solution and pour it into the appropriately labeled cup. Half fill the fifth and sixth cups with water.

3. Place 32 cotton swabs on a clean paper towel. Use a paper bag to collect used swabs.

4. Make a **hypothesis** to describe the distribution of the four taste receptors on the front, back, side, and tip surfaces of the tongue. Write your hypothesis in the space provided.

5. While your partner sits with eyes closed, dip a cotton swab into one of the solutions. Touch the swab to the tip of your partner's tongue. Immediately put the swab into the paper bag. **CAUTION:** *Do not reuse any of the cotton swabs.*

Figure 1.

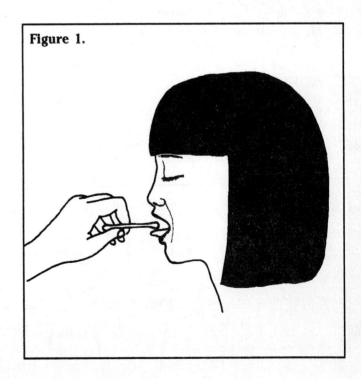

6. Ask you partner to identify the taste. If the identification is correct, ask if the taste was strong, moderate, or weak. Record this response in the proper space in Table 1. If your partner did not taste anything, record "none" in the proper space in Table 1. If your partner responded incorrectly, record this response.

7. Repeat steps 5 and 6 for the four different tastes on the sides, back, front, and tip of your partner's tongue. Use a new swab for each test and throw away each swab after it is used once. Mix up the order of the testing so that your partner cannot guess what taste will be next. Record the results in Table 1. Allow your partner to take a sip of water now and then. This will help wash the test solutions off the tongue.

8. Reverse roles, and repeat steps 5–7. You may reuse the test solutions, but use a new cup of water for the subject to drink, labeled with the person's name.

9. Discard the cups with the taste solutions, but save the two water cups.

10. From your data, use the outline provided in Data and Observations to diagram the distribution of taste receptors on the tongue.

Part B. Determining the Threshold of Taste

1. Use a marking pen to label 5 paper cups full-strength, 1/2-strength, 1/4-strength, 1/8-strength, and 1/16-strength.

2. Measure 10 mL of sweet solution into the cup marked full-strength.

3. Measure 10 mL of sweet solution into the cup marked 1/2-strength. Add 10 mL of water, and stir the mixture. Rinse the graduated cylinder with water. You will dilute this mixture several times. Refer to Figure 2 as you follow the directions for diluting the solution.

4. Measure 10 mL of the 1/2-strength solution into the cup marked 1/4-strength. Add 10 mL of water, and stir the mixture. Rinse the cylinder with water.

5. Measure 10 mL of the 1/4-strength solution into the cup marked 1/8-strength. Add 10 mL of water, and stir the mixture. Rinse the cylinder with water.

6. Measure 10 mL of the 1/8-strength solution into the cup marked 1/16-strength. Add 10 mL of water, and stir the mixture.

7. Obtain 5 cotton swabs. Use them to test your partner's ability to taste the solutions of different strengths. Test the front of the tongue only. Start with the full-strength solution and then test progressively weaker solutions. Immediately discard each swab, as you did in Part A.

8. Record your partner's ability to taste each solution. If the taste was recognized, mark a "+" in the appropriate space in Table 2. If the taste was not recognized, mark a "–" in the appropriate space in Table 2. Allow your partner to take a sip of water now and then. Use the appropriate cup of water from Part A.

The threshold of taste is the minimum concentration of solution that can be tasted. Your partner's exact threshold is somewhere in between the strength of the weakest solution that could be tasted, and the strongest solution that could not be tasted.

9. Reverse roles, and repeat steps 7 and 8. You may reuse the test solutions, but use the appropriately labeled cup of water from Part A.

Part C. The Effect of Smell on Taste

1. Label three paper cups "potato," apple," or "onion." Place six pieces of each food into the appropriate cup, and cover each cup with plastic wrap. Keep the foods covered during the activity, to contain their odors.

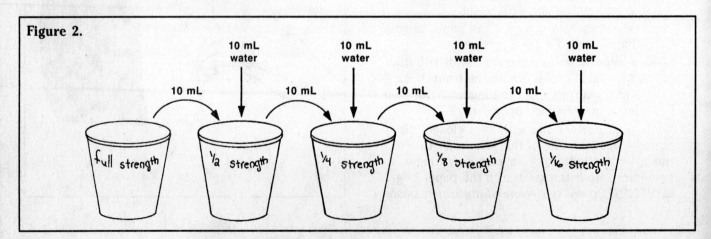

Figure 2.

2. Have your partner sit with **eyes closed,** holding his or her nose. Use a toothpick to spear a piece of one of the foods and place it in your partner's mouth. Your partner should slowly chew and then swallow the food, and then take a sip of water to clear the tongue. Discard the toothpick immediately into the paper bag.

3. Ask your partner to identify the food. If the identification was correct, mark a "+" in the "Taste only" column in Table 3. If the identification was incorrect, mark a "–" in the proper place in Table 3.

4. Repeat steps 2 and 3 with the other two foods. Record all data in Table 3.

5. Have your partner sit with eyes closed, but not holding his or her nose. Use a toothpick to hold a piece of one of the foods beneath your partner's nose. Discard the toothpick.

6. Ask your partner to identify each food. Record the responses under the column labeled "Smell only" in Table 3.

7. Have your partner sit with eyes closed, but not holding his or her nose. Repeat steps 2–4. Record the responses under the column labeled "Taste and smell" in Table 3.

8. Reverse roles, and repeat steps 2–7.

9. Discard all the cups in the bag.

HYPOTHESIS

_____

_____

_____

DATA AND OBSERVATIONS

Table 1.

Name of subject: _____

Areas of Taste Reception				
Location on the tongue	**Taste**			
	Sweet	**Sour**	**Bitter**	**Salty**
Sides				
Front				
Back				
Tip				

Table 2.

Name of subject: _____

Threshold of Taste	
Strength of solution	**Taste**
Full	
1/2	
1/4	
1/8	
1/16	

Table 3.

Combining Smell and Taste			
Food	Taste only	Smell only	Taste and smell
Potato			
Apple			
Onion			

ANALYSIS

1. Describe the distribution of chemoreceptors on the tongue. _____

_____

2. Where was the strongest reception of sweet taste? _____

3. **a.** Where was the strongest reception of bitter taste? _____

   b. How is this related to the fact that many bitter foods taste strongest as they are being swallowed?

   _____

4. Did you and your partner have the same threshold of taste? _____

5. Why was the test in Part B done at the front of the tongue? _____

_____

6. How could you measure more exactly someone's threshold? _____

_____

7. Under which conditions did you and your partner make the most accurate identifications of foods?

_____

8. Why do foods seem to have less taste when you have a cold? _____

_____

CHECKING YOUR HYPOTHESIS

Was your **hypothesis** supported by your data? Why or why not? _____

_____

FURTHER INVESTIGATIONS

1. Repeat Part A using a solution that is a mixture of two of the tastes, such as sweet and salty.
2. Repeat the threshold of taste test using one of the other solutions. Test several areas of the tongue to find out if the threshold of taste or a given taste is the same for all areas of the tongue.

When Does a Chicken Embryo Grow the Fastest?

41–1

LAB

INVESTIGATION

▶ After a sperm has fertilized an egg, the resulting zygote begins to change. The unicellular zygote begins to divide rapidly and develop into a multicellular embryo. After 21 days, a chick hatches from the egg. The chick embryo provides a convenient way to study the changes that take place as an embryo develops, because fertilized eggs can be removed from the incubator at different times during development and examined.

OBJECTIVES

- Hypothesize the rate of growth of an embryo.
- Observe the development of a chick embryo over a period of 72 hours.

- Measure the length of a chick embryo at various stages of development.
- Prepare graphs to show the rate of growth and development of a chick embryo.

MATERIALS

fresh fertilized chicken eggs incubated for
  0 hours, 24 hours, 48 hours, and 72 hours
incubator (38°C)
dissecting probe
small dish or bowl
scissors

forceps
stereomicroscope
metric ruler
graph paper (2 sheets)
absorbent cotton

PROCEDURE

Part A. Egg at 0 Hours Incubation

1. Obtain an egg marked 0 hours.
2. Hold the egg on its side in the palm of one hand. Gently drill a hole into the shell with the pointed dissecting probe.
3. Place the egg into a small dish containing a bed of cotton. Place the tip of the scissors into the hole you drilled and cut an oval in the egg as shown in Figure 1. Use forceps to lift out the pieces of egg shell.
4. Normally the yellow yolk with the embryo will be face up as you open the egg. If not, gently rotate the egg to see if you can locate the blastodisc, a small white spot that surrounds the embryo. It may be necessary to gently pour the contents of the egg into the dish, after removing the cotton.
5. Examine the yellow yolk, which is made of protein and fat. The yolk is enclosed by a membrane.

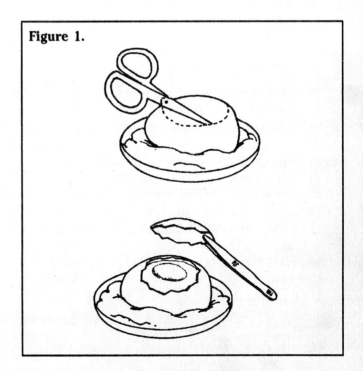

Figure 1.

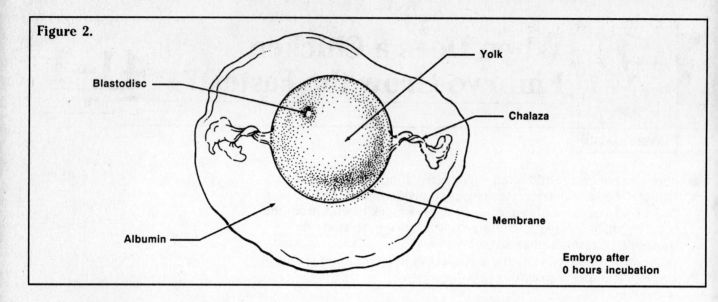

Figure 2.

Blastodisc

Yolk

Chalaza

Membrane

Albumin

Embryo after
0 hours incubation

6. Locate the clear liquid surrounding the yolk. It contains the protein, albumin.

7. Locate the two cordlike structures at each end of the yolk. These are the chalazae, which are made of dense albumin.

8. Examine the blastodisc with a stereomicroscope. The blastodisc contains the developing embryo, a tiny, C-shaped structure.

9. With a metric ruler, measure the length of the embryo in millimeters. Record this measurement in Table 1.

10. Identify the structures shown in Figure 2.

11. Make a **hypothesis** concerning the incubation time period in which the chick will grow the fastest, 0 to 24 hours, 24 to 48 hours, or 48 to 72 hours. Write your hypothesis in the space provided.

Part B. Developing Embryos

1. Obtain an egg marked 24 hours. Open the egg, following the procedure used in Part A.

2. Examine the embryo under a stereomicroscope.

3. Measure the embryo and record the length in millimeters in Table 1.

4. Locate the neural tube, which runs the length of the embryo. This structure will form the spinal cord. Locate the developing brain, an enlargement at one end of the neural tube.

5. Locate the blocks of tissue forming along the neural tube. These are somites. Count and record the somites in the embryo, and compare this number with the number shown in Table 1.

6. Look for the developing heart, a bulge near the neural tube, between the somites and the brain.

7. Identify the neural tube, brain, somites, and heart as shown in Figure 3.

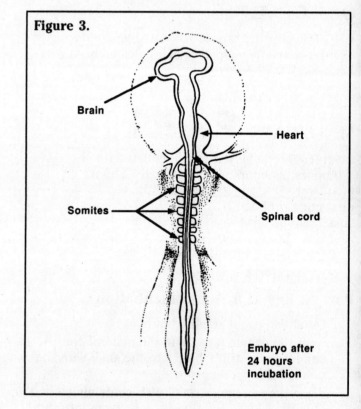

Figure 3.

Brain

Heart

Somites

Spinal cord

Embryo after
24 hours
incubation

8. Repeat steps 1 through 3 with an egg marked 48 hours.

9. Locate the somites and the heart. Count and record the somites, and compare this number with the number shown in Table 1.

10. Notice that the neural tube has formed the spinal cord and the brain. Locate the eye.

11. In Figure 4, notice the brain, spinal cord, eye, heart, and somites.

12. Repeat steps 1 through 3 with an egg marked 72 hours.

13. Find the brain, spinal cord, eye, heart, and somites. Identify these structures in Figure 5.

14. Look for paired bulges at the sides of the body. These bulges are limb buds, and will develop into wings and legs. Look for the tail at the end of the body. Find the limb buds and tail in Figure 5.

15. Dispose of your eggs according to your teacher's directions.

16. Prepare a graph that shows the formation of somites during the first 72 hours of development. Plot time in hours on the horizontal axis. Plot number of somites on the vertical axis.

17. Prepare a graph that shows the growth of the embryo during the first 72 hours of development. Plot time in hours on the horizontal axis. Plot length in millimeters on the vertical axis.

HYPOTHESIS

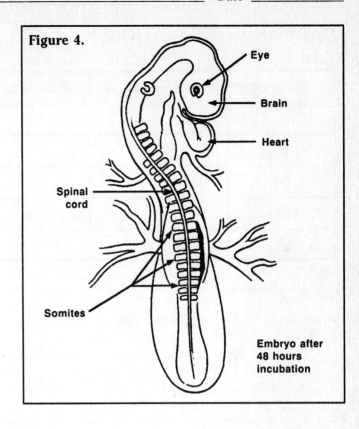

Figure 4.

Embryo after 48 hours incubation

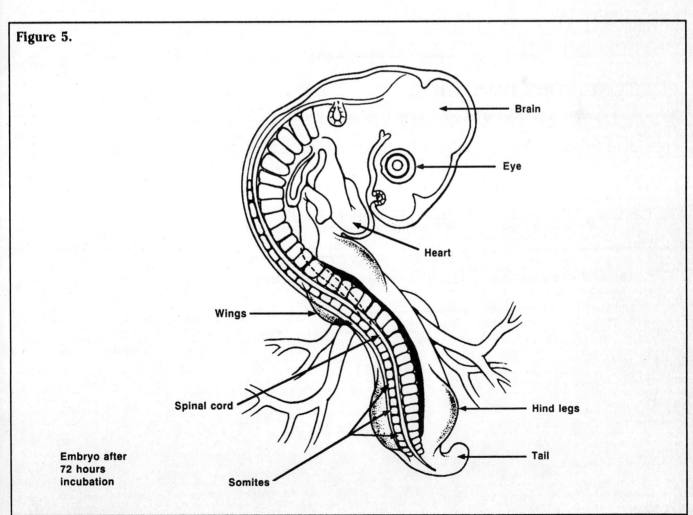

Figure 5.

Embryo after 72 hours incubation

255

DATA AND OBSERVATIONS

Table 1.

Hours of incubation	Length of embryo in mm	Number of pairs of somites	Number of pairs of somites observed
0		0	
24		8	
48		21	
72		36	

ANALYSIS

1. What is the function of the yolk and albumin? _____

_____

2. How does a 72-hour embryo differ from a 24-hour embryo? _____

_____

3. Using the two graphs you made, determine in which time period the embryo grows the fastest. _____

_____

CHECKING YOUR HYPOTHESIS

Was your **hypothesis** supported by your data? Why or why not? _____

_____

FURTHER INVESTIGATIONS

1. Open an unfertilized egg, following the procedure used in Part A. Compare the unfertilized egg with the fertilized egg you studied in Part A.
2. Prepare a "window egg" to follow chick development through hatching. Most college embryology lab books contain instructions for this method.

Human Fetal Growth

41–2
LAB

▶ Complete development of a human fetus takes about 38 weeks. Increases in size and mass are two of the many changes that the fetus undergoes. The increases do not occur at the same rate. Many factors affect the birth size of a human baby, but there is an average mass and an average length standard for each stage of development. The approximate age of a fetus can be determined from its mass and length.

OBJECTIVES

- Calculate the length of a human fetus at various stages of development.
- Graph the length of a developing human fetus.

- Graph the mass of a developing human fetus.
- Determine the period of development during which the greatest changes in mass and in length occur.

MATERIALS

metric ruler

PROCEDURE

Part A. Development of a Human Fetus

1. Examine Figure 1. It shows six stages of a developing human fetus. The stages are shown at 40% of the fetus' natural size.

2. Study the lengths indicated on the diagram of the 38-week fetus. Use these as a guide to measuring the other diagrams.

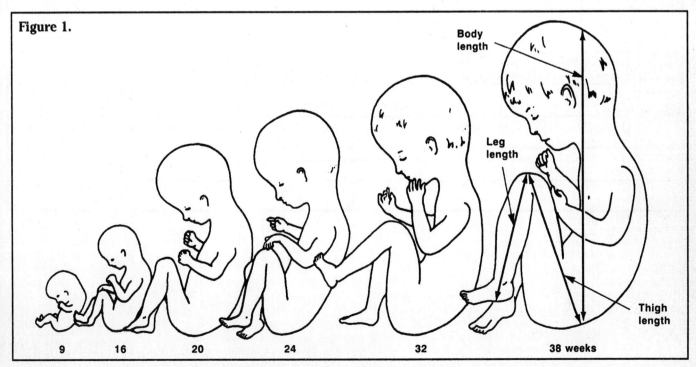

Figure 1.

9   16   20   24   32   38 weeks

Body length

Leg length

Thigh length

3. Measure each length in millimeters. Record your data in the spaces provided in Table 2.
   a. Measure the body length from the rump to the top of the head.
   b. Measure the thigh length from the rump to the knee.
   c. Measure the leg length from the heel to the knee.
4. Add the three measurements for each stage together. Record the total length in the space provided in Table 2.
5. Multiply the total by 2.5 to give a figure that is close to the actual length of the fetus at each stage. Record the actual length in Table 2.

Part B. Graphing the Length of a Developing Fetus

1. The point showing the length (2mm) of the 2-week fetus has been marked on the grid in Figure 2.
2. Using the data in Table 1, mark a point that shows the age and length of each fetal stage.
3. Begin at 0, and connect the points to complete the graph.

Part C. Graphing the Mass of a Developing Fetus

1. Look at the data supplied in Table 1.
2. Mark points on the grid in Figure 3 to show the ages and masses of each fetus.
3. Begin at 0, and connect the points to complete the graph.

Table 1.

Mass of a Developing Fetus			
Time (weeks)	Mass (grams)	Time (weeks)	Mass (grams)
4	0.5	24	650
8	1	28	1100
12	15	32	1700
16	100	36	2400
20	300	38	3300

DATA AND OBSERVATIONS

Table 2.

Lengths of a Developing Fetus					
Age of fetus in weeks	Body length (mm)	Thigh length (mm)	Leg length (mm)	Total length (mm)	Actual length (mm)
2	—	—	—	—	2 mm
9					
16					
20					
24					
32					
38					

Figure 2.

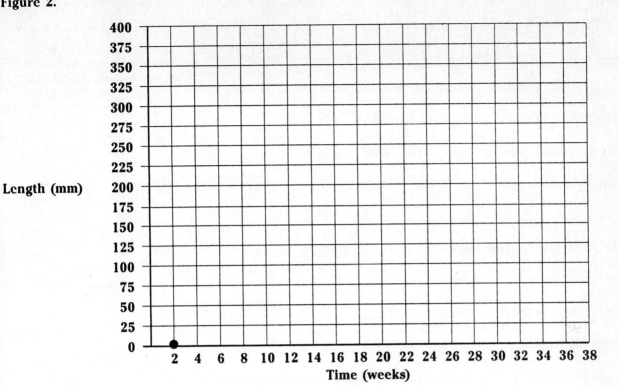

Length of a Developing Fetus

Figure 3.

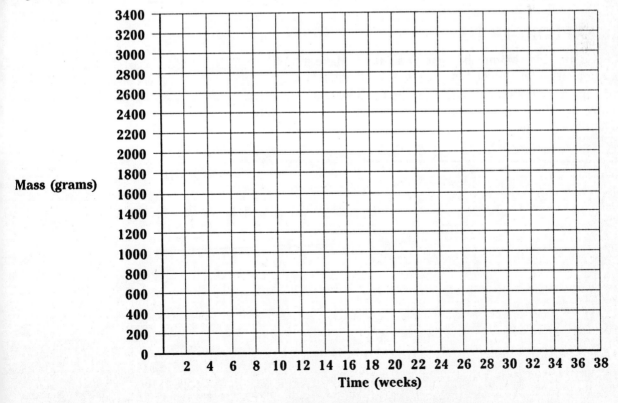

Mass of a Developing Fetus

ANALYSIS

1. What is the actual length of the fetus at week 9? _____

2. How much mass does the fetus gain from 0 to 8 weeks of development? _____

3. Look at Figures 2 and 3 for the halfway point in development at week 19.

   a. Is the fetus half of its full length at this time? _____

   b. Is the fetus half of its full mass at this time? _____

4. Why was the total length of each fetus multiplied by 2.5 to obtain the actual length? _____

   _____

5. Why do you think the length of a fetus increases more rapidly than the mass of a fetus? _____

   _____

6. At what week does the fetus reach

   a. half its full length? _____

   b. half its full mass? _____

7. If a premature baby is born with a mass of

   a. 2200 grams, how old is the fetus? _____

   b. 1800 grams, how old is the fetus? _____

FURTHER EXPLORATIONS

1. A human is a mammal. Mammals come in all sizes. Make a table showing the average birth weights of at least 10 mammals.

2. Prepare a report on the factors affecting the birth weight of a human baby. See if you can find data on how a baby's birth weight might affect its later life.

How Do Bactericides Affect the Growth of Bacteria?

42–1 LAB

INVESTIGATION

In 1928, Sir Alexander Fleming, a British bacteriologist, made a startling discovery. Some mold had fallen accidently onto one of his bacterial cultures. Fleming noticed that the bacteria did not grow near the mold. He experimented with the mold and isolated it. His experiments revealed that the mold, *Penicillium notatum*, produced a substance that stopped the growth of bacteria. Fleming purified the substance from the mold and developed penicillin, the first antibiotic. Antiseptics, disinfectants, and antibiotics are bactericides, substances that inhibit the growth of bacteria. Many bactericides have been developed since Fleming's discovery of penicillin. Different bactericides affect microorganisms to varying degrees. Antibiotics, like penicillin, kill only bacteria, whereas other bactericides can kill other cells. Antiseptics, like hydrogen peroxide, only hinder or prevent the growth of bacteria. Disinfectants, like chlorine bleach, kill bacteria and other organisms. Some bactericides affect only certain kinds of bacteria. In this lab, you will investigate how several bactericides affect the growth of *Escherichia coli*, the common intestinal bacterium, and *Streptococcus lactis*, the bacterium commonly found in milk.

OBJECTIVES

- Compare the effects of different bactericides on the growth of milk and intestinal bacteria.
- Hypothesize how bactericides affect the growth of intestinal and milk bacteria.
- Use sterile techniques for handling bacterial cultures and inoculating agar plates.

MATERIALS

laboratory apron
disinfectant solution (25 mL)
paper towels (2)
sterile cotton swabs (4)
forceps
Bunsen burner
striker
sterile filter paper disks (14)
test tubes (6)

sterile petri dishes containing nutrient agar (2)
sterile petri dishes containing lactose agar (2)
sour milk (25 mL)
Escherichia coli culture
isopropyl alcohol (10 mL)
household bleach (10 mL)
tincture of iodine (10 mL)

3% hydrogen peroxide (10 mL)
antibiotic disks (2)
mouthwash (10 mL)
disinfectant (10 mL)
test-tube rack
masking tape
metric ruler
wax marking pencil

PROCEDURE

CAUTION: *Do not touch eyes, mouth, or any other part of your face while doing this lab.*

Part A. Inoculating the Agar Plates

1. Put on a laboratory apron.

2. Wash your work surface with disinfectant solution, using a paper towel.

3. Place the four petri dishes containing agar medium in front of you. Label each lactose agar dish with an L and each nutrient agar dish with an N. Turn the dishes upside down. Be careful not to open the dishes. Use a wax marking pencil to draw perpendicular lines that divide each dish into four equal sections.

4. Number the four sections of the first lactose agar dish 1 through 4. Number the sections of the second dish 5 through 8, as shown in Figure 1. Number the sections of the two nutrient agar dishes the same way. Turn the dishes right side up.

5. Uncap the sour milk, flame the mouth of the test tube, and dip a sterile swab into the liquid. Continue to hold the swab while you reflame the mouth of the test tube and recap the sour milk. Open one of the lactose agar dishes. **CAUTION:** *Do not let the cotton end of the swab touch the outside of the dish or any other surface. Do not let your fingers touch any sterile surfaces. Use care when working with live bacteria. Avoid spillage on skin, clothing, or work area. Call your teacher immediately if spills occur.* Move quickly, but carefully, when you inoculate the agar plates and later when you introduce the disks. Lift the lid of the petri dish only when necessary. When introducing the disks or inoculating the plates, hold the lid in one hand and the forceps or swab in the other hand. Tip the lid just enough to permit entry of the swab or forceps. Do not allow the forceps or swab to touch the outside of the dish. Replace the lid immediately.

6. Move the swab lightly back and forth over the entire agar surface in a tight s-shaped pattern. Continue to hold the swab and replace the lid.

7. Turn the petri dish one quarter turn, lift the lid, and repeat step 6, as shown in Figure 2. Dispose of the swab as directed by your teacher.

8. Inoculate the second lactose agar petri dish, following steps 5–7 and using a new sterile swab.

9. Inoculate the nutrient agar petri dishes with *Escherichia coli* using the same procedure as in steps 5–7.

Part B. Testing the Effects of Bactericides on the Growth of Bacteria

1. Obtain the six bactericides you will test. Place each in a separate test tube and stand the test tubes in a test-tube rack.

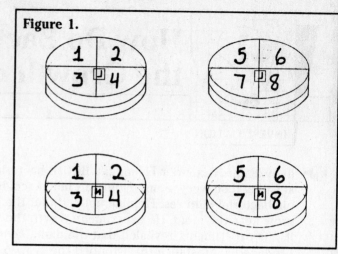

Figure 1.

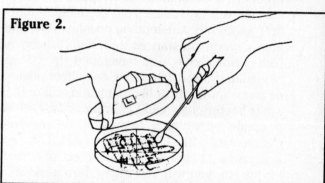

Figure 2.

2. Sterilize the ends of the forceps by holding the end in the flame of a Bunsen burner for 3 seconds. Use the sterile forceps to pick up a sterile untreated filter paper disk.

3. Tip the lid of the first lactose agar dish and place the disk near the outside edge of section 1. Use the end of the forceps to gently press on the disk so it adheres to the agar, as shown in Figure 3. **NOTE:** *Do not let the forceps touch the agar at any time during this procedure.*

4. Sterilize the forceps and use them to pick up an antibiotic disk. Place the disk near the outside edge of section 2. Be sure to record which disk you place in which section.

5. Wash the ends of the forceps with water and dry them.

6. Flame the forceps and use them to pick up another sterile filter paper disk. Dip it into isopropyl alcohol and touch the disk to the side of the test tube to drain off the excess alcohol.

7. Place the alcohol disk near the outside edge of section 3 of the lactose agar dish, as in Figure 4.

8. Repeat steps 5–7 with bleach, iodine, hydrogen peroxide, mouthwash, and disinfectant disks in sections 4 through 8, respectively.

9. Repeat steps 2–8 with the nutrient agar dishes.

Figure 3.

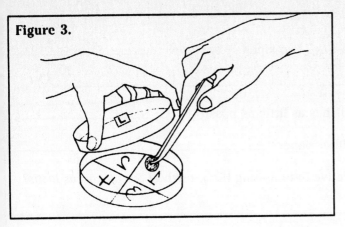

Figure 4.

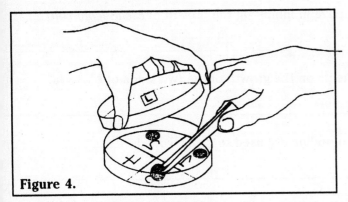

10. Secure the petri dish lids with tape and label them with your name.

11. Make a **hypothesis** to describe how the various bactericides will affect the growth of bacteria. Write your hypothesis in the space provided.

Figure 5.

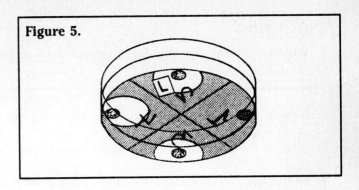

12. Incubate the dishes upside down for 48 hours at 35–37°C.

13. After 48 hours, examine the dishes. DO NOT OPEN THEM. Bacterial growth appears as a cloudy film on top of the agar. Look for zones of inhibition around the disks. These zones are clear areas that indicate an absence of bacterial growth, as shown in Figure 5. Measure the diameter of each zone in millimeters. Record your measurements in Table 1.

14. After completing your observations, dispose of the bacterial cultures as instructed by your teacher. Clean your work surface with disinfectant solution.

HYPOTHESIS

_____

_____

DATA AND OBSERVATIONS

Table 1.

Diameter of Zone of Inhibition (mm)		
Disk	*Streptococcus lactis*	*Escherichia coli*
Untreated		
Antibiotic		
Isopropyl alcohol		
Bleach		
Iodine		
3% hydrogen peroxide		
Mouthwash		
Disinfectant		

ANALYSIS

1. Why are lactose agar and nutrient agar used to grow the two kinds of bacteria? _____

_____

2. Why is it important that you lift the lids from the dishes as little as possible? _____

3. What is the purpose of using an untreated disk of filter paper? _____

4. Of the bactericides used, which were the most effective in inhibiting the growth of *Streptococcus lactis*?

_____

5. Of the bactericides used, which were the most effective in inhibiting the growth of *Escherichia coli*?

_____

6. Why do you think the bactericides had different effects on the growth of the two kinds of bacteria?

_____

7. Why do you think hydrogen peroxide and tincture of iodine are used to clean wounds? _____

_____

8. Did the antibiotic have the same effect on the growth of the two different kinds of bacteria? _____

_____

9. Based on your observations, do you think an antibiotic, such as tetracycline, can effectively help your

immune system fight diseases caused by different types of bacteria? _____

10. Based on your observations, can you conclude which bactericides would kill disease-causing bacteria?

_____

11. Can all bactericides be used to kill bacterial diseases in humans? Explain. _____

_____

CHECKING YOUR HYPOTHESIS

Was your **hypothesis** supported by your data? Why or why not? _____

_____

FURTHER INVESTIGATIONS

1. Design and conduct an experiment to test the effect of salt on the growth of bacteria. Certain bacteria can cause food to spoil. Based on the results of your experiment, suggest a way to prevent food from spoiling.

2. Design and conduct an experiment to test whether or not an antibiotic continues to have the same effect on the growth of bacteria over time. Explain how you think this information affects what antibiotics a doctor might prescribe for you over time.

Testing Water Quality

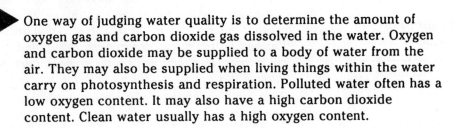

43-1
LAB

EXPLORATION

One way of judging water quality is to determine the amount of oxygen gas and carbon dioxide gas dissolved in the water. Oxygen and carbon dioxide may be supplied to a body of water from the air. They may also be supplied when living things within the water carry on photosynthesis and respiration. Polluted water often has a low oxygen content. It may also have a high carbon dioxide content. Clean water usually has a high oxygen content.

OBJECTIVES

- Treat water samples to determine the presence of oxygen and carbon dioxide in the water.
- Determine the parts per million (ppm) of dissolved oxygen and carbon dioxide in water samples.
- Evaluate whether water samples may be clean or polluted.
- Analyze how the presence of oxygen and carbon dioxide in bodies of water affects life in that water.

MATERIALS

water samples from different sources (2)
small flasks (or beakers) (2)
droppers (7)
masking tape
solution A—manganous sulfate solution
solution B—potassium hydroxide–potassium
   iodide solution

solution C—concentrated sulfuric acid
solution D—starch solution
solution E—sodium thiosulfate solution
safety goggles
laboratory apron
phenolphthalein solution
sodium hydroxide solution

PROCEDURE

CAUTION: *All chemicals used are harmful to skin and clothing. If you spill any chemical, rinse with water and call your teacher immediately.*

Part A. Dissolved Oxygen

1. Obtain 100 mL of two different water samples. Place the samples in small flasks or beakers. If you are to pour samples into the flasks from large containers, pour slowly to avoid making bubbles in the water. If you are collecting samples directly from the source, open the flasks under the water so the flasks fill with water below the surface. Label the flasks with the source of the water. Record in Table 1 where each sample was obtained.

2. With a dropper, add 10 drops of solution A to each water sample. Hold the dropper close to the water surface to avoid splashing. With another dropper, add 10 drops of solution B to each water sample.

3. Gently mix the contents by swirling the flasks. However, be careful to avoid forming bubbles. After swirling, let the flasks stand for one minute. (If you are using seawater as a sample, it must stand for 15 minutes.)

4. With a third dropper, add 15 drops of solution C to each water sample. Gently mix the contents by swirling the flasks. Again, be careful to avoid splashing.

5. With another dropper, while gently swirling each flask, add 5 drops of solution D to the sample. A deep blue color will appear.

6. With a fifth dropper, add solution E one drop at a time to each sample. The number of drops of solution E must be counted. Add drops of solution E until the water sample becomes colorless. Swirl the water samples after the addition of each drop to determine the true color of the solution.

7. Record in Table 1 the number of drops of solution E needed to turn each water sample colorless.

8. The amount of dissolved oxygen in water usually is described in parts per million (ppm)—parts of oxygen per million parts of water. For example, at room temperature, fresh water has 8.4 parts of dissolved oxygen per one million parts of water (8.4 ppm). Convert the drops of solution E to ppm of oxygen in each flask by dividing the numbers of drops of solution E by 20. Express the answers to one decimal place.

9. Record the ppm of dissolved oxygen for each sample tested in Table 1.

Part B. Testing for Carbon Dioxide

1. Obtain 100 mL of each of two different water samples. Place the samples in small flasks or beakers. Label the flasks with the water source. Record the water source in Table 1.

2. With a clean dropper, add 5 drops of phenolphthalein solution to each sample. Mix by gently swirling your flasks. NOTE: *If a light-pink color forms and stays, no carbon dioxide is present. Record this result in Table 1. You are finished testing this water sample. If the pink color forms and then quickly disappears, carbon dioxide gas is present. Go on with the procedure.*

3. With a clean dropper, add sodium hydroxide one drop at a time to each sample. You must count the number of drops of this chemical that you add. Swirl the water samples after the addition of every few drops to determine the true color of the water. Add drops of sodium hydroxide until the water sample becomes light pink and remains pink after swirling.

4. Record the number of drops of sodium hydroxide needed to change the color of the sample.

5. Convert the drops of sodium hydroxide to ppm of carbon dioxide in each flask by multiplying the number of drops used by 5.

6. Record the ppm of carbon dioxide in each sample tested in Table 1.

DATA AND OBSERVATIONS

Table 1.

Results of Oxygen and Carbon Dioxide Tests							
Oxygen results				Carbon dioxide results			
	Water source	Drops of solution E used	Amount of O_2 (ppm)	Water source	Carbon dioxide present?	Drops of sodium hydroxide used	Amount of CO_2 (ppm)
Sample 1							
Sample 2							

ANALYSIS

1. **a.** Which water sample contains more dissolved oxygen? _____

   b. Which contains less dissolved oxygen? _____

2. A lake sample having less than 4 ppm of dissolved oxygen is harmful to most fish.

   a. Which of your samples could have come from a lake that will support most fish? _____

   b. Which samples could not? _____

   c. Explain. _____

   _____

3. Why is oxygen important to organisms living in water? _____

   _____

4. a. Which type of organism (producer or consumer) could provide oxygen to water? _____

   _____

   b. What name is given to the process during which oxygen is produced? _____

   _____

5. The graph in Figure 1 shows the values for dissolved oxygen in a lake at various depths. Does the

   amount of dissolved oxygen increase or decrease with depth? _____

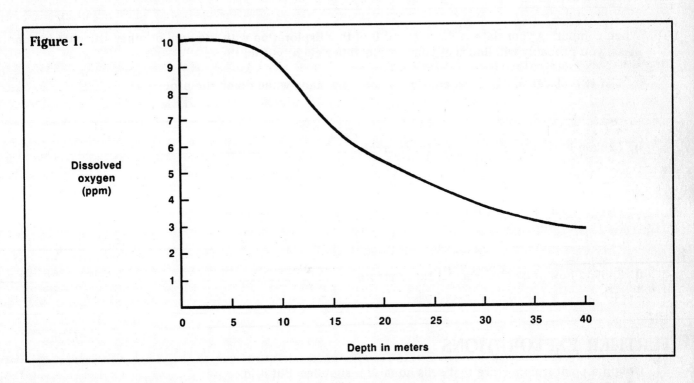

Figure 1.

6. Since light is required for the process during which oxygen is produced, how might this process be

   influenced by changes in water depth? _____

   _____

7. a. Which water sample contains more carbon dioxide? _____

   b. Which water sample contains less carbon dioxide? _____

8. A stream having more than 25 ppm of carbon dioxide is harmful to most bivalves.

   a. Which of your samples could have come from a stream that could support bivalves? _____
   _____

   b. Which samples could not? _____

   c. Explain. _____
   _____

9. Carbon dioxide and light are both required for photosynthesis. Suppose a shallow pond with many plants growing in it was used as a water source for testing.

   a. How might the amount of carbon dioxide in the pond vary from day to night? _____
   _____

   b. Explain. _____
   _____

10. When comparing your data in Parts A and B of this Exploration with the data of other students in your class, you probably will find that your results differ slightly from theirs.

   a. List two places in Part A where errors could be made while doing the procedure. _____
   _____

   b. Explain how you could correct these errors. _____
   _____

   c. List two places in Part B where errors could be made while doing the procedure. _____
   _____

   d. Explain how you could correct these errors. _____
   _____

FURTHER EXPLORATIONS

1. Obtain a plant from a store that sells aquarium supplies. Put it in one of your water samples for a day, keeping it near a window. Determine the ppm of oxygen and carbon dioxide by using the same procedure you used in this Exploration. Compare your results with those obtained before placing the plant in the water. Explain any differences.

2. Put the water sample containing the plant in the dark for 48 hours. Repeat the procedures. Explain any differences in your results.

GLOSSARY

A glossary is the place where you look for the meaning of terms. The glossary lists terms in alphabetical order. In **Biology: The Dynamics of Life,** *Laboratory Manual*, the glossary has been incorporated to save you time in your search for the meaning of a word. At the same time, the glossary gives you extra reference to related topics and gives you a better opportunity for understanding a term in context. Each major vocabulary term that is used in this manual is listed in the glossary in boldface type, followed by its definition. Some terms are followed in parentheses by a guide to pronunciation. Below is a pronunciation key to help you read the terms used in this laboratory manual.

A

abdomen: in mammals, between the thorax and pelvis in mammals; behind the thorax in arthropods

adaptation: a variation in an organism that makes it better able to cope with its environment

AIDS: acquired immune deficiency syndrome

albumin: the protein that makes up the egg white

alcoholic fermentation: the process by which certain yeasts decompose sugars in the absence of oxygen

allele: (uh LEEL) two or more different genes that occupy the same position in homologous chromosomes

allele frequency: the proportion of each allele in the gene pool

alternation of generations: a life-cycle in plants whereby a haploid phase produces gametes that join to form a zygote that germinates to form a diploid phase. Meiosis produces spores that give rise to a new haploid generation

ambulacral grooves: (am byoo LAK rul) the grooves on the ventral side of the starfish rays in which the tube feet are found

ampulla: the sac at the base of a tube foot in certain echinoderms

anaphase: the stage of mitosis in which the sister chromatids or homologous chromosomes move apart and toward the poles of the nucleus

angiosperm: a flowering plant in the division Anthophyta

antenna: sensory organs in many arthropods

anterior: toward the front or head end of an animal's body

anther: pollen-producing structure of a flower

antheridium: a male sex organ in simple plants

antibiotic: a chemical produced by living organisms capable of inhibiting the growth of some bacteria

antibody: a protein produced in response to an antigen

antigen: a substance capable of stimulating a specific immune response

aortic arches: structures that act to keep blood circulating in the earthworm

appendage: any major structure growing out of the body of an organism, such as antennae, bristles, legs

archegonium: a female sex organ in simple plants

arterioles: small connecting vessels branching from arteries into capillaries

artery: blood vessel that carries blood away from the heart

ascus: a membranous sac in ascomycote fungi inside which spores form during sexual reproduction

atria: two anterior chambers of the heart that receive blood from lungs and body organs

auxin: a plant hormone that causes cell elongation

B

bactericide: a substance that destroys bacteria

base: a compound that ionizes to yield hydroxyl ions in water

basidium: microscopic, club-like form that produces basidiospores in some fungi

bast fibers: the fibrous, woody outer layer of the stems of certain plants

biodegradable: the ability of a waste or pollutant to be broken down into chemical nutrients by decomposers

blood: fluid tissue of the body, contained within the circulatory system, carrying oxygen and nutrients throughout and waste materials to excretory channels

brachiopods: solitary, bivalve marine animals

brain: center of control in the body

C

caecum: also cecum; a cavity with only one opening; the intestinal pouch in some mammals containing the microorganisms for digestion of cellulose

Calorie: the amount of heat required to raise the temperature of one kilogram of water one degree Celsius

calyx: the outer whorl of protective sepals in a flower

camouflage: the protective adaptation that allows an organism to blend into its surroundings

canaliculi: small channels in compact bone that carry fluids between the bood vessels and the osteocytes

canines: conical teeth located between the incisors and first bicuspids

capillary: thin-walled vessel of the circulatory and lymphatic systems through which gases and nutrients are exchanged with body cells

capsule: a mucopolysaccharide layer enveloping certain bacteria

carbohydrate: a compound with small molecules, such as in simple sugars, starches and cellulose

carrying capacity: maximum number of organisms in a species that the environment can support

cell membrane: outer layer of proteins and lipids that shapes, protects and regulates what enters and leaves the cell

cell wall: a rigid structure that surrounds the plasma membrane in certain organisms such as plants, bacteria, and fungi

centromere: the point at which sister chromatids are held together

cephalothorax: the anterior section of arachnids composed of a fused head and thorax

cerebellum: one of three main portions of the brain that controls balance, posture and coordination

cerebrum: one of three main portions of the brain that controls conscious activity, memory languages, and the senses

chalazae: the two spiral bands of tissue in an egg, connecting the yolk to the lining membrane

chemotaxis: characteristic orientation of a freely moving living organism relative to a chemical substance

chlorophyll: green pigments found in photosynthetic organisms; traps light in chloroplasts

chloroplast: a plastid in photosynthetic plants that converts light into usable energy

chromatography: separation of complex mixtures though a selectively absorbing medium

chromatophores: pigment-containing structures in the skins of fishes, frogs and other animals

chromosome: a DNA-containing linear structure in cell nuclei

circulatory system: system of structures by which blood and lymph are circulated throughout the body

classification: a systematic arrangement of plants and animals into categories

clitellum: swollen, glandular, saddlelike region in the epidermis of the earthworm

cloaca: the posterior part of the intestinal tract in some invertebrates

coelom: (SEE lum) a true body cavity completely surrounded by tissues that develop from the embryonic mesoderm

collar: an encircling structure suggestive of a collar, immediately above the siphon in the mantle of a squid

colloid: an aggregate of molecules in a finely divided state, dispersed in a medium, which resists sedimentation, diffusion, and filtration

colon: section of the large intestine extending from the cecum to the rectum

colonies: large groups of bacteria descended from a single bacterium

community: the combination of all the populations living and interacting with each other in a given area

conditioned response: new or modified response elicited by a stimulus after conditioning

cones: cylindrical structures borne by certain trees, consisting of cluster of stiff, overlapping, woody scales, between which are the naked ovules

cork cambium: meristematic tissue that produces bark

conditioning: a form of learning by association

conidia: a structure in which a mass of spores are produced by a conidiophore in Ascomycote fungi

consumer: a type of organism that obtains energy from eating other organisms

cork: light, porous elastic outer bark of the cork oak that protects the roots and stems from injury

corolla: the whorl of petals in a flower

cortex: storage tissue that lies between the epidermis and vascular cylinder of a stem or root

crop: the storage chamber in many invertebrates though which food passes from the esophagus into the gizzard

cusps: a prominence on the chewing surface of the tooth

D

deoxyribonucleic acid (DNA): nucleic acid containing deoxyribose as the sugar component and found in the chromatin and chromosomes of cells

density: the degree or measure of degree to which anything is filled or occupied; mass per unit volume of a substance

depth of field: the extent to which one can see clearly under a microscope; foreground or background

development: the sum of changes that take place during the life of an organism

diaphragm: the aperture that controls the amount of light in a microscope

diffusion: particle movement from an area of higher concentration to an area of lower concentration

digestive gland: various endocrine and exocrine glands that secrete digestive enzymes

dihybrid cross: a cross between two individuals that differ in two characteristics

disaccharide: condensation product of two monosaccharides bonding together with the elimination of a water molecule

DNA: see deoxyribonucleic acid

dominant trait: the condition in which only the phenotype determined by one of two genes is expressed in the heterozygous condition

dormant: relatively inactive condition in which some processes are slowed down or suspended

dorsal: located near or on the back of an animal

dorsal aorta: upper artery

dorsal blood vessel: upper blood vessel

duodenum: the part of the small intestine that extends from the lower end of the stomach

E

ecosystem: system that results from the interaction between all the organisms in an area

egg: the female sex cell or gamete

electrophoresis: the movement of particles in an electric field

embryo: an early stage of development in any multicellular organism

endodermis: innermost layer of the cortex; contains a waxy substance that prevents water from moving out of the vascular cylinder of a plant

enzyme: a protein that speed up the rate of a chemical reaction without being permanently changed by the reaction

epidermis: the outer, thin protective layer cells in plants and animals; skin

esophagus: the tubular portion of the alimentary canal that connects the mouth to the stomach

eustachian tube: canal that connects the pharynx to the middle ear through which air pressure is equalized

external nares: external openings in vertebrates for breathing; nostrils

F

fat bodies: found in frogs on the ovary or testis to store fat as nutrition for developing gametes

fatigue: weariness or exhaustion, usually a result of exertion

fetus: the term expressing the embryo after the eighth week of development when all systems are present

field of view: the area that can be seen through a microscope

filaments: the thin, stemlike parts of a stamen in a flower

follicle: a group of cells within the ovary that contains an egg

fossil: the remains, traces, or imprints of an organism preserved in Earth's crust some time during the geologic past

fossil fuel: a hydrocarbon fuel derived from living matter of a previous geologic time

frond: leaf of a fern

funiculus: stalk connecting the ovule with the ovary in a flower

fungus: a unicellular or multicellular eukaryote that has cell walls and obtains food by absorption

G

gamete: a sex cell; formed when the number of chromosomes is halved during meiosis

gametophyte: the form of algae or plants that contains gamete producing organs and produces gametes or sex cells

gastrovascular cavity: cavity in which digestion and circulation take place

gene: a segment of DNA that controls specific hereditary traits.

gene pool: all the genes in a population

genotype: the genetic constitution of an organism; it determines the traits of an organism

geotaxis: the movement of an organism in response to the force of gravity

germination: the development of a seed into a new plant

gill: a respiratory structure that removes oxygen from water

gizzard: in earthworms, a sac with muscular walls and hard particles that grinds soil before it is passed into the intestine; part of the digestive tract of a bird in which food is crushed by muscular action

glottis: the opening that leads to the respiratory system

glucose: a form of sugar

gonad: the organ that produces gametes; testis or ovary

gymnosperm: seed plant having a naked ovule

H

Haversian canals: small channels containing the blood vessels that nourish the osteocytes in the bone

Haversian system: elongated structures that surround the Haversian canal

head: upper, anterior vertebrate extremity, containing the brain or principal ganglia

heart: muscular organ that pumps blood to all parts of the body

hermaphrodite: a condition in which both male and female reproductive organs are present in an individual

holdfast: structure resembling a root that serves as an anchor for many algae

heterozygous: condition in which two homologous chromosomes have different alleles for a trait

homozygous: condition in which two homologous chromosomes have identical alleles for the same trait

hormone: the chemical secreted by an endocrine gland that brings about an effect in a specific tissue or organ; also produces growth and responses in plants

hypothesis: a possible solution to a problem based on all currently known facts; a prediction that must be testable

I

incomplete dominance: the condition resulting when two alleles produce three phenotypes instead of two; neither allele of a pair is completely dominant

inheritance: the process of genetic transmission of characteristics

ink sac: extends from the intestine, near the siphon in a squid; it is used for defense

innate behavior: behavior that is inherited

intestine: the portion of the alimentary canal extending from the stomach to the anus

invertebrate: an animal without a backbone

K

karyotype: picture of paired human chromosomes arranged by size; used to identify chromosomal abnormalities

key: a table for identifying organisms

kidneys: organs that filter liquid waste from the blood

L

landfill: a method of rehabilitating land in which garbage and trash are buried in low-lying ground

lens: transparent structure in the eye that helps focus images

lichens: the mutualistic relationship of a green organism and a fungus, being so complete that a genus and species name is assigned them as though they were a single organism

ligament: the fibrous tissue that connects bone to bone

lipid: an organic compound made by cells for long-term energy storage

liver: an organ that secretes bile and acts in formation of blood and metabolism; breaks down substances such as alcohol and drugs

lung: spongy, saclike organ where breathing occurs

lymphocyte: white blood cell found in lymph nodes that function in the body's defense system

M

macroorganism: a plant or animal large enough to be seen with the naked eye

madreporite: (muh DREH puh rite) a delicately perforated sieve plate at the distal end of the stone canal in echinoderms

mantle: a thin membrane that surrounds the digestive, excretory and reproductive organisms of mollusks, and secretes a shell in shelled species

mantle cavity: a space formed by the mantle which hangs down over the back and sides of the body of the mollusk

maxillary teeth: small teeth, conical, in the upper jaw of a frog that aid in holding prey

medulla: the inner layer of an organ such as a kidney

medulla oblongata: the portion of the brain stem that controls involuntary activities

megaspore: reproductive cell in plants that gives rise to the egg

meiosis: the process of cell division that results in the formation of haploid gametes by reducing the number of chromosomes by one half through two divisions of the nucleus

metaphase: phase in mitosis during which sister chromatids become attached by the centromere to the spindle fibers

microscope: optical instrument that uses lenses to produce magnified images of small objects

microspore: reproductive cell in plants that gives rise to sperm

molars: teeth with a broad crown for grinding

motile: having the ability to move

mouth: oral cavity; body opening through which an animal takes in food

mutualism: symbiotic relationship in which both species derive benefit

mycelium: the mass of hyphae that comprises a fungus growth

N

natural selection: the theory that a mechanism for change in populations occurs when organisms with characteristics most favorable for survival in a particular environment are able to pass these traits on to offspring

nematocyst: in cnidarians, a capsule containing a sharp barb that delivers a poison for obtaining food or self defense on tentacles

nictitating membrane: third eyelid in frogs that keeps the eyeball moist

nosepiece: the part of a microscope, often rotatable, to which one or more objective lenses are attached

nostril: either of two openings of the nose

nucleic acid: a large complex macromolecule that stores information in the form of a code; DNA, RNA

nucleotides: the individual monomers that link together to form a nucleic acid; made up of a nitrogen base, a sugar, and a phosphate group

nutrients: carbohydrates, fats, proteins, vitamins, and minerals

O

objective: the lens system in a microscope that is closest to the object

organelle: a membrane-bound cell structure that performs one or more functions

osmosis: the diffusion of water through a semipermeable membrane

osteocytes: bone cells that produce the system of canals within the Haversian system

ovary: the basal portion of a pistil in a seed plant; the organ in an animal that produces eggs

ovule: a structure in the ovary of a seed plant

P

paleontologist: scientist who looks for and studies fossils

pancreas: a gland behind the stomach that secretes digestive enzymes and produces insulin

pedicellaria: (ped uh sehl AH ree uh) a small grasping structure on an echinoderm

pen: the chitinous internal shell of a squid

petal: leaflike flower part

pharynx: the upper expanded portion of the digestive tube between the esophagus and the mouth and nasal cavities

phenotype: the physical appearance of an organism as opposed to its genotype

phloem: (FLOH em) a complex food-conducting vascular tissue in higher plants

photosynthesis: the building up of chemical sustances under the influence of light

phototaxis: the movement of an organism in response to a source of light

pistil: the female structure of a flower

plasmolysis: the loss of water resulting in a drop in turgor pressure in plant cells

posterior: the back surface of the body; toward the tail

pollen grain: the sperm-producing organ of a seed plant

population: all organisms of one species in a specific area

predation: the capturing of prey as a means to maintaining life

premolars: one of eight bicuspid teeth, behind the canines and in front of the molars

primates: mammals such as humans, monkeys, and apes

producer: an organism that generates its own food; an autotroph

prophase: the initial stage of mitotic or meiotic cell division

protein: complex nitrogen-containing organic compounds of high molecular weight that have amino acids as their basic structural units

prothallus: the heart-shaped gametophyte stage of a fern; develops from a protonema

Punnett square: (PUN ut) a tool used to predict the possible offspring of crosses between different genotypes

pyloric sphincter: the muscular valve that controls the passage of food out of the stomach

pyrenoids: small protein bodies that store starch in chloroplasts

R

radial canal: in starfish, one of five pipelike structures that delivers water to the tube feet

radula: a snail's tongue-like organ with rows of teeth used for scraping food

rays: stiff narrow rods that support fins on fish; the arms of an echinoderm

recessive trait: a hereditary characteristic not expressed in the presence of another characteristic

rectum: the last section of the digestive system

reducing sugars: monosaccharides and some disaccharides that yield a positive Benedict's test

respiration: the process by which food molecules in a cell are broken down to release energy; also refers to the process of breathing

response: the reaction of an organism to a stimulus

retina: a photoreceptive layer of cells in the eye

ribosomal RNA: (rRNA) comprises part of the ribosomes

root hair: single-celled hairlike outgrowth of the roots epidermis; absorbs nutrients from soil

S

seed: a plant structure protecting the plant embryo

segment: a clearly differentiated subdivision of an organism or part

seminal vesicles: two glands located beneath the bladder that release a sugary fluid into the vas deferens

sepal: one of the leaflike parts of a flower that compose the calyx; protects the flower in the young bud stage

septa: the walls separating the chambers of the heart

sessile: an organism that remains permanently attached to a surface for its entire adult life

setae: in earthworms, pairs of bristles that aid in locomotion

skeleton: system of bones and joints that provides protection and support and which allows body movement

small intestine: the part of the intestine between the outlet of the stomach and the large intestine; where most digestion and nutrient absorption takes place

solvent: a substance that dissolves another substance

sperm: the male germ cell or gamete

spinal cord: the bundle of nerves of the central nervous system that extends down the back.

sporangium: (spuh RAN jee um) a cell in which asexual spores are produced

spore: a cell from which a new organism is produced without fertilization by another cell

sporophyte: the spore-producing phase in algae and plant life cycles

stamen: the male reproductive structure of a flower

stigma: the rough or sticky surface of the pistil to which pollen grains will stick in flowering plants

stimulus: the condition that provokes a reaction from an organism

stomach: a muscular pouch-like organ of the digestive system

stomata: small pores in the surfaces of leaves

style: the usual slender part of a pistil

swimmerets: appendages on the abdomen of a crustacean

T

taxonomy: the science, laws, or principles of classification

telophase: the final phase of mitosis in which the chromosomes of daughter cells are grouped in new nuclei

tentacle: in a cnidarian, the long structures surrounding the mouth used for obtaining food

testes: male reproductive organ of animals

titration: a process by which the concentration of a substance is determined by comparing it to a standard solution of known concentration

trachea: the main tubelike structure through which air passes to and from the lungs

transpiration: the process through which plants lose water vapor

tube feet: footlike extensions of the radial canals of and controlled by the water vascular system in echinoderms; capable of powerful grasping

tympanic membrane: a circular structure in a frog's head that responds to air or water vibrations and transmits them to the inner ear and to the brain.

U

unicellular organism: a single-celled organism in which all life functions are carried out

V

vein: blood vessel that carries blood toward the heart

ventral: near the undersurface of an animal body

ventricles: the two chambers of the heart from which blood circulates to the lungs and body

W

water vascular system: an internal closed system of reservoirs and ducts containing a watery fluid in echinoderms that controls movement, respiration, and food capture

X

xylem: the principal water-conducting tissue and the chief supporting tissue of higher plants

Z

zygote: a single cell that is the result of fertilization

Using the Balance

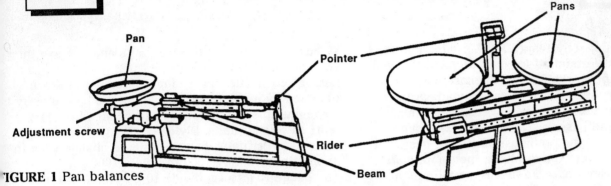

FIGURE 1 Pan balances

Although the balance you use may look somewhat different from the balance pictured in Figure 1, all beam balances require similar steps to find an unknown mass.

Follow these steps when using a beam balance.

1. Slide all riders back to the zero point. Check to see that the pointer swings freely along the scale. The beam should swing an equal distance above and below the zero point. Use the adjustment screw to obtain an equal swing of the beams. You should "zero" the balance each time you use it.

2. Never put a hot object directly on the pan. Air currents developing around the hot object may cause massing errors.

3. Never pour chemicals directly on the balance pan. Dry chemicals should be placed on paper or in a glass container. Liquid chemicals should be massed in glass containers.

4. Place the object to be massed on the pan and move the riders along the beams beginning with the largest mass first. If the beams are notched, make sure all riders are in a notch before you take a reading. Remember, the swing should be an equal distance above and below the zero point on the scale.

5. The mass of the object will be the sum of the masses indicated on the beams, as shown in Figures 2 and 3. Subtract the mass of the container from the total mass.

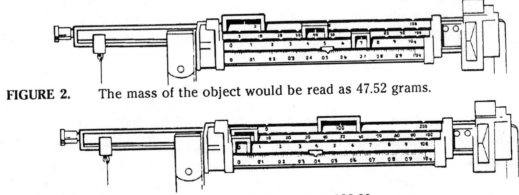

FIGURE 2.   The mass of the object would be read as 47.52 grams.

FIGURE 3. The mass of the object would be read as 100.39 grams.

Measuring Volume

The surface of liquids when viewed in glass containers is curved. This curved surface is called the meniscus. Most of the liquids you will be using form a meniscus that curves down in the middle. Read the volume of these liquids from the bottom of the meniscus, as shown in Figure 4. This measurement gives the most precise volume because the liquids tend to creep up the sides of glass containers.

If you are using plastic graduated cylinders and no meniscus is noticeable, read the volume from the level of the liquid.

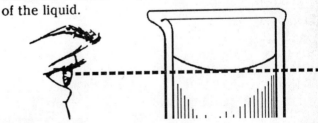

FIGURE 4.

Measuring in SI

The International System (SI) of Measurement is accepted as the standard for measurement throughout most of the world. Four of the base units in SI are the meter, liter, kilogram and second. Other frequently used SI units are degrees, joules, newtons, watts, and pascals. The size of a unit can be determined from the prefix used with the base unit name. For example: *kilo* means a thousand; *milli* means a thousandth; *micro* means a millionth; and *centi* means a hundreth. The tables below give the standard symbols for these SI units and some of their equivalents.

Larger and smaller units of measurement in SI are obtained by multiplying or dividing the base unit by some multiple of ten. Multiply to change from larger units to smaller units. Divide to change from smaller units to larger units. For example, to change 1 km to m, multiply 1 km by 1000 to obtain 1000m.

To change 10 g to kg, divide 10 g by 1000 to obtain 0.01kg.

TABLE 1

COMMON SI UNITS			
Measurement	Unit	Symbol	Equivalents
Length	1 millimeter	mm	1000 micrometers (μm)
	1 centimeter	cm	10 millimeters (mm)
	1 meter	m	100 centimeters (cm)
	1 kilometer	km	1000 meters (m)
Volume	1 milliliter	mL	1 cubic centimeter (cm^3 or cc)
	1 liter	L	1000 milliliters(mL)
Mass	1 gram	g	1000 milligrams (mg)
	1 kilogram	kg	1000 grams (g)
	1 ton	t	1000 kilograms (kg)=1 metric ton
Time	1 second	s	
Area	1 square meter	m^2	10 000 square centimeters (cm^2)
	1 square kilometer	km^2	1 000 000 square meters (m^2)
	1 hectare	ha	10 000 square meters (m^2)
Temperature	1 Kelvin	K	1 degree Celsius (°C)